Experimental Methodology

SIXTH EDITION

Experimental Methodology

LARRY B. CHRISTENSEN
Texas A&M University

ALLYN AND BACON
Boston London Toronto Sydney Tokyo Singapore

Series Editorial Assistant: Marnie Greenhut
Cover Administrator: Linda Dickinson
Manufacturing Buyer: Megan Cochran
Editorial-Production Service: Lifland et al., Bookmakers

Copyright © 1994, 1991, 1988, 1985, 1980, 1975 by Allyn and Bacon
A Division of Simon Schuster, Inc.
160 Gould Street
Needham Heights, MA 02194

Library of Congress Cataloging-in-Publication Data
Christensen, Larry B., 1941–
    Experimental methodology // Larry B. Christensen.—6th ed.
        p.   cm.
    Includes bibliographical references and index.
    ISBN 0-205-15506-5
    1. Psychology, Experimental. 2. Psychology—Experiments.
3. Experimental design.   I. Title.
BF181.C48 1993   93-31421
150'.724—dc20   CIP

Printed in the United States of America
10 9 8 7 6 5 4 3 2   98 97 96 95

Contents

CHAPTER 3   The Experimental Research Approach   73

CHAPTER 4   Problem Identification and Hypothesis Formation   97

CHAPTER 5   Ethics   121

CHAPTER 6   Variables Used in Experimentation   173

CHAPTER 7   Control in Experimentation   211

CHAPTER 8   Techniques for Achieving Constancy   243

CHAPTER 15   The Research Report   477

Preface

Goals of the Textbook

When I wrote the first edition of *Experimental Methodology*, my goal was to provide an introduction to the basic principles of psychological research. I wanted to create a textbook that explained the key principles of the research process in as clear and straightforward a manner as possible, using examples to illustrate each of the various points. This goal has guided all of the subsequent editions, including the current one. In explaining the research process, I elected to focus attention on methodological issues as opposed to providing a balanced presentation of methodology and statistical analysis. This decision was reached for two reasons. First, virtually every psychology department offers a separate statistics course, so including extensive material on statistics in a methodology course represents duplication of effort (although a good argument can be made for incorporating statistics in a research methods course). Second, and more important, focusing attention on methodology allows for more extensive coverage of methodological issues involved in the design of experiments and thus provides the student with a better foundation in the research process.

Organization of the Textbook

In presenting the basic principles of research, I chose to organize the textbook so as to follow the steps involved in the research process. After a discussion of the nature of science, the two basic research processes—descriptive and experimental research—are introduced, to provide the student with an appreciation of both processes. Identification of a research question is then covered at some length, because students frequently have difficulty developing a good researchable idea.

Ethics of research are taken up next, because consideration of ethical issues is an integral part of research study. Once a research problem has been developed, ethical issues must be kept in mind as the study is conducted and the results are reported. Control issues are then discussed, to give the student an appreciation of the necessity of exercising control as well as information on how to implement the needed control. Actual designs, both multisubject and single-subject, are presented, to provide the student with knowledge of the various approaches that can be used to test hypotheses. Following the chapters on design are chapters that focus attention on data collection and statistical analysis, as well as on communication of the research results in a research report.

The Sixth Edition

Although the sixth edition has maintained the same goal and basic orientation as the prior editions, a number of changes have been made that not only update the presentation but also provide for more complete coverage of the research process. Following are some of the important changes in the sixth edition:

1. Gender issues relating to the research process have been incorporated in appropriate chapters. Gender issues are increasingly being addressed in psychological research, as the impact of gender on research results becomes more evident. Students need to be aware, for example, of how the researcher's gender may influence the topic he or she selects.

2. The section on survey methodology has been expanded. Several reviewers recommended additional coverage of this method, because it is frequently used by researchers.

3. Coverage of scientific misconduct has been extended substantially. Unfortunately, scientific misconduct is

on the rise, making it vital that psychology departments help students guard against such totally unacceptable behavior.

4. Discussion of informed consent has been added.

5. The section on animal ethics has been extensively revised to incorporate the new guidelines recently proposed by the American Psychology Association.

6. The chapter on statistics has been expanded to include an example of an ANOVA summary table as well as discussion of two-way factorial ANOVA.

7. The new sample research report is a recently published study on nightmares, providing interesting information and serving as an excellent example of a research report.

8. Information has been added on several other topics, including the types of variables used in research and how to take a random sample from a population. A description of an actual interaction effect has been incorporated in the section that discusses the nature of interactions.

Acknowledgments

The current version of *Experimental Methodology* is, in part, the product of the excellent input I have received from many of my colleagues and students. Their comments have improved the quality and educational value of the final product. Without this continuous input I am sure this textbook would not have been as successful or as well received as it has been. In an effort to continue to improve the quality and instructional value of the book, I invite students and faculty to send me their candid comments—both positive and negative—as well as suggestions for alterations and additions of material I may have omitted. I can assure you that each comment will be taken seriously and I will do my best to include your suggestions in the next edition. You can send your comments to me at the Department of Psychology, Texas A&M University, College Station, TX 77843.

Finally, I would like to thank the staff at Allyn and Bacon for their careful attention to detail in the handling of the current version. Seeing a manuscript through the publication process is a difficult task requiring the coordinated efforts of many individuals and personalities. Much of the thanks for making sure the task was completed smoothly goes to my editor, Laura Pearson.

L. B. C.

CHAPTER 1

What Is Science?

LEARNING OBJECTIVES

1. To gain an understanding of the nature of the scientific method.
2. To understand how the scientific method differs from other methods of acquiring information.
3. To learn the unique characteristics of the scientific method and understand why each of these characteristics is necessary.
4. To understand the sort of characteristics that typify the person who is adept at pursuing the scientific enterprise.
5. To gain an appreciation of the role gender plays in the accumulation of scientific information.

On May 9, 1982, the Los Angeles Times *ran an article in its Sunday Supplement under the headline "The Devil Made Me Do It." According to this article, some religious groups were claiming that several rock bands and recording companies were putting satanic messages on their records. The religious groups claimed that these messages could be heard only when the records were played backwards, but that the listener could subconsciously decipher the meaning of the messages when the record was played in the normal manner. Further, these groups asserted, listeners would accept the satanic message uncritically and would act accordingly.*

Readers of this article probably wondered whether this was true, and whether they had been listening to and accepting satanic messages. The article also stated that these messages occasionally praised the devil and that several states had introduced legislation to require certain records to carry a label stating that they contained a satanic message. This seemed to give the assertion some legitimacy—surely legislators would not introduce a bill unless the claim had some truth to it.

The question is whether this accusation is true. Did several rock bands, in conjunction with their recording companies, introduce satanic messages on the records, or were the religious groups making a false accusation? In interviews, the recording companies and the bands flatly denied any such attempt. Their accusers, however, insisted that they had heard the messages. Thus, the information obtained from the religious groups and from the bands and recording companies is contradictory. Is it possible for both sides to be correct? Is it possible that no satanic messages were included on the records but that members of the religious groups heard *such messages? The persons who stated that they had heard the messages were most likely sincere individuals who actually believed that they had heard the messages. If anyone tried to deny that satanic messages actually existed on the records, these individuals would probably strongly oppose any such suggestion. After all, they had heard the messages, and this would seem to make their claim irrefutable.*

But direct experience is only one of the many ways in which we gain information. In fact, psychological research has shown that people are likely to accept and even to experience suggestions made by others with prestige or authority if the situation they are in is unfamiliar or vague. This seems to be exactly what happened in this instance. Very few people ever play records backwards; if they do, the sounds they hear are unfamiliar and unclear. This was an ideal situation for suggestion to operate, particularly since the suggestion was coming from esteemed religious leaders.

In order to determine most accurately whether suggestion could have led to the perception of hearing satanic messages, we must conduct a scientific study. Few people, however, understand the nature of a scientific investigation or the need to conduct such an investigation in a situation like this. Scientists are usually conceptualized as people in white coats who work in a laboratory, conducting experiments on complex theories that are far beyond the comprehension of the average individual. The actual process by which scientists uncover the mysteries of the universe totally eludes most people. It is as if the scientific process were encompassed in a shroud of secrecy and could be revealed only to the scientist.

Science, however, is not a mysterious phenomenon. Rather, it is a very logical and rigorous method for attempting to gather facts. This chapter is designed to remove the mystery surrounding science and to acquaint you with the scientific process. This chapter will cover not only the meaning of science but also the unique characteristics that distinguish the scientific method from other methods of gathering information.

INTRODUCTION

In our daily lives, we continually encounter problems and questions relating to behavior. For example, one person may have a tremendous fear of taking tests. Others may have problems with alcoholism or drug abuse, or problems in their marriage. People who encounter such problems typically want to eliminate them but often lack the knowledge or ability to handle the problems themselves. Consequently, they seek out professionals, such as psychologists, to help them to remediate such difficulties.

Other people may enlist the assistance of professionals in understanding the behavior of others. For example, salespeople differ greatly in their ability to sell merchandise. One car sales-person may be capable of selling twice as many cars as another salesperson can. If the sales manager could discover why such differences in ability exist, he or she might be able to develop either better training programs or more effective criteria for selecting the sales force.

In an attempt to gain information about behavior, people turn to the field of psychology. As you should know by now, a great deal of information about the behavior of organisms has been accumulated. We have knowledge that enables us to treat disorders such as "test anxiety." Similarly, we have identified many of the variables influencing persuasion and aggression. Although we know a great deal about the behavior of humans and infrahumans, there is still much to be learned. For exam-ple, we have an inadequate understanding of childhood autism and leadership ability. In order to learn more about such behav-iors, we must engage in scientific research because this is the only way in which we can fill the gaps in our knowledge. However, the ability to understand and engage in the research process does not come easily; it is definitely not an ability that comes from taking introductory or abnormal psychology. These content courses give little insight into the way in which psychological facts and data are acquired. They state implicitly or explicitly that such facts and data have been acquired from scientific research, but the nature of the scientific research process itself remains elusive.

In order to learn about the scientific research process, one needs more direct instruction. The course in which you are now enrolled is aimed at providing you with information about the way in which the scientific research process is con-ducted. Some students may object that such a course is not necessary for their education because they have no intention of becoming research psychologists. But there are a number of very good reasons for all students to study experimental meth-odology. First, at some time in the future you may be asked to conduct a study (such as a community survey) on some issue. Second, virtually all the material you are required to learn in your science courses is based on knowledge acquired from the scientific method, so you should be familiar with the method. Third, we are all continually bombarded by the results of sci-entific research, and we need experimental tools to determine which research outcomes are conclusive. For example, saccha-rin has been demonstrated to cause cancer in laboratory ani-mals, yet there are many people who consume saccharin and do not contract cancer. You as a consumer must be able to resolve these discrepancies in order to decide whether or not you are going to eat foods containing saccharin.

Similarly, television commercials often present what appears to be a scientific test in order to convince us of the superiority of one product over another. Several years ago the manufacturers of Schlitz beer were concerned with the decline in the sales of their product. In an effort to reverse this decline, the company conducted a live "challenge" on television in which devotees of another brand were challenged to see if they could distinguish their preferred brand from Schlitz. This live demonstration consistently showed that about 50 percent of these beer drinkers chose Schlitz over their preferred brand as the better tasting beer. On the surface, this challenge seems to reveal that Schlitz is an excellent beer because so many people chose it. If you had some knowledge of research design and statistics, however, you would be able to see that this contest did not prove anything about the superiority of Schlitz over other beers because the challenge was conducted on live television, in the midst of a lot of noise and commotion. Such distractions would minimize a person's ability to distinguish one beer from another. If there were enough distractions that people could not distinguish one beer from another, they would probably select one beer about the same number of times as the other. This is exactly what happened, since Schlitz and the other brand were *each* picked by about 50 percent of the people. From this example, you can see that an understanding of the scientific research process induces a way of thinking that will enable you to evaluate critically the information with which you are confronted. Given that our society is constantly becoming more complex and we are having to rely more and more on scientific evidence, our ability to evaluate the evidence intelligently becomes increasingly important.

METHODS OF ACQUIRING KNOWLEDGE

There are many procedures by which we obtain information about a given phenomenon or situation. We acquire a great deal of information from the events we experience as we go through life. Experts also provide us with much information.

Helmstadter (1970) has posited that there are at least six different approaches to acquiring knowledge, only one of which is the scientific method. In order to enable you to gain an appreciation of the rigor and accuracy that is achieved by the scientific method, we will begin by taking a look at the five unscientific approaches to acquiring knowledge and then look at the scientific method. You should be able to see that each successive approach represents a more acceptable means of acquiring knowledge.

Tenacity

Tenacity
A method of acquiring knowledge based on superstition or habit

The first approach can be labeled **tenacity,** defined in *Webster's Third New International Dictionary* as "the quality or state of holding fast." This approach to acquiring knowledge seems to boil down to the acquisition and persistence of superstitions, because superstitions represent beliefs that are reacted to as if they were fact. Habit, or what might be labeled the principle of longevity, also illustrates tenacity at work. Habit leads us to continue believing something we have always believed.

Here is an example of tenacity based on superstition. Jerry Glanville took over as head coach of the Houston Oilers in 1986. The Oilers had been one of the worst teams in professional football, but during the 1986 preseason games something happened. The Oilers won all four of their games, something they had not done for about a decade. During this preseason series, Jerry Glanville was wearing black pants, a black shirt, and white shoes. When the regular series was to begin, however, he planned to change from his white shoes to another color. His assistant coaches refused to allow this change because, they said, they were winning, and they did not want to change anything that might affect their continued winning.

The assumption seemed to be that the white shoes somehow enhanced the probability of the Oilers' winning football games. The Oilers proceeded to beat the Green Bay Packers in their first regular season game, which probably reinforced this superstitious assumption. But they lost the next five or six games, demonstrating the fallacy of using this approach to acquire information.

The principle of longevity can be seen in statements such as "You can't teach an old dog new tricks." In general, the more frequently we are exposed to such statements, the more we tend to believe them. Social psychologists have identified a similar process operating in attitude formation; they call it **mere exposure.** The more we are exposed to something or the more familiar it becomes, the more we like it. Politicians are very aware of this principle and discuss it in terms of name recognition. When running for office, a politician will plaster his or her name all over town, repeatedly exposing the public to it without ever mentioning campaign issues. This repeated exposure can engender in voters a more positive attitude toward the politician and a belief that he or she is the best candidate for the position.

Mere exposure
The development of a positive attitude toward something as a function of increased familiarity with it

Although tenacity is a method of acquiring knowledge, it has two problems that diminish its value. First, knowledge that is acquired through mere exposure may be inaccurate. Everyone has heard that old dogs can't learn new tricks, but in fact the elderly can and do learn. They may be more resistant, but they learn. Second, tenacity does not provide a mechanism for correcting erroneous superstitions and habits in the face of evidence to the contrary.

Intuition

Intuition
An approach to acquiring knowledge that is not based on reasoning or inferring

Intuition is the second approach to acquiring knowledge. *Webster's Third New International Dictionary* defines intuition as "the act or process of coming to direct knowledge or certainty without reasoning or inferring." Psychics such as Edgar Cayce seem to have derived their knowledge from intuition. The predictions and descriptions made by psychics are not based on any known reasoning or inferring process; therefore, such knowledge must be intuitive.

The problem with the intuitive approach is that it does not provide a mechanism for separating accurate from inaccurate knowledge. This does not mean that knowledge acquired from psychics is undesirable or inappropriate—only that it is not scientific. (In fact, it has been suggested that intuitive thought takes place in the right hemisphere of the brain and logical thinking occurs in the left hemisphere. Consequently, instruction in the scientific process may involve exercising and using only the left hemisphere.)

Authority

Authority
A basis for acceptance of information because it is acquired from a highly respected source

Authority as an approach to acquiring knowledge represents an acceptance of information or facts stated by another because that person is a highly respected source. For example, authority exists within the various religions. A religion typically has a sacred text, tribunal, person, or some combination of these that represents the facts, which are considered indisputable and final. This example is not meant to be critical of religions, but only to demonstrate that the authority approach to gaining knowledge differs from the scientific approach. Another example comes from the political-social arena. On July 4, 1936, the Central Committee of the Communist Party of the Soviet Union issued a "Decree Against Pedology" (Woodworth and Sheehan, 1964), which, among other things, outlawed the use of standardized tests in schools. Since no one had the right to question such a decree, the need to eliminate standardized tests had to be accepted as fact. The problem with the authority approach is that the information or facts stated by the authority may be inaccurate.

The authority approach should not be confused with our increasing dependence on experts for information. Experts do transmit scientific knowledge, and they usually base their opinions on scientific knowledge. The distinction between the authority approach and an appeal to an expert is that the authority approach dictates that we accept whatever is decreed, whereas the appeal to an expert leaves us free to accept or reject whatever the expert says.

Rationalism

Rationalism
The acquisition of knowledge through reasoning

A fourth approach to gaining knowledge is **rationalism.** This approach uses reasoning to arrive at knowledge and assumes that valid knowledge is acquired if the correct reasoning process is used. During the sixteenth century, rationalism was assumed to be the dominant mode by which one could arrive at truth. In fact, it was believed that knowledge derived from reason was just as valid as, and often superior to, knowledge gained from observation. The following anecdote represents an extreme example of the rationalistic approach to acquiring knowledge.

> In the year of our Lord 1432, there arose a grievous quarrel among the brethren over the number of teeth in the mouth of a horse. For thirteen days the disputation raged without ceasing. All the ancient books and chronicles were fetched out, and wonderful and ponderous erudition, such as was never before

heard of in this region, was made manifest. At the beginning of the fourteenth day, a youthful friar of goodly bearing asked his learned superiors for permission to add a word, and straightway, to the wonderment of the disputants, whose deep wisdom he sore vexed, he beseeched them to unbend in a manner coarse and unheard-of, and to look in the open mouth of a horse and find answer to their questionings. At this, their dignity being grievously hurt, they waxed exceedingly wroth; and joining in a mighty uproar, they flew upon him and smote his hip and thigh, and cast him out forthwith. For, said they, surely Satan hath tempted this bold neophyte to declare unholy and unheard-of ways of finding truth contrary to all the teachings of the fathers. After many days of grievous strife the dove of peace sat on the assembly, and they as one man, declaring the problem to be an everlasting mystery because of the grievous dearth of historical and theological evidence thereof, so ordered the same writ down. (Francis Bacon, quoted in Mees, 1934, p. 17)

This quote should clearly illustrate the danger of relying solely on rationalism for acquiring knowledge. Rationalism, or reasoning, does not necessarily reflect reality and frequently does not provide accurate information. For example, it is not unusual for two well-meaning and honest individuals to use rationalism to reach different conclusions. Undoubtedly both conclusions are not correct—possibly neither is correct.

This does not mean that science does not use reasoning or rationalism. In fact, reasoning is a vital element in the scientific process, but the two are not synonymous. In the scientific process, reasoning is used to arrive at hypotheses, which are then tested for validity using the scientific method.

Empiricism

Empiricism
The acquisition of knowledge through experience

The fifth and final unscientific approach to gaining knowledge is through **empiricism.** This approach says, "If I have experienced something, then it is valid and true." Therefore, any facts that concur with experience are accepted, and those that do not are rejected. This is exactly the approach that was used by the members of religious groups who stated that satanic messages were included on some records. These individuals had played the records backwards and had heard messages such as "Oh Satan, move in our voices." Since these individuals had actually listened to the records and heard the messages, this information seemed to be irrefutable.

Although this approach is very appealing and has much to recommend it, several dangers exist if it is used alone. Our perceptions are affected by a number of variables. Research has demonstrated that variables such as past experiences and our motivations at the time of perceiving can drastically alter what we see. Research has also revealed that our memory for events does not remain constant. Not only do we tend to forget things, but at times an actual distortion of memory may take place. Stratton's classic experiment on the inversion of the retinal image epitomizes the alteration that can take place in our experience of the world. Stratton (1897) designed a set of lenses, or glasses, that would turn the world upside down. Anyone who put on the glasses would see objects on the left when they were really on the right, and vice versa. Likewise, objects that were above the observer would appear to be below. Consequently, once a person put on the glasses, the world would appear totally opposite to reality. When Stratton first put these glasses on, his movements were very confused and uncoordinated because objects were not as they appeared. He had to remind himself that if he saw something on his right it was really on his left. He had to use a trial-and-error process whenever he reached for something. After about three days, however, something astonishing happened. His movements became more skilled, and his confusion began to disappear. By the eighth day, the world no longer appeared upside down. Rather, the world appeared normal, and objects that were actually above him now appeared to be above him. Similarly, objects on the right appeared on the right and objects on the left appeared on the left. Somehow, the brain had compensated for the distortion produced by the glasses. After eight days of wearing the glasses, Stratton took them off. Up to now, Stratton's brain had been compensating for the distortion produced by the glasses, a distortion that no longer existed once he took the glasses off. But his brain was still compensating, so the world again appeared reversed. It took about another eight days for the brain to compensate for the absence of the distortion previously produced by the glasses and once again to see the world right side up. This experiment dramatically illustrates that the things we see are not necessarily true.

Empiricism is a vital element in science, but in science empiricism refers to the collection of data through the use of the scientific method, not to the personal experience of an event.

SCIENCE

The best method for acquiring knowledge is the scientific method, because the information it yields is based as much as possible on reality. Through the scientific method investigators attempt to acquire information that is devoid of personal beliefs, perceptions, biases, values, attitudes, and emotions. This is accomplished by empirically testing ideas and beliefs according to a specific testing procedure that is open to public inspection. The knowledge attained is dependable because it is ultimately based on objectively observed evidence.

To understand the scientific method, we must first take a look at the definition of science. Most philosophers of science define it as a *process* or *method*—a method for generating a body of knowledge. **Science,** therefore, represents a *logic of inquiry,* or a specific method to be followed in solving problems and thus acquiring a body of knowledge. This method can be broken down into a series of five steps, which are presented here as a prelude to the remainder of the book, the primary focus of which is a detailed discussion of each of these steps.

Science
A method or logic of inquiry

Identifying the Problem and Forming a Hypothesis

The beginning point of any scientific inquiry involves identifying a problem, which is actually a simple process. All one has to do is look around, and numerous problems that need solutions come readily to mind. Child abuse, cancer, alcoholism, and crime are just a few of the more apparent problems. However, it is not enough just to identify a problem. Before a problem can be investigated, it must be refined and narrowed so that it is researchable. Thorne and Himelstein (1984), for example, were interested in the concern expressed by the various religious groups over the possibility that some records conveyed satanic messages. These investigators were also aware that the accused bands and recording companies totally denied the charges. Consequently, a practical problem existed. Thorne and Himelstein narrowed and refined this problem by asking whether suggestion could account for the apparent perception of satanic messages on the part of the various religious groups.

Once the problem has been stated in researchable terms, hypotheses are formulated that express the expected or predicted relationships between the variables. These hypotheses must in turn be stated in such a way that they are testable and capable of being refuted. Thorne and Himelstein hypothesized that suggestion might account for the apparent perception of satanic messages, particularly when the suggestion was presented by someone with prestige or authority.

Designing the Experiment

The stage of actually designing the experiment is crucial and demands a tremendous amount of preparation on the part of the experimenter to ensure that the hypotheses stated are those actually tested. Proper controls over extraneous variables have to be established, and the experimental variable as well as the response variable must be specified. These procedures are extremely important because this stage represents the outline to be followed in conducting the experiment. This scheme is constructed to overcome the difficulties that would otherwise distort the results and to help make sure the data are properly analyzed and interpreted.

In designing their study of the effect of suggestion on the perception of satanic messages, Thorne and Himelstein randomly assigned research participants to one of three groups. Members of each group were given a different suggestion about what they would hear. They then listened to three rock-and-roll records played backwards. After hearing the records, the subjects recorded what they had heard.

Conducting the Experiment

After the experiment has been designed, the researchers must make a number of very important decisions regarding the actual conduct of the experiment. Before any data are collected, they must decide what subjects are to be used, what instructions are necessary, and what equipment and materials are needed. Actually, this involves filling in the outline set forth in the design stage of the experiment. After these decisions have been made, the experimenters are then ready to collect the data, precisely following the prescribed procedure and accurately recording responses made by the subjects. For some studies, this involves little more than plugging in electronic equipment. Other studies are much more demanding, since the experimenters must interact with the subjects and record the subjects' responses. In many experiments, debriefing or postinterviews must be conducted with the subjects to determine their reaction to the experiment and to eliminate any undesirable influence the experiment may have created.

In the Thorne and Himelstein experiment, a decision was made to use introductory psychology students, probably because they were most accessible. The experimenters then obtained the necessary equipment—in this case, only a cassette recorder—and formulated the instructional sets to be administered to the subjects. These instructional sets included one in which the subjects were asked only to listen to the recordings and record any word or words that they heard, one in which the subjects were told that satanic messages had been recorded and that their task was to detect these messages, and one in which the subjects were merely told to listen to the cassette and record their reactions. The experiment was then conducted by following the procedure outlined in the design stage.

Testing the Hypothesis

After the data have been collected, the experimenters must analyze and interpret the data to determine whether the stated hypotheses have been supported. With the advent of the computer and statistical software packages, investigators have been spared the task of making the necessary computations. But even though the computer is a marvelous piece of machinery, it will do only what it is told to do. The investigators still must decide on the appropriate statistical analyses. After the statistical analyses have been conducted, the investigators must interpret the results and specify exactly what they mean.

Thorne and Himelstein, in analyzing the data from their study, first computed the percentage of individuals in each group who reported hearing words relating to Satan. These percentages, which appear in Figure 1.1, reveal that of the individuals who were requested only to listen to the recording and record their reactions, only 5 percent stated that they heard satanic messages. However, 18 percent of the individuals who were requested to record any word or words and 41 percent of the individuals who were told that satanic messages had been recorded stated that they heard satanic messages.

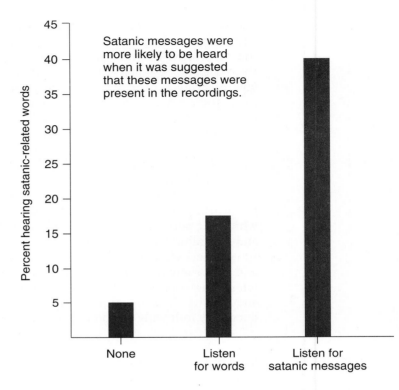

Satanic messages were
more likely to be heard
when it was suggested
that these messages were
present in the recordings.

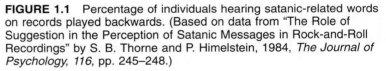

FIGURE 1.1 Percentage of individuals hearing satanic-related words
on records played backwards. (Based on data from "The Role of
Suggestion in the Perception of Satanic Messages in Rock-and-Roll
Recordings" by S. B. Thorne and P. Himelstein, 1984, *The Journal of
Psychology, 116*, pp. 245–248.)

Thorne and Himelstein then statistically analyzed these percentages and found that the individuals who were told that
satanic messages were placed on the recording actually heard
significantly more satanic messages. These researchers interpreted this to mean that suggestion can play a large role in
whether or not one hears satanic messages.

Communicating the Research Results

After the data have been analyzed, the scientists want to communicate the results to others. Communication most frequently takes place through the professional journals in a field.

Consequently, the scientists must write a research report stating how the research was conducted and what was found. Thorne and Himelstein, after completing their research, wrote a research report and were successful in having it published in *The Journal of Psychology*. Consequently, the results of the experiment are now available for anyone to read.

REALITY AND THE SCIENTIFIC RESEARCH PROCESS

The five steps just described represent the *logical* analysis of the scientific research process. They suggest that research flows in an orderly process, from the first step to the last. A logical analysis such as this one has didactic value but is not an accurate representation of actual practice.

Selltiz, Jahoda, Deutsch, and Cook (1959, p. 9) identified two ways in which the actual research process differs from the model presented. First, the sequence of activities suggested is almost never followed. Instead, the researcher's experience allows him or her to consider all five steps at once. Definition of the problem and statement of the hypotheses in part determine both the design and the conduct of the experiment. The design of the experiment in turn has a great deal to say about the method to be followed in collecting the data and the way the results are analyzed. The point is that there is a tremendous interaction between the various components. One activity cannot be carried out without affecting another. Second, the research process involves many activities not included in published studies or in the model presented. Frequently, equipment breaks down, subjects do not show up at the designated time and place, experimenters have to be trained, and cooperation has to be obtained not only from subjects (assuming human subjects are used) but also from administrative personnel. These are only a few of the additional activities that may be required as a study gets under way.

The veteran researcher moves back and forth through the research steps to accomplish the goal of obtaining scientific knowledge. On the other hand, the student, who is a novice at research, may perform better if the logical sequence of steps is followed. As a researcher gains more experience, the need for rigid adherence to these steps declines. Skinner's 1956 talk, "A

[handwritten margin notes: Interaction of steps / Unforeseen problems]

Case History in Scientific Method," depicts the informal nature of the scientific process as used by the seasoned researcher. This informality is depicted in his description of the "unformalized principles of scientific practice." These principles, derived from his own experience, are as follows:

1. When you run into something interesting, drop everything else and study it.
2. Some ways of doing research are easier than others. (The now famous *Skinner box* resulted from the fact that Skinner, in his early maze studies, saw no reason why he always had to retrieve the rat at the end of the runway or why the rat could not deliver its own reinforcement.)
3. Some people are lucky. (Skinner's construction of the cumulative recorder and observation of other aspects of the rate of responding were a result of not throwing away an apparently useless appendage of the discarded apparatus from which he built his first food magazine.)
4. An apparatus sometimes breaks down. (The jamming of the food magazine resulted in an extinction curve, which he later investigated.)
5. Never underestimate serendipity—the art of finding one thing while looking for another. (This principle led Skinner to investigate schedules of reinforcement.)

You should not assume that Skinner, because he elaborated on the informality of the scientific process, was lax or fuzzy in his approach to experimentation. Quite the contrary. He advocated more careful control and thought about experimentation and more rigorous definition of terms and quantification of research findings.

ADVANTAGE OF THE SCIENTIFIC METHOD

The scientific method has been contrasted with five other methods of acquiring knowledge, and it has been stated that the scientific method is the preferred method. If you closely scrutinize the six methods of acquiring knowledge, you should be able to see why. Science relies on data obtained through systematic empirical observation. The scientific method specifies that we obtain our observations through a particular systematic logic of inquiry. This logic is established in order to allow us to obtain **objective observations.** In other words, the scientific method enables us to make observations that are independent of opinion, bias, and prejudice. Such is not the case with the other five methods. Empiricism, though based on experience, does not provide a means for eliminating the possibility that our experience is biased in some manner. Similarly, rationalism, intuition, authority, and tenacity do not preclude the existence of prejudice. In order for us to uncover the basic laws of behavior, we must acquire data that are devoid of such bias, and the only way we can do so is through the scientific method.

> *Objective observation*
> Observation that is independent of opinion or bias

The scientific method is also better than the other methods of attaining knowledge because it enables us to establish the superiority of one belief over another. For example, two people may experience different results from taking a vitamin supplement. One person may suddenly become more energetic, whereas another may feel no difference at all. Based on their experiences, the two people would hold different beliefs regarding the vitamin supplement. Which belief is correct? Or are they both correct? Only through the systematic logic specified by the scientific method can we ultimately weed out fact from fiction.

RELATIONSHIP BETWEEN SCIENCE AND THE OTHER METHODS OF ACQUIRING KNOWLEDGE

The scientific method is considered to be superior to the other methods of acquiring knowledge because it allows us to make objective observations of phenomena and thereby allows us to establish the superiority of one belief over the other. However, the other methods of acquiring knowledge are used in the process of implementing the scientific method. Tenacity, for example, permeates the scientific approach when a scientist persists in believing in an idea, a hypothesis, or the results of research in the face of criticism from colleagues. Garcia's (1981) research on conditioned taste aversion, for example, was severely criticized initially as being incompetently conducted and was rejected for publication. However, Garcia believed in his ideas and the quality of his research. His tenacity eventually led to the publication of his research and the demonstration of conditioned taste aversion as a robust psychological phenomenon. If Garcia had not held on to his belief that he was correct and that his research was competently conducted, he would probably have dropped this line of investigation and the discovery and communication of the taste aversion phenomenon would have been delayed.

Tenacity also operates in science when a favorite theory is maintained in the face of conflicting evidence or in the absence of much scientific evidence. Freud's psychoanalytic theory, for example, is taught in virtually every introductory psychology course. However, there is little scientific evidence supporting such concepts as the superego or the id. Festinger's theory of cognitive dissonance (1957) has received a lot of scientific support. However, some of its propositions have not been supported. In spite of this lack of support, it is still taught and used.

The use of intuition in science is probably seen most readily in the process of forming hypotheses. Although most scientific hypotheses are derived from prior research, some hypotheses arise from "hunches." You may, for example, think that women are better at assessing the quality of a relationship than are men. This belief may have been derived from things others told you, your own experience, or any of a variety of other factors. Somehow you put together prior experience and any of a variety of other sources of information to arrive at this belief. If someone asked you why you held this belief, you probably could not identify the relevant factors—you might instead say it was based on your intuition. From a scientific perspective, this intuition could be molded into a hypothesis and tested. A scientific study could be designed to determine whether women are better at assessing the quality of a relationship than are men.

Authority plays a role in virtually all stages of the scientific process. In the beginning stages of the scientific process, when the problem is being identified and the hypothesis is being formed, a scientist may consult someone who is considered "the" authority in the area to assess the probability that the hypothesis is one that is testable and addresses an important research question. Virtually every area of endeavor has a leading proponent who is considered the authority or expert on a given topic. This is the person who has the most information on a given topic.

Although authority plays an integral part in the development of hypotheses, it is not without its problems. A person who is perceived as an authority can be incorrect. For example, Key (1980) has been a major proponent of the claim that advertisers resort to "subliminal advertising" to influence public buying and has been perceived by some as being the authority on this topic. He has stated, for instance, that implicitly sexual associations in advertisements enhance memorability. Fortunately for science, such claims by authority figures are subject to assessment through the use of the scientific method. The claims made by Key (1980) are readily testable and were tested by Vokey and Read (1985) in their study of subliminal messages. Vokey and Read scientifically demonstrated that Key's claims were unfounded.

Authority is also used in the design stage of a study. If you are unsure of how to design a study to test a specific variable, you may call someone who is considered an authority in the area and get his or her input. Similarly, if you have collected data on a given topic and you are not sure how to interpret the data or how they fit with the other data in the field, you may consult with someone who is considered an authority in the area and obtain input.

Rationalism is probably one of the more obvious approaches to gaining knowledge that is also used in science. Scientists make use of the reasoning process not only to derive hypotheses but also to identify the manner in which these hypotheses are to be tested. Vokey and Read (1985), for example, set out to test the hypothesis that meaning could be derived from sentences played backwards. In one of their studies they used the fact that the breaks and pauses that occur in speech are typically syllabic breaks rather than word breaks. Vokey and Read reasoned that if subjects could decipher the meaning of a sentence played backwards, their assessment of the number of words in the sentence should approximate the actual number of words rather than the number of syllables. However, if subjects were not able to decipher meaning from the sentences played backwards, they would base their word count on the number of syllables—which is what the researchers found. Consequently, these researchers concluded that meaning cannot be derived from sentences played backwards.

Empiricism is probably the most obvious approach that is used in science. Science is based on observation, and empiricism refers to the observation of a given phenomenon. The scientific studies investigating the satanic messages that supposedly existed when certain records were played backwards made use of the same empirical observations as did the unscientific approach. Greenwald (mentioned in Vokey & Read, 1985), for example, played records backwards and asked people to hear for themselves the satanic messages that appeared on the records. In doing so Greenwald relied on empiricism to convince the listeners that satanic messages were actually on the records. Scientific studies such as those conducted by Vokey and Read (1985) and Thorne and Himelstein (1984) make use of the same type of data. These studies also ask people to identify what they hear on records played backwards. The difference is the degree of objectivity that is imposed on the observation. Greenwald proposed to the listeners that the source of the messages was Satan or an evil-minded producer, thereby generating an expectation of the type of message that might exist on the records. In science researchers avoid setting up such an expectation unless the purpose of the study is to test such an expectation. Vokey and Read (1985), for example, used religious material as well as a meaningless passage and asked subjects to try to identify messages. These subjects were not, however, informed of the probable source of the message. Interestingly, Vokey and Read discovered that messages were identified in both meaningless and religious passages played backwards and subjects found that some of these messages had satanic suggestions.

CHARACTERISTICS OF THE SCIENTIFIC APPROACH

Science has been defined as a logic of inquiry, and it has been specified that this method of acquiring information is superior to other methods because it allows us to obtain knowledge that is free of bias and opinion. In order to produce such objective knowledge, the process must possess certain characteristics that, although necessary to science, are not limited to the realm of science. We will now look at the three most important characteristics of science: control, operational definition, and replication.

Control

Perhaps the single most important element in scientific methodology, *control* refers to eliminating the influence of any extraneous variable, that could affect observations. Control is important because it enables scientists to identify the causes of their observations. Experiments are conducted in an attempt to answer certain questions, such as why something happens, what causes some event, or under what conditions an event occurs. In order to provide unambiguous answers to such questions, experimenters must use control. Marx and Hillix (1973, p. 8) present an example of how control is necessary in answering a practical question.

> A farmer with both hounds and chickens might find that at least one of his four dogs is sucking eggs. If it were impractical to keep his dogs locked away from the chicken house permanently, he would want to find the culprit so that it could be sold to a friend who has no chickens, or to an enemy who does. The experiment could be run in just two nights by locking up one pair of hounds the first night and observing whether eggs were broken; if so, one additional dog would be locked up the second night, and the results observed. If none were broken, the two dogs originally released would be locked up with one of the others, and the results observed. Whatever the outcome, the guilty dog would be isolated. A careful farmer would, of course, check negative results by giving the guilty party a positive opportunity to demonstrate his presumed skill, and he would check positive results by making sure that only one dog was an egg sucker.[1]

In this example, the farmer, in the final analysis, controlled for the simultaneous influence of all dogs by releasing only one dog at a time in order to isolate the egg sucker. To answer questions in psychology, we also must eliminate the simultaneous influence of many variables in order to isolate the cause of an effect. Controlled inquiry is an absolutely essential proc-

---

[1]From *Systems and Theories in Psychology* by M. H. Marx and W. A. Hillix. Copyright © 1973 by McGraw-Hill, Inc. Used by permission of McGraw-Hill Co.

ess in science because without it the cause of an effect could not be isolated. The observed effect could be due to any one or a combination of the uncontrolled variables. The following historical example shows the necessity of control in arriving at causative relationships.

In the 1930s Norman Maier presented a paper at a meeting of the American Association for the Advancement of Science. In this paper (Maier, 1938) he illustrated a technique for producing abnormal behavior in rats by giving them a discrimination problem that had no solution. Shortly thereafter, other investigators examined the procedure used by Maier and became interested in one of its components. Maier had found that the rats, when confronted with the insoluble problem, normally refused to jump off the testing platform. To induce jumping behavior, experimenters directed a blast of hot air at an animal. Shortly thereafter, Morgan and Morgan (1939) duplicated Maier's results by simply exposing rats to the high-pitched tones of the hot air blast that Maier had used to make his rats leave the jumping stand. The significant point made in the Morgan and Morgan study was that conflict and elaborate discrimination training were not necessary to generate the abnormal behavior. This led to a controversy regarding the role of frustration in fixation that lasted for a number of years (Maier, 1949).

The important point for our purposes is that the potential effect of the auditory stimulus was not controlled. Therefore, one could not conclude from Maier's 1938 study that the insoluble discrimination problem produced the abnormal behavior, since the noise also could have been the culprit. Exercise of control over such variables is essential in science.

Operational Definition

The principle of operational definition was originally set forth by Bridgman (1927) and was incorporated into psychology shortly thereafter. **Operationism** means that terms must be defined by the steps or operations used to measure them. Such definition is necessary to eliminate confusion in communication. Consider the statement "Hunger causes one to perceive food-related objects selectively." One might ask, "What is meant by hunger?" Stating that hunger refers to being starved only adds to the confusion. However, stating that hunger refers to eight hours of food deprivation communicates a clear idea.

Operationism
The definition of terms by the operations used to attain or measure them

Now others can understand what you mean by hunger and can, if they so desire, generate the same condition. Setting down an operational definition forces one to identify the empirical references, and so ambiguity is minimized.

Consider a more difficult statement. Assume that you just had an encounter with a car salesperson whom you thought did an excellent job. How would you operationally define an excellent salesperson as opposed to a poor one? First, the empirical referents that correspond to this term must be identified. As Figure 1.2 reveals, these referents might consist of selling many cars, pointing out a car's good features, helping the customer to find financing, and complimenting the customer on an excellent choice. Once such behaviors have been identified, meaning can be communicated with minimal ambiguity. This does not mean, as Stevens (1939) has pointed out, that operationism is a panacea. It is merely a statement of one of the requirements of science that must eventually be used if science is to provide communicable knowledge.

One of the early criticisms of operationism was that its demands were too strict. If everything had to be defined operationally, one could never begin the investigation of a problem. Critics were concerned that it would be virtually impossible to formulate a problem concerning the functional relationships among events. Instead of stating a relationship between hunger and selective perception, one would have to talk about the relationship between number of hours of food deprivation and inaccurate description of ambiguous stimuli presented ta-

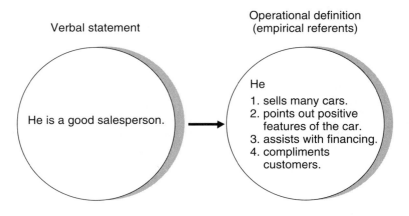

FIGURE 1.2 Example of an operational definition of a good car salesperson.

chistoscopically—that is, for very brief time intervals. Operationism, however, does not preclude verbal concepts and higher-level abstraction. Problems and questions are originally formulated in nonoperational terms—that is, scientists begin by using terms such as *hunger, anxiety,* and *frustration.* Once the problem is formulated, then the terms must be operationally defined. Verbal concepts are, therefore, admissible and useful in the pursuit of science.

Another criticism leveled at operationism was that each operational definition completely specified the meaning of the term. Any change in the set of operations would specify a new concept, which would lead to a multiplicity of concepts. Such a notion suggests that there is no overlap among the operations—that, for example, there is no relationship among three different operational measures (responses to a questionnaire, galvanic skin response [GSR] readings, and amount of urination and defecation by rats in an open-field situation) of a concept such as anxiety or that they are not concerned with the same thing. Stevens addresses this issue and notes that this "process of generalization proceeds on the basis of the notion of classes" (Stevens, 1939, p. 233). A verbal concept such as *anxiety* merely represents the name or symbol that has been given to that class of events. Each of the many operational definitions that can be given to a verbal concept merely represents a member of that class; consequently, the various operational definitions of a verbal concept are concerned with the same thing.

Replication

Replication
The reproduction of the results of a study

A third requirement of science is that the observations made must be replicable. As a characteristic of science, **replication** refers to the reproduction of the results obtained from a study. In other words, the data obtained in an experiment must be reliable—the same results must be found if the study is repeated. The need for science to have such a requirement is quite obvious, since the goal of science is to obtain knowledge about the world in a scientific manner. If observations are not repeatable, our descriptions and explanations are unreliable and therefore useless.

Reproducibility of observations can be investigated by making intergroup, intersubject, or intrasubject observations.

Intergroup observations involve attempting to duplicate the results on another group of subjects; intersubject observations involve assessing the reliability of observations on other individual subjects; and intrasubject observations involve attempting to duplicate the results with the same subject on different occasions.

Whenever we use one or more of these procedures to obtain evidence of the replicability of our results, we will obtain one of two possible outcomes. We will either replicate or fail to replicate the results of a previous study. If we replicate the results of a previous study, it gives us additional assurance that the results are reliable. Failure to replicate the results of a previous study, however, can be interpreted in several ways, because there are several possible reasons why it might occur. The first and most obvious possibility is that the results of the prior study were due entirely to chance, which means that the phenomenon that was previously identified did not really exist. If the phenomenon did not exist, it obviously cannot be reproduced in a replication study. The second reason that one may not be able to replicate the results of a study is more subtle. The replication experiment might have altered some seemingly nonsignificant element of the experiment, and this element in turn may have produced an altered response on the part of the subjects. For example, Gardner (1978) found that subjects' performance in the presence of environmental noise could be altered by either telling or not telling the subjects that they were free to discontinue participation in the study at any time. Most investigators would not consider such a minor alteration to have a significant impact on the subjects. However, psychological experiments can be affected by seemingly minor influences, and thus such apparently minor influences can be the cause of a failure to replicate. Consequently, whenever you conduct a replication experiment, you must remember that the experiment is just that—a replication. Therefore, you must conduct the experiment in exactly the same manner as it was previously performed. Even if you replicate every detail exactly, however, there is one element that in many instances cannot be replicated. That element is the experimenter—unless, of course, you are the one who conducted both the original and the replication experiment. More commonly, however, the replication is conducted by a different experimenter at a different location using different subjects. These differences must be taken into consideration when interpreting a failure to replicate.

Although replication is accepted as a characteristic of science, Campbell and Jackson (1979) have pointed out that an inconsistency exists between the acceptance of this characteristic and the researchers' behavioral commitment to replication research. Few researchers are conducting replication research, primarily because it is difficult to publish such studies. Also, it seems as though most researchers believe that well-designed and well-controlled studies can be replicated and, therefore, that replication research is not as important as original research. The one exception is in the field of parapsychology, where numerous replication studies are conducted because it is very difficult to attain replication. Lack of commitment to conducting replication studies does not, however, diminish the role of replication as one of the salient characteristics of science. Only through replication can we have any confidence that the results of our studies are valid and reliable.

OBJECTIVES OF SCIENCE

Ultimately, the objective of science is to understand the world in which we live. This goal pervades all scientific disciplines, but to say that the objective of science is understanding is rather nebulous. Ordinary people as well as scientists demand understanding. There is, however, a difference between the level of understanding referred to by the scientist and that referred to by the ordinary person. Understanding on the part of the nonscientist usually consists of the ability to provide some explanation, however crude, for the occurrence of a phenomenon. Most people, for example, do not totally understand the operation of the internal combustion engine. Some are satisfied with knowing that it requires turning a key in the ignition switch and simultaneously depressing the accelerator. Others are not satisfied until they acquire additional information. For the ordinary person, understanding—or knowing the reasons—ceases when curiosity rests.

Science is not satisfied with such a superficial criterion; it demands a detailed examination of a phenomenon. Only when a phenomenon is accurately described and explained—

and therefore predictable and in most cases capable of being controlled—will science say that it is understood. Consequently, scientific understanding requires four specific objectives: description, explanation, prediction, and control.

Description

Description
The portrayal of a
situation or phenomenon

The first objective of science, the process of **description,** requires one to portray the phenomenon accurately, to identify the variables that exist, and then to determine the degree to which they exist. For example, Piaget's theory of child development arose from his detailed observations and descriptions of his own child. Any new area of study usually begins with the descriptive process because it identifies the variables that exist. Only after we have some knowledge of which variables exist can we begin to explain why they exist. For example, we would not be able to explain the existence of separation anxiety (an infant's crying and visual searching behavior when the mother departs) if we had not first identified this behavior. Scientific knowledge typically begins with description.

Explanation

Explanation
Determination of the
cause of a given
phenomenon

The second objective is the **explanation** of the phenomenon, and this requires knowledge of why the phenomenon exists or what causes it. Therefore we must be able to identify the antecedent conditions that result in the occurrence of the phenomenon. Assume that the behavior connoting separation anxiety existed when an infant was handled by few adults other than its parents and that it did not exist when the infant was handled by and left with many adults other than parents. We would conclude that one of the antecedent conditions of this behavior was frequency of handling by adults other than parents. Note that frequency was only *one* of the antecedents. Scientists are cautious and conservative individuals, recognizing that most phenomena are multidetermined and that new evidence may necessitate replacing an old explanation with a better one. As the research process proceeds, we acquire more and more knowledge concerning the causes of phenomena. With this increasing knowledge comes the ability to predict and possibly control what happens.

Prediction

Prediction, the third objective of science, refers to the ability to anticipate an event prior to its actual occurrence. We can, for example, predict very accurately when an eclipse will occur. Making this kind of accurate prediction requires knowledge of the antecedent conditions that produce such a phenomenon. It requires knowledge of the movement of the moon and the earth and of the fact that the earth, moon, and sun must be in a particular relationship for an eclipse to occur. If we knew the combination of variables that resulted in academic success, we could then predict accurately who would succeed academically. To the extent that we cannot accurately predict a phenomenon, we have a gap in our understanding of it.

Control

The fourth objective of science is **control,** which refers to the manipulation of the *conditions that determine a phenomenon.* Control, in this sense, means knowledge of the causes or antecedent conditions of a phenomenon. When the antecedent conditions are known, they can be manipulated so as to produce the desired phenomenon. Psychologists, therefore, indirectly influence behavior by directly controlling the variables that, in turn, influence behavior.

It has often been said that the psychologist is interested in the control of behavior. Books such as *Walden Two* (Skinner, 1948) and *Beyond Freedom and Dignity* (Skinner, 1971) have promoted this kind of statement because readers of these books note that Skinner was explaining or laying out a scheme for making people behave in a certain way. Thus they conclude that psychologists destroy people's free will and have the ability to control the behavior of others. The point that most nonscientists miss is that Skinner's concern was to identify the antecedent variables that generate the behavior in question. Skinner believed that behavior is determined and that the psychologist's task is to isolate the antecedent conditions.

Once psychologists understand the conditions that produce a behavior, the behavior can be controlled by either allowing or not allowing the conditions to exist. Consider the hypothesis that frustration leads to aggression. If we knew that this hypothesis were completely correct, we could control aggressive behavior by allowing or not allowing a person to be frustrated. Control, then, refers to the manipulation of conditions that produce a phenomenon, not of the phenomenon itself.

At this point, it seems appropriate to provide some additional insight into the concept of control. So far, control has been discussed in two slightly different ways. In the discussion of the characteristics of the scientific approach, control was referred to in terms of holding constant or eliminating the influence of extraneous variables in an experiment. In the present discussion, control refers to the antecedent conditions determining a behavior. Boring (1954) noted that the word *control* has three meanings. First, control refers to a check or verification in terms of a comparison. Second, it refers to a restraint—keeping conditions constant or eliminating the influence of extraneous conditions from the experiment. Third, control refers to guidance or direction in the sense of producing an exact change or a specific behavior. The second and third meanings identified by Boring are those used in this book so far. Since all of these meanings will be used at various times, it would be to your advantage to memorize them.

BASIC ASSUMPTIONS UNDERLYING SCIENCE

In order for scientists to have confidence in the capacity of scientific inquiry to achieve a solution to questions and problems, they must accept one basic axiom about the nature of the objects, events, and things with which they work. Scientists must believe that there is uniformity in nature; otherwise, there can be no science. Skinner (1953, p. 13), for example, stated that science is "a search for order, for uniformities, for lawful relations among the events in nature." If there were no uniformity in nature, there could be no understanding, explanation, or knowledge about nature. Without uniformity, we could not develop theories, laws, or facts. Implicit in the assumption of uniformity is the notion of

Determinism
The belief that behavior is caused by specific events

determinism—the belief that there are causes, or determinants, of behavior. In our efforts to uncover the uniform laws of behavior, we attempt to identify the variables that are linked together. We construct experiments that attempt to identify the effects produced by given events; in this way, we try to establish the determinants of events. Once we have determined the events that produce a given behavior or set of behaviors, we have uncovered the uniformity of nature.

Although a belief in the uniformity of nature is the basic assumption underlying science, several axioms must exist to enable the scientist to uncover the deterministic relationships inherent in this uniformity. These axioms refer to the reality, rationality, regularity, and discoverability of events in nature.

Reality in Nature

Reality in nature
The assumption that the things we see, hear, feel, and taste are real and have substance

One of the assumptions that we all make as we go through our daily lives is that the things we see, hear, feel, and taste are real and have substance. We assume that other people, objects, or events such as marriage or divorce are not just creations of our imagination. This **reality in nature** represents an assumption underlying science. This does not mean that creations of our imagination don't exist. Remember that Stratton (1897) devised a pair of glasses that turned the world upside down. After wearing them for eight days Stratton found that the world had righted itself somehow and no longer appeared upside down. This suggests that regardless of how the world really appears, we want to see it a particular way and our mind will eventually create that image. Science, however, represents an investigation of the uniformity of nature and not our own perceptions of it. This means that science makes the assumption that there is an underlying reality and attempts to uncover this reality.

Rationality

Rationality
The assumption that there is a rational basis for the events that occur in nature and they can be understood through the use of logical thinking

For the scientific study of behavior to be successful the events that occur in nature must be rational. This means that there must be some rational basis for their existence and that the events must be understood through use of logical thinking. This requirement of **rationality** logically follows from the basic assumption of the uniformity of nature. For uniformity and determinism to exist in nature there must be some logical and rational basis for the existence of all events. For example, a depressed person's ruminations and negative outlook on life may seem unreasonable and illogical to friends and relatives.

However, the scientist studying depression focuses on these negative cognitions (in addition to other events), attempts to understand them through logic and rationality, and derives experiments to verify the logical and rational basis of depression.

Regularity

Regularity
The assumption that events in nature follow the same laws and occur the same way at all times and places

Regularity refers to the fact that events in nature follow the same laws and occur the same way at all times and places. Depressed people, for example, consistently view the world, themselves, and the future very pessimistically. Without this regularity it would be impossible to identify any underlying causes of behavior because a lack of regularity would suggest that there is little uniformity in nature. If there is uniformity in nature, there must be regularity. If regularity exists, then we can conduct studies to identify these regularities and in this way uncover the uniformity of nature.

Discoverability

Discoverability
The assumption that it is possible to discover the uniformities that exist in nature

Scientists believe not only that there is uniformity in nature but also that there is **discoverability**—that is, it is possible to discover this uniformity. This does not mean that the task of discovering the uniformities that exist in nature will be simple. Nature is very reluctant to reveal its secrets. It takes little imagination to provide examples of the difficulty in discovering the relationships that nature holds. Scientists have been working on discovering the cause and cure for cancer for decades. Although significant progress has been made, we still do not know the exact cause of all forms of cancer or the contributors to the development of cancer. Similarly, a complete cure for cancer still does not exist. An intensive effort is also taking place within the scientific community to identify the cause and cure for AIDS. However, scientists have yet to uncover nature's secrets in this arena.

The intensive effort that has existed to uncover the cause of diseases such as cancer and AIDS or, within the field of psychology, disorders such as schizophrenia, depression, or numerous other behavioral aberrations reveals one of the basic processes of science. Science is similar to putting a puzzle together. When putting a puzzle together you have all the pieces in front of you and then try to put them together to get the overall picture. Science includes the difficult task of first discovering the pieces of the puzzle. Each study conducted on a given problem has the potential of uncovering a piece of the puzzle. Only when each of these pieces has been discovered is it possible for someone to put them together to enable us to see the total picture. Consequently, discoverability in science incorporates two components. The first is discovery of the pieces of the puzzle, and the second is putting the pieces together or discover the nature of the total picture.

METHOD VERSUS TECHNIQUE

Technique
The specific manner in which the scientific method is implemented

Up to this point, I have focused attention on the scientific method as a logic of inquiry. This logic of inquiry must be distinguished from **technique,** which refers to the specific manner in which the scientific method is implemented. Many people confuse methodology with technique. They think that psychology uses a scientific method that is somehow different from the method used by other sciences. Actually, all sciences are characterized by a common logic of inquiry in their quest for knowledge. Psychology merely uses different techniques in applying the basic scientific method. The techniques used in applying the scientific method vary with the nature of the subject matter, the nature of the specific problem, and the stage of inquiry.

Variation in technique as a function of subject matter can be illustrated by contrasting various fields of inquiry as well as by contrasting the techniques used in the various areas of psychology. Consider, for example, the different observational techniques used by the astronomer, the biologist, and the psychologist doing research on small groups. The astronomer uses a telescope and, more recently, interplanetary probes; the biologist uses the microscope; and the small-group researcher may use a one-way mirror to unobtrusively observe subjects' interactions. The scientists in these various fields are all using the same scientific method, the key aspect being controlled inquiry, but the techniques they use in implementing this method differ. The techniques used in different fields do not necessarily vary any more than the techniques used in the different areas of psychology. The physiological psychologist might use stimulation electrodes in investigating cortical processes, whereas the learning psychologist uses reinforcement techniques.

Variation in technique as a function of the nature of the problem can be illustrated by contrasting two studies, one conducted by Hayes and Cone (1981) and the other conducted by Yoburn, Cohen, and Campagnoni (1981). Table 1.1 shows that Hayes and Cone (1981) investigated the influence of feedback on electrical consumption, whereas Yoburn et al. (1981) investigated the influence of interrupted access to food on attack behavior. Hayes and Cone used a feedback technique that was administered to residential consumers; Yoburn et al. used a reinforcement technique as well as a specific technique

TABLE 1.1 Contrasting Techniques Used by Two Studies Employing the Scientific Method

Scientific Method, or Logic of Inquiry	Hayes and Cone (1981) Study	Yoburn, Cohen, and Campagnoni (1981) Study
Identify problem and form hypothesis	Is residential consumption of electricity decreased through monthly feedback of changes in kilowatt-hours and cost over the same month in previous years?	Will interrupted access to food induce pigeons to attack?
Design experiment	Residential consumers were randomly assigned to feedback and no-feedback groups.	Pigeons were given preliminary training in obtaining food under a specific reinforcement schedule and then tested to see if they would attack an image projected on a screen.
Conduct experiment	Consumers either received or did not receive feedback and information on change in cost and kilowatt-hours used.	Pigeons were given one type of reinforcement; an image was presented in the twenty-first session. The rate of attack on the image was recorded.
Test hypothesis	Changes in electrical consumption for feedback and no-feedback groups were compared, and feedback groups revealed reduction in electrical consumption.	Changes in rate of attack when image was present and absent were compared. Attack on the image increased following an intermittent food schedule.
Write research report	Research was published in the *Journal of Applied Behavior Analysis.*	Research was published in the *Journal of Experimental Analysis of Behavior.*

for projecting an image on a screen. Different techniques had to be used because the two studies investigated different problems. But as Table 1.1 indicates, both studies followed the logic of inquiry inherent in the scientific method. What is perhaps not so apparent is that both studies also employed other characteristics of science, such as control. For example, Hayes and Cone controlled for differences that may exist in residential consumers' use of electricity by randomly assigning the consumers to either a feedback or a no-feedback group.

The point that one has reached in the research process also dictates the technique used. If one is at the data collection stage, then questionnaires, verbal responses, bar pressing, or any of numerous other techniques could be used. The technique used is determined by the demands of the study. Likewise, if one is at the data analysis stage, the appropriate statistical technique has to be selected.

The wide variances that exist among different fields (and even among different areas within a field) are, therefore, primarily a function of the diverse techniques used in applying the same scientific method.

THE ROLE OF THEORY IN SCIENCE

Use of the scientific method in making objective observations is absolutely essential to the accumulation of a highly reliable set of facts. Accumulating such a body of facts, however, is not sufficient to answer many of the riddles of human nature. For example, scientific research has revealed that individuals who are paid less than someone else for doing the same job get angry and upset. Similarly, couples break up if one of them always demands his or her way. Employees who are paid more than they think is appropriate for a job will tend to work harder and perhaps do more than is absolutely necessary to get the job done. Once facts such as these have been accumulated through the use of the scientific method, they must somehow be integrated and summarized to provide a more adequate explanation of behavior. This is one of the roles theory plays in the scientific enterprise. Equity theory, for example, summarized and integrated a large portion of the data related to the notion of fairness and justice to provide a more adequate explanation of interpersonal interactions. Theories are not created just to summarize and integrate existing data, however. A good theory must also suggest new hypotheses that are capable of being tested empirically. Consequently, a theory must have the capacity to guide research as well as to summarize the results of previous research. This means that there is a constant interaction between theory and empirical observation, as illustrated in Figure 1.3. From this figure you can see that theory is originally based on empirical observations obtained from using the scientific method; once the theory has been generated, it must direct future research. The outcome of the future research then feeds back and determines the usefulness of the theory. If the predictions of the theory are confirmed by the scientific method, evidence exists that the theory is useful in accounting for a given phenomenon. If the predictions are refuted by the scientific method, the theory has been demonstrated to be inaccurate and must either be revised so as to account for the experimental data or be thrown out.

From this discussion you can see that theory generation is a valuable part of the scientific enterprise. But generating a theory is not synonymous with science; rather, it is a valued activity within the field of science because it integrates and summarizes scientific facts and thus allows us to arrive at a

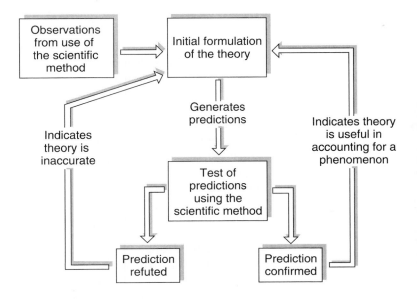

FIGURE 1.3 Illustration of the relationship between theory and research.

more adequate explanation of a given phenomenon. Another reason theory is valued within the scientific enterprise is because it suggests research studies that might otherwise be overlooked.

THE ROLE OF THE SCIENTIST IN SCIENCE

One very significant component in the scientific approach is the scientist—the individual who employs the scientific approach and who ultimately makes science possible. Is the scientist just any person, or does he or she possess special characteristics? As might be expected, certain characteristics are necessary. A scientist is any individual who rigorously employs the scientific method in the pursuit of knowledge. Nature's secrets are revealed reluctantly, though. The scientist must actively search and probe nature to uncover orderly relationships, so he or she must be curious, patient, objective, and tolerant of change.

Curiosity

The scientist's goal is the pursuit of knowledge, the uncovering of the laws of nature. Scientists attempt to answer questions: What? When? Why? How? Under what conditions? With what restriction? These questions are the starting point of scientific investigation, and they continue to be asked throughout the study. To ask these questions, the scientist must be inquisitive, must exhibit curiosity, and must never think that the ultimate solution has been reached. If these questions ever cease, then so does the scientific process.

Garcia, for example, was involved in radiobiological research for about a decade. This research convinced him that several of the classical principles of conditioning had some specific limitations. One of these limitations, he felt, was that unconditioned stimuli had a selective effect on what was learned. He then proceeded to demonstrate this limitation (Garcia, 1981). Such an example illustrates that the scientist must maintain an open mind, never becoming rigid in orientation or in method of experimentation. Such rigidity could cause him or her to become blinded and incapable of capitalizing on, or even seeing, unusual events. This relates to Skinner's "fifth unformalized principle of scientific practice . . . serendipity—the art of finding one thing while looking for another" (1972, p. 112). If they weren't inquisitive and open to new and different phenomena, scientists would never make the accidental discoveries that periodically occur.

Patience

The reluctance of nature to reveal secrets is seen in the slow progress made in scientific inquiry. When individuals read or hear of significant advances in some field of scientific inquiry, they marvel at the scientists' ability and think of the excitement and pleasure that must have surrounded the discovery. Although moments of excitement and pleasure do occur, most people do not realize the many months or years of tedious, painstaking work that go into such an achievement. Many failures usually precede a success, so the scientist must be extremely patient and must be satisfied with rewards that are few and far between. Note the many years of effort that have gone into cancer research; many advances have been made, but a cure is still not available.

Objectivity

One of the prerequisites of scientific inquiry is objectivity. Ideally, the scientist's personal wishes and attitudes should not affect his or her observations. Realistically, however, perfect objectivity cannot be attained, as scientists are only human. No matter how severe the attempt to eliminate bias, the scientist still has certain desires that may influence the research being conducted. Rosenthal (1966) has repeatedly demonstrated that experimenter expectancies can influence the results of experiments.

Change

Scientific investigation necessitates change. The scientist is always devising new methods and new techniques for investigating phenomena. This process typically results in change.

When a particular approach to a problem fails, a new approach must be devised, which also necessitates change. When change no longer exists, the scientific process ceases because we then continue to rely on and accept old facts and old ways of doing things. If we are no longer asking questions, we are assuming that we have solved all the problems. Change does not necessitate abandoning all past facts and methods; it merely means the scientist must be critical of the past and constantly alert to facts or techniques that may represent an improvement.

Despite the need for the scientist to accept change as part of the scientific process, it seems that new ideas are resisted if they do not somehow fit in with current knowledge. Polanyi (1963), for example, relayed his own experience of the reaction to his theory of the absorption (adhesion) of gases on solids following its publication in 1914. He was chastised by Albert Einstein for showing a "total disregard" for what was then known about the structure of matter. Polanyi, however, was later proved to be correct. Garcia (1981) experienced a similar reaction to his research on taste aversion. Some of his manuscripts were returned with caustic and gratuitous personal insults. Garcia noted that in one instance a reviewer stated that his "manuscript would not have been acceptable as a learning paper in [the reviewer's] learning class." However, as Garcia also pointed out, these comments typically surfaced when he and his colleagues sent in their first paper on a given research topic; the next research manuscript they submitted on the same topic, designed to answer the same research question, was typically accepted. Such a phenomenon seems to illustrate that greater familiarity with a point of view increases the probability of its acceptance, as long as the research is conducted competently.

GENDER AND THE MAKING OF A SCIENTIST

A scientist is any individual who rigorously employs the scientific method in the pursuit of knowledge. Ideally, gender should not be an issue with regard to the development of a scientist. However, data gathered over the past several decades indicate that this ideal situation does not exist. For example, females are less likely to become scientists than are males. In 1979 about 21 percent of the Ph.D.s in the fields of science and engineering were granted to women. By 1989 this figure had risen to 28 percent. However, only 5 percent of the tenured jobs in universities and colleges were held by women in 1979, and this figure had risen to only 7 percent by 1989 (Gibbons, 1992). Clearly the percentage of women being granted tenure was inconsistent with the rise in the percentage of women receiving Ph.D.s. Because most research is conducted at colleges and universities, this inconsistency means that the vast proportion of scientists are males. This proportion has changed little, although the percentage of women receiving training in the scientific method has increased.

Idealistically and theoretically, the gender of the scientist should have no bearing on the knowledge base that is acquired from use of the scientific method, because objectivity is one of the prerequisites of scientific inquiry. In spite of this prerequisite training, it is impossible for any scientist to be totally objective. Attitudes, values, and opinions influence observations and interpretations of these observations. In recent years, some female scientists have pointed out that the domination of the field of science by males has created a biased knowledge base. Matlin (1993) and Unger and Crawford (1992) as well as others have pointed out that a gender bias can enter the research process at virtually any stage, from the selection of the research question and formulation of the research design to the communication of the findings. These possibilities for bias will be discussed throughout the remainder of this textbook.

SUMMARY

There are at least six different approaches to acquiring knowledge. Five of these approaches—tenacity, intuition, authority, rationalism, and empiricism—are considered to be unscientific. Tenacity, or the state of holding fast, represents knowledge acquired through superstition. Intuition refers to knowledge acquired in the absence of any reasoning or inferring. Authority represents knowledge acquired from a highly respected source of information. Rationalism refers to the acquisition of knowledge through correct reasoning. Empiricism represents the acquisition of knowledge from experience.

The sixth and best approach to acquiring knowledge is science, which is a logic of inquiry requiring that a specific method be followed. The five different activities that make up the scientific method are (1) identifying the problem and formulating the hypotheses, (2) designing the experiment, (3) conducting the experiment, (4) testing the hypotheses, and (5) writing the research report. The advantage of the scientific method is that it enables us to make objective observations and establish the superiority of one belief over another.

The scientific method as a logic of inquiry has certain rules or characteristics. *Control* is the most important characteristic because it enables the scientist to identify causation; without control, it would be impossible to identify the cause of a given effect. A second characteristic of the scientific method is *operational definition,* which refers to the fact that terms must be defined by the steps or operations used to measure them. Defining terms operationally is necessary to eliminate confusion in meaning and communication. The third characteristic of the scientific method is *replication.* The scientific observations that are made must be able to be repeated. If these characteristics are not satisfied, the results of an investigation are useless because they are not reliable.

Science has certain objectives that it strives to achieve in attempting to reach the ultimate goal of providing us with an understanding of the world in which we live. The first objective is *description,* or portraying an accurate picture of the phenomenon under study. The second objective is *explanation,* or determining why a phenomenon exists or what causes it. The third objective is *prediction,* or the anticipation of an event prior to its occurrence. The fourth and last objective is *control,* in the sense of being able to manipulate the antecedent conditions that determine the occurrence of a given phenomenon. In order to pursue these goals, the scientist must believe that there is uniformity in nature.

Science is a logic of inquiry—a method for generating a body of knowledge. As such, science must be distinguished from *technique,* which represents the specific manner in which one implements the scientific method. The various fields that use the scientific method for generating data employ a wide variety of techniques. Although these techniques differ across fields and across areas within given fields of study, the logic of inquiry used is identical.

Once knowledge has been gained through use of the scientific method, it must somehow be integrated and summarized. This is the role that theory plays in the scientific process. Theory should not only assume the role of integrating and summarizing scientific facts, but also suggest new ideas and additional research studies. Consequently, there is always a constant interaction between theory and the scientific process.

In attempting to gain knowledge through use of the scientific method, the scientist must implement this methodology. Any individual who rigorously employs the scientific method is a scientist. Nature is reluctant to reveal its secrets, however, so the successful scientist must be curious enough to ask questions and patient enough to gain the answers. The scientist must also be objective so as not to bias the data and must accept change in the form of new techniques and facts.

The gender of the scientist should have no bearing on the scientific data accumulated. However, the attitudes, opinions, and values of the scientist can influence the research process at any stage from formulation of the research question to the interpretation of the data. To avoid the development of a data base that precludes specific gender and cultural viewpoints, a more diverse group of investigators must engage in science.

STUDY QUESTIONS

1. Identify the different methods by which we acquire knowledge and the unique characteristic(s) of each of these methods.
2. Identify the steps that are followed in gaining knowledge through the use of the scientific method.
3. What advantage does the scientific method have over the other methods of acquiring knowledge?
4. Identify the important characteristics of the scientific method, and explain why each of these characteristics is important.
5. Identify the four objectives of science, and explain what each is attempting to accomplish.
6. What is the basic assumption underlying all science, and why is it important that this assumption is valid?
7. Why do people tend to confuse technique with method, and why is it important to keep these two concepts separate?

8. What role does theory play in the scientific enterprise?
9. What are the important characteristics a scientist must possess, and why are these characteristics important?
10. Why do scientific theories constantly change, whereas other theories (such as religious theories of creation) remain constant? What does this imply about the scientific method?

KEY TERMS AND CONCEPTS

Tenacity   Description
Mere exposure   Explanation
Intuition   Prediction
Authority   Control
Rationalism   Determinism
Empiricism   Reality in nature
Science   Rationality
Objective observation   Regularity
Operationism   Discoverability
Replication   Technique

CHAPTER 2

Descriptive Research Approaches

LEARNING OBJECTIVES

1. To understand the basic characteristics of the descriptive research approach.
2. To become familiar with the different types of descriptive research approaches.
3. To become aware of the differences among the various descriptive research approaches.
4. To learn when the different descriptive research approaches should be used.

Tuesday, January 28, 1986 was to be a momentous occasion for the United States and the U.S. space program. Christa McAuliffe, a high school social studies teacher from New Hampshire, was to accompany the crew of the Challenger *space shuttle on their flight. Amid all the fanfare surrounding the selection of Christa McAuliffe and the events leading up to the launch, no one even considered the possibility that doom awaited the space shuttle crew. After all, the National Aeronautics and Space Administration (NASA) had just completed twenty-four successful space shuttle launches.*

At T-minus-three-hours, the Challenger *was in a holding period. During this time the countdown proceeded while the astronauts boarded the orbiter and the spacecraft received its final inspection for ice and other potentially damaging debris that could be blasted up at it during ignition. The exterior temperature of the external tank was also checked for unusual cold spots that could indicate flawed insulation, because a temperature of 29 degrees Fahrenheit was predicted for launch time. All signs indicated no abnormalities. After two subsequent reviews also suggested that the* Challenger *was ready for launch, at T-minus-nine-minutes the decision was made to proceed with the launch. At T-minus-zero-minutes the orbiter's main engines ignited with an ear-splitting roar. The shuttle rocked in a tight arc as its engines strained against the launch pad. Then, as the shuttle steadied, the two solid rocket boosters burst into life and the* Challenger *broke free of the launch pad and thundered into the sky. Another perfect launch.*

Thirty-five seconds into the flight, the Challenger's *solid fuel rockets were pumping out maximum power, causing the shuttle, like all its predecessors, to undergo severe aerodynamic stress. The shuttle's engines were, as planned, automatically throttled back to 65 percent thrust to prevent the shuttle from tearing itself apart. Seventeen seconds later the throttles were again wide open. All systems looked normal, and mission control in Houston radioed: "Challenger, go at throttle up." Pilot Michael J. Smith replied, "Roger, go at throttle up."*

Moments later the orbiter was engulfed in a fireball that raced the length of the spacecraft, followed by a titanic explosion that consumed the Challenger *and its crew.*

Within about five minutes the news reached the White House. White House communications director Patrick

Buchanan rushed into the Oval Office and announced: "Sir, the shuttle's blown up." President Reagan looked stricken. "Isn't that the one with the teacher on it?" he asked. Back in Houston, a wave of panic swept through NASA. The engineers began sifting through the tons of data sent back from Challenger's *computers. Film taken by news agencies was impounded to provide a photographic record of the explosion, and Coast Guard and NASA teams began to search the Atlantic Ocean for debris.*

Almost immediately, speculation began. One of the solid rocket boosters might have somehow burned through its casing and operated like a blowtorch, igniting the external fuel tank. Or perhaps the explosives in Challenger's *self-destruct system were set off by a stray radio signal. Or maybe it was sabotage. These were only a few of the possibilities that could have led to the explosion and the death of the astronauts.*

With so many possibilities, how can one identify the most likely cause? In a situation like this, the only option is to use the descriptive research approach. *Investigators were restricted to the data that had already been collected or the data relayed back from the* Challenger's *computers and the photographs of the explosion. These are the data that the investigating team pored over for weeks before they constructed a description of the events preceding the launch that seemed to be most directly related to the explosion.*

This chapter is designed to acquaint you with the basic characteristics of the descriptive research approach—the approach used by investigators of the Challenger *explosion.*

INTRODUCTION

Descriptive research approach
A technique that provides a description or picture of a particular situation, event, or set of events

The primary characteristic of the **descriptive research approach** is that it represents an attempt to provide an accurate description or picture of a particular situation or phenomenon. This approach does not try to ferret out cause-and-effect relationships. Instead, it attempts to identify variables that exist in a given situation and, at times, to describe the relationship that exists between these variables. Therefore, the descriptive approach is widely used and is of great importance. We see the outcome of the descriptive approach whenever the results of Gallup polls or other surveys are reported. Helmstadter (1970, p. 65) has even gone so far as to say that the "descriptive approaches are the most widely used . . . research methods."

When initially investigating a new area, scientists use the descriptive method to identify existing factors and relationships among them. Such knowledge is used to formulate hypotheses to be subjected to experimental investigation. Also, the descriptive method is frequently used to describe the status of a situation once a solution, suggested by experimental analyses, has been put into effect. Here the descriptive method can provide input regarding the effectiveness of the proposed solution, as well as hypotheses about how a more effective solution could be reached. Thus the descriptive method is useful in both the initial and the final stage of investigation into a given area.

SECONDARY RECORDS

Secondary records
Any recorded observation, measure, or physical trace made by others

Secondary records consist of the records of observations, measures, or some type of physical trace made by others. Within our society, such records include birth records, census data, graffiti, and even the photographs and computer data obtained from the launch of the *Challenger.* Virtually any observation or measure that has been recorded qualifies as a secondary record. Once the observation or measure has been recorded, it can be used later on to describe a given situation. For example, census data are used to describe changes in population growth or divorce rates.

Secondary records can also be used to describe the relationship between variables. Following the explosion of the *Challenger,* for example, a broad inquiry was immediately opened to try to identify the cause of the disaster. Note, however, that the only data available to the investigating team were the secondary records—the computer data and photographs of the launch. The investigating team pored over these data in an attempt to identify the variables that had led to the explosion. This investigation resulted in the description of a variety of events, any of which could have resulted in the explosion. Secondary records revealed that an unusual plume occurred near the right solid rocket booster during the *Challenger's* last seconds, and that one of the boosters had lost power just ten seconds before the explosion. This led to speculation that the right solid rocket booster might have burned through its casing, becoming a blowtorch that burned through the thin wall of the external tank. Such a scenario gained in credibility because the photographs of the launch clearly depicted the external tank as the source of the huge fireball. But this description of the loss of power and the external tank's being engulfed in flames did not verify that the solid rocket booster had indeed burned through its casing to create the blowtorch effect. This was only speculation, although the investigating team used data from other secondary records to describe a situation that would suggest that this was the case.

Secondary records also revealed that NASA had, several years earlier, shaved a fraction of an inch off the steel casings of the solid rocket booster to save weight. This would have weakened the casing and made it more likely to burn through at a faulty weld or other weak spot. Again, however, this was only speculation based on a description of the events that had occurred. As the investigating team continued to study the available secondary records, they described a series of interactions between the engineers at Morton Thiokol, the company that had built the solid booster rockets, and NASA executives that seemed to represent the most likely set of events leading up to the explosion of the *Challenger*. Because the temperature at launch time was forecast to be only 29 degrees, the Morton Thiokol engineers were extremely concerned that the booster rocket seals might fail. In fact, they officially recommended that the launch be delayed until warmer weather. NASA executives disagreed, however, and actually coerced the engineers to issue a recommendation to launch. This description of the engineers' concern over the performance of the seals in the solid booster rocket, along with photographs revealing that the initial flame shot out of the booster at the point of one of the seals, resulted in the most likely explanation of the cause of the explosion.

The advantages and disadvantages of using secondary records are evident from the attempt to identify the cause of the *Challenger* explosion. The primary advantage of using secondary records is that they are not subject to a reactivity effect.

Reactivity effect
An alteration in response created by the knowledge that one is part of an investigation or that one's responses will be evaluated by others

The term **reactivity effect** refers to the potentially biasing effect of knowing that one is part of a study or that the data being collected will be scrutinized in the future. To illustrate this potential bias, consider the reactivity effect that could exist in a study investigating the efficacy of a weight control program. Participants would naturally be aware that the study was investigating their weight and that they would be required to weigh themselves at different intervals. This knowledge could motivate participants to eat less and thus to lose more weight than if they had followed the same diet guidelines but had not participated in the study. Most studies using secondary records have little potential for a reactivity effect because the data are typically recorded outside the context of a research study. The last thing NASA officials expected was that the *Challenger* would explode, despite the Morton Thiokol engineers' grave concern over the performance of the booster rocket seals. Consequently, the officials probably did not alter the letters, memoranda, or other secondary records revealing their interactions with the engineers at Morton Thiokol.

Selective deposit
Selective recording of data

The main disadvantage of using secondary records is the problem of selective deposit and selective survival. **Selective deposit** refers to the recording or depositing of only certain types of data. The data that NASA officials chose to record during the launch consisted almost entirely of information transmitted by the *Challenger's* computers. If the news media had not been photographing the launch, there would have been no record of the puff of black smoke that appeared at the site of the seal during initial ignition, nor any record of the actual explosion. Without these records it would have been much more difficult to discover the probable cause of the explosion.

Selective survival
Selective retention of data

Selective survival refers to survival over time of only certain records. Someone must decide what records are to be maintained. In the case of the *Challenger* disaster, will all the computer data, all photographs of the launch, and all correspondence between NASA and Morton Thiokol be retained, or will only certain segments of this data base be kept? Archaeologists must constantly contend with the problem of selective survival because their data base is made up of the artifacts that have survived the passage of time. Because objects made of clay survive whereas those made of wood do not, archaeologists have an incomplete data base. Despite these disadvantages, secondary records can provide a rich and meaningful data base that can give us information we could not attain in any other manner.

NATURALISTIC OBSERVATION

Naturalistic observation
A descriptive research
technique for
unobtrusively collecting
data on naturally
occurring behavior

Naturalistic observation is a technique that enables the investigator to collect data on naturally occurring behavior. Ebbesen and Haney (1973), for example, were interested in determining the relationship between the proportion of drivers who turned in front of an oncoming car and the risk of a collision with that car. The investigators naturally hypothesized that as the risk of a collision increased, the proportion of drivers turning would decrease. To obtain data to test the hypothesis, the researchers situated an unobtrusive observer in a parking lot next to the T-shaped intersection selected for study. Results of the study supported the hypothesis. However, it was observed that males took significantly greater risks than females and that risk taking increased if drivers had to wait in a line of cars before being allowed to turn, particularly if they had no passengers with them.

A unique characteristic of naturalistic observation is the unobtrusiveness of the observer. Rather than taking an active part in the experiment, the observer must remain completely aloof in order to record natural behavior. If subjects in Ebbesen and Haney's study had known they were being observed, their behavior probably would not have been the same. A second, related characteristic is the lack of artificiality of the situation. The subjects are left in their natural environment so as to eliminate any artificial influence that might be caused by bringing them out of their natural habitat.

For certain types of studies, these characteristics are crucial. If a research project were directed at answering the question of what baboons do during the day, naturalistic observation would be the technique to use. Such research would also generate hypotheses that could be tested with field or laboratory experimentation. If we observe that baboons fight when conditions *a*, *b*, *c*, and *d* exist and we want to know why they fight, we could conduct an experiment to determine this. Condition *a* would be presented without *b*, *c*, or *d*, and we would observe whether fighting occurred. Then condition *b* would be presented without *a*, *c*, or *d*, and so forth, until all combinations of conditions had been presented. We have now moved from observation to experimentation.

Obviously, naturalistic observation is necessary in studying issues that are not amenable to experimentation. (It is not experimentally feasible, for example, to study suicide.)

Although there are many positive components of naturalistic observation, it also has a number of constraints. Naturalistic observation is excellent for obtaining an accurate description, but causes of behavior are almost impossible to isolate using this method. Any given behavior could be produced by a number of agents operating independently or in combination, and observation does not provide any means of sorting these agents out. In no way could the Ebbesen and Haney study have isolated why males take more risks than females. Also, the observational approach is very time consuming. Observers in the Ebbesen and Haney study spent about a month observing drivers at selected time intervals between 10:00 A.M. and 5:00 P.M. to collect data for just one portion of the study. These are just some of the difficulties encountered in such a study.

CASE STUDY

Case study
An intensive description and analysis of an individual, organization, or event

A **case study** is an intensive description and analysis of a single individual, organization, or event, based on information obtained from a variety of sources such as interviews, documents, test results, and archival records. Exhibit 2.1 summarizes a case study of autocastration. Such an intensive description of an event has both advantages and disadvantages. Probably the biggest advantage of the case study is that it provides a fertile breeding ground for ideas and hypotheses. For example, in the case study in Exhibit 2.1, the decline in sexual tension following castration suggests a decrease in the male hormone testosterone. The validity of this hypothesis could be tested using experimental methodology.

Case studies are important in describing rare events, such as autocastration. They are also useful in providing a counter-instance of a universally accepted principle. For example, the motor theory of speech stated that the ability to decode the speech signals of others depended on the listener's ability to speak—if a person could not speak, he or she would not be able to understand what others were saying. In 1962, however, Lenneberg described an eight-year-old boy who lacked the motor skills for speech but could clearly understand others. This one case was all that was needed to refute the basic tenet of the motor theory.

EXHIBIT 2.1 A Case Study of Autocastration

Meyer and Osborne (1982) described a case study of a twenty-nine-year-old male who castrated himself with a kitchen knife while immersed in the ocean because he thought the cool water would act as an anesthetic. He then returned home and handed his testicles to his mother, an act that he thought would return to her the life she had given him at birth. Subsequent in-patient psychiatric treatment revealed that this man had been emotionally disturbed during most of his childhood. When he was 17, he withdrew from social activities and was diagnosed as suffering from psychotic depression. Visual hallucinations were frequent, and he had the persistent delusion that he was draining his brain of nuclear material when he masturbated. During this time he frequented prostitutes and engaged in homosexual activities. These sexual exploits increased his feelings of guilt, anxiety, and depression. He considered suicide but chose castration instead because it would destroy the object of his guilt. The autocastration was interpreted by the subject's therapist as a substitute for suicide. The act was performed under sustained and mounting sexual tension. The male hormone testosterone was suggested as the possible cause of this tension because the man's sex drive and anxiety decreased sharply after he castrated himself.

Probably the biggest disadvantage of the case study is that the cause of a specific event cannot be identified with any degree of certainty. The case study presented in Exhibit 2.1 concludes that autocastration was a substitute for suicide and that the decline in sexual tension was due to the decline in testosterone resulting from castration. Although these interpretations may be correct, it is also possible that they are wrong. The decline in the man's sexual tension may have been due to temporary decline in guilt, resulting from a belief that the object of his guilt, his testicles, had been destroyed.

Another problem with case studies is the difficulty of generalizing to other individuals or situations. It was suggested in Exhibit 2.1 that the autocastration was a substitute for suicide. Even if this hypothesis were true, it would be inappropriate to conclude that autocastration is always performed for this reason. One can generalize from a single case only if there is no variability within the population. If there is variability—if, for example, autocastration is performed for many reasons (as a way of assuaging sexual guilt, as a substitute for suicide, as a result of a psychotic delusion, in an attempt to alter sexual identity)—then it is inappropriate to generalize from a single case.

CORRELATIONAL STUDY

Correlational study
A study that seeks to describe the degree of relationship that exists between two measured variables

In its simplest form, a **correlational study** consists of measuring two variables and then determining the degree of relationship that exists between them. Consequently, a correlational study can be incorporated into other descriptive research approaches. A relatively old study commonly cited in introductory and developmental texts is the study by Conrad and Jones (1940) of the relationship between the IQ scores of parents and those of their offspring. To accomplish the goals of this study, Conrad and Jones measured the IQs of the parents and correlated them with their children's IQs. In this way, a descriptive index was obtained that accurately and quantitatively portrayed the relationship between these two variables. As you can see, correlational studies do not make any attempt to manipulate the variables of concern, but simply measure them in their natural state.

The correlational approach enables us to accomplish the goals of prediction. If a reliable relationship is found between two variables, then we not only have described the relationship between these two variables but also have gained the ability to predict one variable from a knowledge of the other variable. Sears, Whiting, Nowlis, and Sears (1953), for example, found that there is a positive relationship between severity of weaning and later psychological adjustment problems. Knowledge of this relationship enables one to predict a child's psychological adjustment when one knows only the severity with which the child was weaned.

The weakness of the correlational approach is apparent in the Sears et al. study. Given the study's results, some individuals would say that severity of weaning was the agent causing later psychological maladjustment. But because of the so-called third variable problem, such an inference is not justified. The *third variable problem* refers to the fact that two variables may be correlated not because they are causally related but because some third variable caused both of them. As Figure 2.1 illustrates, the degree of severity with which a child is weaned and that child's later psychological adjustment may both be influenced by the parents' child-rearing skills. If the parents are unskilled, they may both wean their child in a severe manner and inflict verbal or physical abuse, leading the child to become psychologically maladjusted later. If the parents are skilled at child-rearing, weaning may take place with little or no trauma and the child may have a healthy relationship with

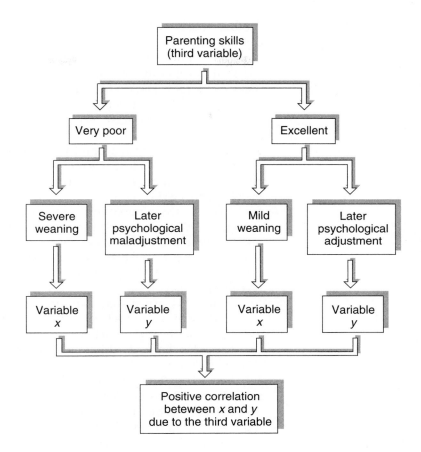

FIGURE 2.1   Illustration of the third variable problem in correlation.

his or her parents, leading to stable psychological adjustment. In such a situation, although a correlation would exist between severity of weaning and later psychological adjustment, these variables would not be causally related. Rather, both would be caused by an underlying third variable: the parents' parenting skills.

To further emphasize the third variable problem, let us look at a correlational study conducted by Sternglass and Bell (1979). As you may know, Scholastic Aptitude Test (SAT) scores demonstrated a consistent decline from the 1960s to the early 1980s. This decline had been attributed to multiple causes, including curricular changes, grade inflation, and broken homes. Sternglass and Bell reasoned that, because the decline in SAT test scores was so widespread, the cause of the decline must be an equally widespread environmental phenomenon. Since the decline began in 1962–1963 and most of the students would have been seventeen to eighteen years old when they took the SATs, this meant they would have been born around 1945. Sternglass and Bell also made the assumption that the causal agent was something that began at birth and gradually exerted its effect over time. One variable that fit this criterion—it began around 1945 and had a gradual effect—was fallout from nuclear weapons testing, which began in 1945 and gradually increased until atmospheric testing was banned in 1959 and 1963.

To test their hypothesis, Sternglass and Bell correlated students' SAT scores with the tonnage of nuclear testing that was conducted in Nevada during the year they were born and found a negative correlation between these two variables. In other words, the SAT scores declined as the amount of nuclear testing increased.

Does this prove that radiation from the nuclear testing caused the decline in SAT scores? Certainly not. All this study did was describe the relationship that existed between these two variables, both of which could have been caused by some third variable.

Similarly, Benbow and Stanley (1980) revealed that males score higher than females on the mathematics portion of the SAT (SAT-M). This does not mean that being male causes an increase in mathematical ability; it simply describes the relationship that exists between gender and SAT-M test scores. There are some rather complex correlational procedures that do give evidence of causation, but the two-variable cases just presented do not.

The fallacy of assuming causation is not inherent in the correlational study but is merely a tendency on the part of the people who use the study results. If the purpose of an investigation is to describe the degree of relationship that exists between variables, then this approach is the appropriate one to use.

EX POST FACTO STUDY

An **ex post facto study** is a study in which the variable or variables of interest to the investigator are not subject to direct

Ex post facto study
A study comparing the
effects of two or more
variables where the
variables being
manipulated are not
under the experimenter's
control
manipulation but must be chosen after the fact. The investigator begins with two or more groups of subjects that already differ according to one variable (such as sex, age, prior experience, or internal state) and then records their behavior to determine whether they respond differently in a common situation. These and other characteristics of ex post facto research are illustrated by a study by Atkeson, Calhoun, Resick, and Ellis (1982), who investigated the incidence, severity, and duration of the depressive symptoms of rape victims. Using several measures of depression, these researchers compared rape victims over time with a group of females who had not been raped. As Figure 2.2 illustrates, the results show that rape victims initially exhibited quite a bit of depression but that their depression declined to an approximately normal level over time.

If you look at Figure 2.3, you can see that the Atkeson et al. study has the appearance of a field experiment by virtue of the fact that the experience of being raped was

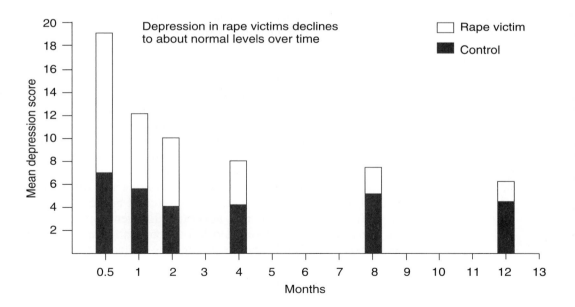

FIGURE 2.2 Mean depression scores for rape victims and control subjects. (Based on data from "Victims of Rape: Repeated Assessment of Depressive Symptoms" by B. M. Atkeson, K. S. Calhoun, P. A. Resick, and E. M. Ellis, 1982, *Journal of Consulting and Clinical Psychology, 50,* pp. 96–102.)

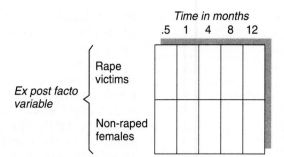

FIGURE 2.3 Graphic illustration of the design of the Atkeson et al. (1982) study.

varied and then the consequence of having been raped was assessed across time. Consequently, the investigation compared the degree of depression experienced by rape victims with that of control, or nonraped, females. However, the nature of the manipulation of the experience of rape determines the ex post facto nature of the study. The experimenters did not have control over who was and who was not raped. Instead, the females came to the study with their prior differential experiences and, therefore, *assigned themselves* to one of the two groups. So the variable of interest was out of the experimenter's control, which is the distinguishing characteristic of an ex post facto study. The process of self-selection that determines the ex post facto nature of the research also is responsible for its weakness. Subjects who make up the different groups because of some self-selected characteristic or experience may also possess other characteristics extraneous to the research problem. It may be one of these other characteristics and not the variable supposedly being manipulated in the study that produced the observed difference. Atkeson et al. recognized this limitation of the study and, instead of concluding that the experience of being raped was the causative factor of the observed differences, discussed the possible contribution of other factors (such as repeated assessment of degree of depression, the subjects' knowledge of the fact that they were in a psychological study, the stress on the poorer victims created by living in an area with a high crime rate, and the age of the research participants). As can be seen from this example, ex post facto studies resemble correlational studies in that the obtained relationships may have been produced by variables other than those investigated.

LONGITUDINAL STUDY AND CROSS-SECTIONAL STUDY

Longitudinal study
A developmental field study that repeatedly measures the same characteristics in a single sample of individuals at selected time intervals

Cross-sectional study
A developmental field study that measures the same characteristics in representative samples of individuals at different age levels

Longitudinal and cross-sectional studies investigate developmental changes that take place over time. The approaches used by these two basic techniques are somewhat different, however. The **longitudinal study** involves choosing a single group of subjects and measuring them repeatedly at selected time intervals to note changes that occur over time in the specified characteristics. For example, Brown, Cayden, and Bellugi-Klima (1969) studied the language development of children by systematically recording the verbalizations and language productions of three children for almost two years. On the other hand, a **cross-sectional study** identifies representative samples of individuals at specific age levels and notes the changes in the selected characteristics of subjects in these different age groups. Liebert, Odem, Hill, and Huff (1969) took this approach in their study of language development. They identified three relatively large groups of children at three different age levels and then observed and recorded the differences among these groups.

The longitudinal and cross-sectional descriptive approaches to developmental research have frequently been used in the past, and there has been much discussion about the relative advantages and disadvantages of each technique. One significant point is that these two techniques have not always generated similar results. The classic example of this discrepancy in results is in data regarding the development of intelligence during adulthood. As seen in Figure 2.4, cross-sectional studies have suggested that adult intelligence begins to decline around the age of thirty, whereas longitudinal studies show an increase or no change in intellectual performance until the age of fifty or sixty (Baltes, Reese, and Nesselroade, 1977). This difference has been attributed to what is called an *age-cohort effect.* In other words, longitudinal studies follow just one group or cohort of individuals over time, so all individuals within this cohort are experiencing similar environmental events. However, cross-sectional studies investigate a number of different groups of individuals or different cohorts. Because of changes in environmental events, these cohorts have not been exposed to similar experiences. For example, members of a forty-year-old cohort would not have been exposed to video

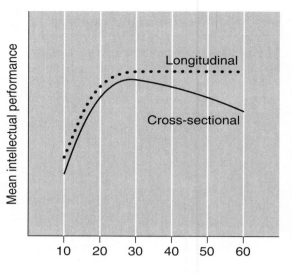

FIGURE 2.4   Change in intellectual performance as a function of the longitudinal versus the cross-sectional method. (From *Life-Span Developmental Psychology: Introduction to Research Methods* by P. B. Baltes, H. W. Reese, and J. R. Nesselroade. Copyright © 1977 by Wadsworth Publishing Company, Monterey, California.)

games or computers when they were ten years old, but a group of eleven-year-olds would have. Such differences are confounded with actual age differences in cross-sectional studies.

I have presented longitudinal and cross-sectional studies as descriptive research approaches because most studies using either of these methods are in fact descriptive in nature. It is possible, however, to conduct an experimental longitudinal or cross-sectional study. There is currently a great deal of emphasis on prevention research. In a number of recent studies, a prevention technique has been used and then the individuals to whom the technique was administered followed to determine, for example, whether they became drug abusers. Such research requires an experimental longitudinal approach.

THE SURVEY

Survey
A field study in which an interview technique is used to gather data on a given state of affairs in a representative sample of the population

The **survey** is a widely used descriptive research technique. It is often defined as a method of collecting standardized information by interviewing a representative sample of some population. In other words, the survey represents a probe into a given state of affairs that exists at a given time. Therefore direct contact must be made with the individuals whose characteristics, behaviors, or attitudes are relevant to the investigation.

Probably the most widely known surveys are those conducted by the Gallup organization. Gallup polls are frequently conducted to survey the voting public's opinions on such issues as the popularity of the president or a given policy, or to determine the percentage of individuals who may be expected to vote for a given candidate at election time. Surveys are initially conducted to answer the questions "how many" and "how much." But collection of frequency data is only a preliminary phase of the research in many studies. Studies often want to answer the questions "who" and "why." Who votes for the Republican candidate, and who votes for the Democrat? Why do people buy a certain make of car or brand of product? Such information helps us to understand why a particular phenomenon occurred and increases our ability to predict what will happen.

For example, Table 2.1 illustrates the results of a Gallup poll (1980) concerning the issue of handguns. One of the questions asked in the survey was "Do you think there should or should not be a law which would forbid the possession of pistols and revolvers except by police and other authorized persons?" The responses to this question reveal that only 31 percent of the individuals polled would favor such a law. However, opinions varied according to the background characteristics of the respondents. A greater percentage of Catholics, Easterners, and individuals living in large cities favored such a law (as opposed to Protestants, Southerners, and those living in rural communities). The responses to this one question provide answers to the questions of *who* and *how many* are in favor of a law restricting gun ownership.

The survey is a technique that is applicable to a wide range of problems. It is also a technique that is deceptively easy to use. The unsophisticated researcher may think that all that is needed is to construct a number of questions addressing the issue on which information is desired and then get a number of people to respond to these questions. Superficially, this is essentially all there is to the survey. However, completing these seemingly simple steps requires a lot of thought and work. Without this thought and work, the questions asked will elicit unreliable answers. For example, in 1936 the periodical *Literary Digest* set out to predict the winner of the presidential election. A sample of subjects was selected from telephone

TABLE 2.1 Response Obtained from a Gallup Poll to the Question "Do you think there should or should not be a law which would forbid the possession of pistols and revolvers except by police and other authorized persons?"

	Yes, Should	No, Should Not	No Opinion
National	31%	65%	4%
Sex			
Male	25	71	4
Female	36	59	5
Race			
White	32	65	3
Nonwhite	22	67	11
Education			
College	33	64	3
High school	29	66	5
Grade school	32	63	5
Region			
East	46	50	4
Midwest	32	62	6
South	20	77	3
West	23	73	4
Age			
Total under 30 years	30	65	5
18–24 years	28	66	6
25–29 years	33	65	2
30–49 years	29	67	4
50 years & older	33	63	4
Income			
$25,000 & over	32	66	2
$20,000–$24,999	33	66	1
$15,000–$19,999	32	60	8
$10,000–$14,999	27	67	6
$5,000–$9,999	30	65	5
Under $5,000	29	68	3
Political affiliation			
Republican	33	65	2
Democrat	32	64	4
Independent	26	69	5
Religion			
Protestant	24	72	4
Catholic	44	53	3
Occupation			
Professional & business	33	64	3
Clerical & sales	38	59	3
Manual workers	25	70	5
Non-labor force	32	64	4

(continued)

TABLE 2.1 *(continued)*

	Yes, Should	No, Should Not	No Opinion
City size			
1,000,000 & over	46	47	7
500,000–999,999	35	58	7
50,000–499,999	37	61	2
2,500–49,999	22	76	2
Under 2,500; rural	18	78	4
Labor union families	28	65	7
Non–labor union families	31	65	4

Source: Gallup Opinion Index: Political, Social, and Economic Trends, Report No. 174, January 1980, p. 29.

directories and car registrations. This sample was surveyed to determine whom they would vote for in the 1936 election. Based on the results of this survey, the *Literary Digest* predicted that Alfred Landon would defeat Franklin Roosevelt by a wide margin. However, if you know your U.S. presidents, you know that Franklin Roosevelt won the election.

Why was such an error made? If you think about the climate that existed in 1936, you will realize that the United States was in the middle of the Great Depression. During this time many individuals did not have phones or cars simply because they couldn't afford them. Consequently, selecting a sample of individuals from telephone directories and car registrations provided a biased sample of individuals, which underrepresented the less affluent segment of the population and in doing so underestimated Roosevelt's popularity. To collect accurate data using the survey technique such errors must be eliminated. To give you an appreciation of the effort involved in conducting a good survey, several of the major types of decisions that must be made will be discussed briefly.

Methods of Data Collection

Face-to-face method
A survey method involving a personal interview, often conducted in the interviewee's home

There are a variety of methods for collecting survey data. The most popular methods are face-to-face, telephone, and mail. Each has its own set of advantages and disadvantages in terms of issues such as cost and response rate. The **face-to-face method,** as the name suggests, is a person-to-person interview, which typically involves going to the interviewee's home and obtaining responses to the survey by conducting a personal interview. This technique has the advantages of allowing the interviewer to clear up any ambiguities in the question asked or to probe for further clarification if the interviewee provides an inadequate answer, and it generally gives a higher completion rate and more complete information. However, this technique is the most expensive. It is also possible that the interviewer may bias the responses. For example, an interviewer may (either consciously or unconsciously) spend more time with and probe more effectively an attractive interviewee of the opposite sex and thus bias the results.

Telephone method
A survey method involving a telephone interview

With the **telephone method,** as the name suggests, the survey is conducted by means of a telephone interview. This method is about half as expensive as the face-to-face interview (Groves & Kahn, 1979), and some data (Rogers, 1976) demonstrate that the information collected is comparable to that obtained in a face-to-face interview. This seems to be particularly true with the use of random-digit dialing. In some areas 20 to 40 percent of customers elect not to list their phone numbers in telephone directories (Rich, 1977). Such unlisted numbers are accessible with the **random-digit dialing method.** In this method telephone numbers are dialed through use of a random process, usually by a computer, which means that unlisted numbers are just as accessible as listed numbers. If a survey researcher has access to a computer-assisted telephone interview (CATI) system, the interviewer's questions are prompted on the computer screen and the interviewee's responses are input directly into the computer for analysis.

Random-digit dialing method
A survey method that involves dialing telephone numbers composed through a random process

Mail method
A survey conducted by sending a questionnaire through the mail

The **mail method,** as the name suggests, involves sending questionnaires to interviewees through the mail and asking them to return the completed questionnaires, typically in envelopes provided by the organization conducting the survey. The primary advantage of this technique is its low cost. You can send a questionnaire anywhere in the world for the price of postage. However, the disadvantage is that most questionnaires are never returned. The return rate is typically 20 to 30 percent for the initial mailing (Nederhof, 1985), although the rate can be increased by use of techniques such as follow-up letters reminding the respondent of the survey and enclosing another copy of the questionnaire.

Questionnaire Construction

In addition to deciding on the mode of data collection, it is necessary to construct a number of questions that will provide an answer to the research question. In constructing these questions it is imperative that the researcher have an explicitly identified research question. An explicit research question states what the researcher wishes to know. For example, Veroff, Douvan, and Kulka (1981) conducted a survey "to determine in some detail how American people cope with problems of adjustment which arise in their lives." Given such a specific research question, a number of items directed at assessing coping mechanisms of the American public can be written.

Although specification of the research question is an essential beginning point for writing survey questions, there are a number of other factors that must be considered. Several of these factors are discussed below. More detailed discussions of these factors are found in Babbie (1990) and Sudman and Bradburn (1982).

Open-ended questions Questions that enable respondents to answer in any way they please

Closed-ended questions Questions that require respondents to choose from a limited number of predetermined responses

Type of Question There are two types of questions that can be used in a survey: open-ended or closed-ended questions. An **open-ended question** enables respondents to answer in any way they please, whereas a **closed-ended question** requires respondents to choose from a limited number of predetermined responses. For example, if you wanted to find out what people do when they feel depressed, you could ask an open-ended question such as "What do you do when you feel depressed?" Alternatively, you could ask a closed-ended question such as

What do you do when you feel depressed?
a. Eat
b. Sleep
c. Exercise
d. Talk to a close friend
e. Cry

The open-ended question requires the respondent to come up with the answer. Such a question is valuable when the researcher needs to know what people are thinking or when the dimensions of a variable are not well defined. However, the responses to an open-ended question must be coded and categorized, which takes time. Also, sometimes the responses given don't make sense and can't be categorized.

The closed-ended question requires the respondent to select one of the alternative answers given. Generally, closed-ended questions are appropriate when the dimensions of a variable are known. In such an instance the alternative responses can be specified and the respondent can select among these alternatives.

Question Wording Once the type of question has been specified, the actual questions must be written. There are a number of pitfalls that must be avoided in writing the questions. **"Double-barreled" questions,** or questions that ask two things, must be avoided. A question such as "Do you agree that the president should focus his primary attention on the economy and foreign affairs?" includes two different issues: the economy and foreign affairs. Each issue may elicit a different attitude, and combining them into one question makes it unclear which attitude or opinion is being assessed. Leading and ambiguous questions must also be avoided. For example, a congressman used the following item in one of his mail surveys:

Double-barreled question
A question that covers two different issues at once

> A site in Ellis county, in your congressional district, is one of seven national finalists for the superconducting super collider (SCC) project. During this time of budget restraint, do you support programs vital to the future growth of our country such as the SCC?

This item is ambiguous—it is not clear whether it is trying to assess an attitude toward the SCC or toward the programs vital to the growth of the country. To identify the ambiguous or unclear items it is necessary to pilot test the questionnaire by administering it to a small sample of individuals who are instructed to identify any difficulties they have in responding to each item.

Ordering of the Questions The ordering, or sequencing, of the questions must also be considered. When the questionnaire includes both positive and negative items, it is generally better to ask the positive questions first. Similarly, the more important and interesting questions should come first to capture the attention of the respondent. Roberson and Sundstrom (1990) found that placing the important questions first and demographic questions (age, gender, etc.) last in an employee attitude survey resulted in the highest return rates.

Questionnaire Length There are many significant questions that can be asked in any survey, but every data-gathering instrument has an optimal length for the population to which it is being administered. After a certain point, the respondents' interest and cooperation diminish. The survey researcher must therefore ensure that the questionnaire is not too long, even though some important questions may have to be sacrificed. It is impossible to specify the optimum length of any survey questionnaire because length is partially dependent on the topic and the method of data collection. As a general rule telephone interviews should be no longer than 15 minutes. However, a face-to-face interview may consume more time without making the interviewee feel uncomfortable.

Obtaining Subjects After the survey questionnaire is constructed, it must be administered to a group of individuals to obtain a set of responses that will provide an answer to the research question. There are many ways in which a researcher can select the subjects who will be given the survey questionnaire. Most research projects involve selecting a sample of subjects from a population of interest. A **population** refers to all of the events, things, or individuals to be represented, and a **sample** refers to any number of individuals less than the population. However, the manner in which this sample of subjects is selected depends on the goals of the research project. If the research question focuses on identifying the relationship between variables, then a haphazard sampling technique will probably be used. However, if it is important that the results obtained from the sample mirror the results that would have been obtained if the total population had been included, then a random sampling technique must be used.

Population
All of the events, things, or individuals to be represented

Sample
Any number of individuals less than the population

    Haphazard sampling is a nonprobability sampling technique whereby the sample of subjects selected is based on convenience and includes individuals who are readily available. For example, a significant amount of psychological research is conducted using introductory psychology students as subjects because these students are conveniently available to researchers. The obvious advantage of using the haphazard sampling technique is that subjects can be obtained without spending a great deal of time or money. However, researchers usually want the results of their studies to generalize to "people in general" or at least to "a college student population." Making such a generalization from a sample of college stu-

Haphazard sampling
A nonprobability sampling technique whereby the sample of subjects selected is based on convenience

dents is hazardous because the sample is composed of students volunteering for the study who have elected to take introductory psychology during the semester in which the study was conducted.

Random sampling
Sampling in which every member of the population has an equal chance of being selected for the study

When the research question requires an accurate depiction of the general population, a **random sampling** technique must be used. When this technique is used, every member of the population has an equal chance of being selected for the study. (This technique is discussed in more detail in Chapter 8.) The advantage of this technique is that it provides a sample of subjects whose responses represent those of the general population. For example, during presidential campaigns polls are frequently conducted to test the pulse of the voting public. These polls are taken to determine the popularity of the candidates as well as the influence of various issues, such as prior drug use, on public opinions of candidates.

When a true random sample of subjects is obtained from the population, the results can be amazingly accurate. For example, in 1976 a *New York Times*–CBS poll correctly predicted that 51.1 percent of the voters would vote for Jimmy Carter and 48.9 percent would vote for Gerald Ford (Converse and Traugott, 1986). The prediction was made using a sample of less than 2,000 individuals selected from almost eighty million voters. This perfectly accurate prediction was unusual, but it illustrates the accuracy with which the population responses can be predicted from a sample of just a few individuals, if these individuals are selected randomly. In virtually all polls such as this there is sampling error, or error that arises from the fact that the sample of subjects selected does not provide a perfectly accurate representation of the population. However, this error is typically quite small and much smaller than it would be if any other sampling technique were used.

PARTICIPANT OBSERVATION

Participant observation
Observation and recording of the behavior of others while actively participating with them in a situation

Participant observation is a descriptive research approach whereby the behavior of others is recorded in a context in which the researcher is an active participant. For example, Marquart (1983) wanted to analyze the formal and informal mechanisms of social control that existed within one of the prison units of the Texas Department of Corrections (TDC). To accomplish this goal, Marquart managed (with some difficulty) to get hired as a prison guard at a maximum-security unit. During the next eighteen months, while performing the routine duties of a guard, Marquart observed the behavior of other prison guards and of the inmates. As he searched for weapons, broke up fights between inmates, and patrolled the cell blocks, showers, and dining halls, he was also observing the informal structure that existed.

Although the Johnson unit where Marquart worked housed hard-core habitual criminals, it also had a reputation for strict discipline, control, and order. This order was achieved through the use of an informal system of control in addition to such formal control techniques as withdrawal of privileges and restriction of inmates to their cells. One component of the informal system involved using inmates to control other inmates. These inmates, known as building tenders and turnkeys, were made responsible for helping the guards maintain discipline and control within a given section of the prison unit. They were typically the brightest, biggest, and meanest inmates in the prison, who would loyally assist the warden and prison guards in exchange for a variety of rewards, such as having open cells while other inmates' cells were shut. Marquart used the participant observation method to describe a control mechanism that had evolved over a number of years and was being used effectively to exercise control in a potentially explosive environment.

Marquart's study of the TDC's informal control system illustrates the characteristics of participant observation as a research approach. Participant observation can typically be used in situations that otherwise might be closed to scientific investigation. Marquart, for example, had to earn the trust of the inmates before they would reveal many of their control techniques. Suspicion and paranoia run rampant in a prison environment; it would be virtually impossible for an unknown researcher to walk into a prison and expect not only the inmates but also the guards to divulge their informal system of control.

Although participant observation may allow the researcher to describe a situation that might otherwise be closed to investigation, it also involves a significant risk of introducing various biases into the description. With this approach, the researcher plays two roles: researcher and participant. Even as the researcher is trying to make objective observations, he or she must also interact with those being observed. This dual role maximizes the chances for the observer to lose objectivity and allow personal biases to enter into the description. For example, Kirkham (1975) noted that as he worked as a police officer in the slums, he began to take on the attitudes and behavior of the officers with whom he worked. But the researcher is also constantly aware that he or she is just that, a researcher, and this may influence the way others interact in his or her presence, particularly if they are aware of the researcher's purpose. Marquart, for example, told the prison guards and inmates of his intentions in hopes that this would make them more likely to reveal the nature of the control system. At one point in the study, however, this approach backfired, when the rumor spread that Marquart was a nephew of the director of the TDC. Until Marquart managed to dispel this rumor, few inmates would talk to him.

Such limitations of participant observation may threaten the accuracy of a researcher's description of a given situation. A conscientious investigator, however, can minimize such bias and use this research approach to describe the behavior of others in a situation that may not be amenable to any other method of investigation.

META ANALYSIS

Meta analysis
A quantitative technique that is used to integrate and describe a large number of studies

Meta analysis is a term that was introduced by Glass (1976) to describe a quantitative approach that could be used to integrate and describe the results of a large number of studies. In the field of psychology, we rarely find a single study that provides a complete answer to a research question. Not only is human behavior too complex to be explained within the framework of a single study, but the ability to control the research environment, the subject sample, and the procedures used may vary from study to study. This means that a number of studies must be conducted in order to obtain a sufficiently definitive answer. Within the field of psychology, there has been a proliferation of studies investigating such research questions as the effectiveness of psychotherapy, leadership characteristics, and causes of obesity. These studies typically use different sets of definitions, subject samples, variables, and procedures; their conclusions may even be different. At some point, however, these studies need to be integrated and described. Prior to Glass's (1976) article, the traditional approach to accomplishing this task involved collecting the studies conducted on a given topic, categorizing them in some manner (for example, those that were methodologically sound versus those that contained obvious control problems), and then reaching a conclusion based on the proportion of studies that suggested a given outcome. Such an approach, however, maximizes the opportunity for the introduction of subjective judgments, preferences, and biases, which become obvious when two reviews of the same literature arrive at different conclusions. Meta analysis gets around this problem because it involves the analysis of analyses or the use of a variety of quantitative techniques to analyze the results of the studies in question.

There are actually two broad classes of meta-analytic techniques, and each provides an answer to a different research question. One technique involves determining if an overall significant result will emerge if the findings from many studies are combined in a statistical fashion. For example, assume that four different studies were conducted investigating the effectiveness of psychotherapy in treating depression. In two studies, a significant effect was not found; the probability (p) values attained in these studies were .10 and .08, respectively (a p value of .05 or less is necessary to conclude that a result is statistically significant). The other two studies indicated that psychotherapy was effective in treating depression, because p values of .04 and .001, respectively, were attained. It would be valuable to combine the studies to determine whether an overall significant effect would emerge. Meta analysis can be used to combine the p values of these four studies to yield one overall p value that would be representative of the combined effect of all four studies.

The other meta-analytic technique deals with the size of the treatment effect produced in the study. For example, if you conduct a study investigating the impact of some therapeutic intervention on depression, you want to know whether the treatment program was effective in reducing depression. You also want to know how much effect the treatment program had on depression—how large the effect was. Different studies investigating the same intervention technique would arrive at different estimates of the size of the effect. Consequently, it would be advantageous to be able to combine these effect sizes in order to come up with one overall effect size. This second meta-analytic technique enables the researcher to combine the results of several studies and arrive at a combined estimate of the size of the effect.

To illustrate the use of meta analysis, let us look at the research conducted on lunar effects. For centuries, a significant proportion of the population has believed that the moon influences behavior. The typical assumption is that a variety of behaviors—suicides, admissions to mental institutions, homicides—are more likely to occur during a full moon. Between 1961 and 1984 at least forty studies were conducted that investigated the lunar hypothesis—a surprisingly large number of studies given that most researchers are skeptical about the value of such investigations. However, the lunar hypothesis has a great deal of popular appeal and, if supported, would become a legitimate and valuable area for research. To determine whether the lunar hypothesis had any support, Rotton and Kelly (1985) collected all the studies between 1978 and 1984 that investigated the lunar hypothesis and then conducted a meta analysis of the results.

Rotton and Kelly first used the technique of combining the p values of these studies. The results of this meta analysis revealed that none of the combined p values were significant for studies that focused on activities taking place under a full moon or a new moon. Rotton and Kelly then focused on a meta analysis of the effect size computed for each of the studies identified. This analysis showed that the combined size of the effect of the moon is extremely small—so small, in fact, as to be considered negligible. Consequently, the results of this meta analysis indicated that the lunar hypothesis has no substance when the results of many studies are considered together.

SUMMARY

In applying descriptive approaches to gaining scientific knowledge, researchers attempt to paint a picture of a particular phenomenon. There are nine basic research approaches used in attaining this objective:

In studies based on *secondary records*, current or previously gathered data are used to describe a given situation.

In *naturalistic observation,* the researcher unobtrusively watches and records naturally occurring behavior.

A *case study* involves the intensive description of an individual, organization, or event.

A *correlational study* describes the degree of relationship between two variables.

An *ex post facto study* describes the relationship between a given behavior and a variable on which groups of subjects naturally differ, such as skin color.

Longitudinal and cross-sectional studies are of the developmental type and describe changes that take place across time.

A *survey* describes a given state of affairs through use of the interview technique.

In *participant observation,* the researcher actively participates with and observes the behavior of others.

Meta analysis is a research approach involving the quantitative analysis of the results of many studies in order to integrate and describe the overall picture presented by these studies.

STUDY QUESTIONS

1. What is the primary objective of a descriptive research approach?
2. Identify the type of situation in which you would use each of the various descriptive research approaches.
3. Identify the advantages and disadvantages of each of the descriptive research approaches.
4. What are the similarities and differences among the various descriptive research approaches?
5. What is the primary way in which meta analysis differs from the other descriptive research approaches?

KEY TERMS AND CONCEPTS

Descriptive research   Selective deposit

Secondary records   Selective survival

Reactivity effect   Naturalistic observation

Case study

Correlational study

Third variable problem

Ex post facto study

Longitudinal study

Cross-sectional study

Age-cohort effect

Survey

Face-to-face method

Telephone method

Random-digit dialing method

Mail method

Open-ended question

Closed-ended question

Double-barreled question

Population

Sample

Haphazard sampling

Random sampling

Participant observation

Meta analysis

CHAPTER 3

The Experimental Research Approach

LEARNING OBJECTIVES

1. To understand the basic characteristics of the experimental research approach.
2. To understand the different settings in which experimental research is conducted and the advantages and disadvantages associated with each setting.
3. To gain some understanding of the nature of causation and appreciate the difficulty of trying to identify causal relations.

At 3:20 A.M. on March 13, 1964, Kitty Genovese parked her car in the lot adjacent to the railroad station and started to walk to her apartment, about 100 feet away. As she proceeded to the entrance, she noticed a man at the far end of the parking lot. Kitty stopped and then nervously headed up the street toward Lefferts Boulevard, where there was a police call box. When she got to the street light, the man grabbed her. She screamed, and lights went on in the ten-story apartment house where she lived. Windows slid open and voices broke the early morning stillness. Kitty screamed: "Oh, my God, he stabbed me! Please help me! Please help me!" A man hollered from one of the upper windows: "Let that girl alone!" The assailant looked up at him, shrugged, and walked down the street toward a car.

Kitty struggled to her feet as the lights went out in the apartment house. The killer immediately returned and stabbed Kitty again as she screamed: "I'm dying! I'm dying!" Again, windows opened and lights came on. This time the assailant got into his car and drove away, while Kitty again struggled to her feet. But the killer returned a third time. By now Kitty had made her way to the rear entrance to the building. When the killer found her there, slumped on the floor at the foot of the stairs, he stabbed her a third time—the final, fatal blow.

This is a true incident abstracted from police reports. Many explanations were offered for the neighbors' failure to intervene in this obvious emergency. One sociologist attributed the lack of intervention to a disaster syndrome. A psychoanalyst talked about the effect of living in a megalopolis. All of these explanations, however, represented only speculations by these individuals. How, then, do we explain the lack of intervention? There must be some set of variables that makes people more or less likely to intervene in such a situation. Perhaps it is the potential danger involved in the situation, or maybe it is the number of other people present, who could potentially intervene.

Such variables represent factors that not only might affect bystander intervention but also actually existed in the Kitty Genovese killing. The only way to determine whether such variables do in fact affect a person's tendency to intervene is to conduct a psychological experiment, or to use the experimental research approach. This chapter will provide a basic orientation to the experimental research approach. Since this approach attempts to identify cause-and-effect relationships, the chapter will also discuss the meaning of causation.

INTRODUCTION

The **experimental research approach** is the technique designed to ferret out cause-and-effect relationships. This research method enables us to identify causal relationships because it allows us to observe, under controlled conditions, the effects of systematically changing one or more variables. Because of its ability to identify causation, the experimental approach has come to represent the prototype of the scientific method for solving problems. But before we turn to this research approach, which will be the focus of attention for the remainder of this book, you must have some grasp of the concept of causation.

CAUSATION

Causation is one of those terms that people frequently use but often don't really understand. People ask questions like "What causes cancer?" "What causes a person to murder someone else?" "What causes a man to beat his wife?" What do they really mean? Common sense suggests that causality refers to a condition in which one event—the cause—generates another event—the effect; however, causality is much more complex.

When individuals discuss the effects of events, they tend to use the words *cause* and *effect* rather informally. People are likely to assume that manipulation is implicit in the concept of causation. If we manipulate or do something, we expect something else to happen. If something does happen, the thing or event we manipulate is called the *cause* and what happens is called the *effect.* For example, if we spank a child for coloring on a wall and then observe that he no longer colors on the wall, we assume that the spanking caused the child to stop the coloring. But this casual interpretation of the word *causation* does not take into consideration the necessity of ruling out alternative explanations. Perhaps the child actually stopped coloring because he got tired of coloring or because his attention was diverted to other interests. The problems associated with the use of the word *causation* have been studied by philosophers for years.

John Stuart Mill (1874), a British philosopher, set forth canons that could be used to experimentally identify causality. These canons form the basis of many of the approaches currently used.

Method of agreement
The identification of the
common element in
several instances of an
event

Method of difference
The identification of the
different effects
produced by variation in
only one event

*Joint methods of
agreement and difference*
The combination of the
methods of agreement
and difference to
identify causation

1. The first canon is the **method of agreement,** by which one identifies causality by observing the element common to several instances of an event. This canon can be illustrated by the frequently cited case of the man who wanted to find out scientifically why he got drunk. He drank rye and water on the first night and became drunk. On the second night, he drank scotch and water and became drunk again. On the third night, he got drunk on bourbon and water. He therefore decided that the water was the cause of his getting drunk because it was the common element each time. This method, as you can see, is inadequate for unequivocally identifying causation because many significant variables—such as the alcohol in the rye, scotch, and bourbon—may be overlooked.

2. The second canon is the **method of difference,** by which one attempts to identify causality by observing the different effects produced in two situations that are alike in all respects except one. The method of difference is the approach taken in many psychological experiments. In an experiment designed to test the effect of a drug on reaction time, the drug is given to one group of subjects while a placebo is given to another group of matched subjects. If the reaction time of the drug-taking group differs significantly from that of the control group, the difference is usually attributed to the drug (the causal agent). This method provides the basis for a great deal of work in psychology aimed at identifying causality.

3. The third canon set forth by Mill is the **joint methods of agreement and difference.** This method is exactly what its name implies. The method of agreement is first used to observe common elements, which are then formulated as hypotheses to be tested by the method of difference. In the case of the man who wanted to find out why he got drunk, the common element, water, should have been formulated as a hypothesis to be tested by the method of difference. Using the method of difference, researchers would give one group of subjects water and a matched group another liquid (such as straight bourbon). Naturally, the group drinking only water would not get drunk, indicating that the wrong variable had been identified even though it was a common element.

Method of concomitant variation
The identification of parallel changes in two variables

4. The fourth canon is the **method of concomitant variation.** This method states that a variable is either a cause or an effect, or else is connected through some factor of causation, if variation in the variable results in a parallel variation in another variable. Plutchik (1974) interprets this canon to be an extension of the method of difference in that, rather than just using two equated groups in an experiment, the researchers use three or more, with each group receiving a different amount of the variable under study. In the previously cited drug example, rather than just a placebo group and a drug group, one placebo and several drug groups could be used, with each drug group receiving a different amount of the drug. Reaction times could then be observed to determine if variation in the quantity of the drug results in a parallel variation in reaction time. If this parallel variation is found, then the drug is interpreted as being the cause of the variation in reaction time.

Some writers interpret this canon as including correlation studies. One is on extremely shaky ground in attempting to infer causative relationships from correlational studies, as many correlational studies simply describe the degree of relationship. Identification of causation requires direct manipulation of the variables of interest. However, recent work is making strides in enabling causation to be inferred from correlational studies.

From looking at the works of people such as Mill, one gets the idea that we have a fairly adequate grasp of what causation is and how to obtain evidence of it. This belief is reinforced when we see that many studies of causation are based on Mill's canons. Such philosophizing and experimentation have not completely clarified the meaning of the word *cause,* however.

Exhibit 3.1, which presents Morison's (1960, pp. 193–194) review of the history of attempts to find the cause of malaria, illustrates the ambiguity of this word. Essentially, the method of agreement was used first to hypothesize that the bad air in the lowlands caused malaria, because a common element among people living on top of the hills was better air. Subsequent investigation using the method of difference revealed that only individuals with the malaria parasite in their blood

---

EXHIBIT 3.1 Morison's Discussion of the History of Attempts to Find the Cause of Malaria

---

Whatever the reason, medical men have found it congenial to assume that they could find something called *The Cause* of a particular disease. If one looks at the history of any particular disease, one finds that the notion of its cause has varied with the state of the art. In general, the procedure has been to select as *The Cause* that element in the situation which one could do the most about. In many cases it turned out that, if one could take away this element or reduce its influence, the disease simply disappeared or was reduced in severity. This was certainly desirable, and it seemed sensible enough to say that one had got at the cause of the condition. Thus in ancient and medieval times malaria, as its name implied, was thought to be due to the bad air of the lowlands. As a result, towns were built on the tops of hills, as one notices in much of Italy today. The disease did not disappear, but its incidence and severity were reduced to a level consistent with productive community life.

At this stage it seemed reasonable enough to regard bad air as the cause of malaria, but soon the introduction of quinine to Europe from South America suggested another approach. Apparently quinine acted on some situation within the patient to relieve and often to cure him completely. Toward the end of the last century the malaria parasite was discovered in the blood of patients suffering from the disease. The effectiveness of quinine was explained by its ability to eliminate this parasite from the blood. The parasite now became *The Cause,* and those who could afford the cost of quinine and were reasonably regular in their habits were enabled to escape the most serious ravages of the disease. It did not disappear as a public health problem, however, and further study was given to the chain of causality. These studies were shortly rewarded by the discovery that the parasite was transmitted by certain species of mosquitoes. For practical purposes *The Cause* of epidemic malaria became the Mosquito, and attention was directed to control of its activities.

Entertainingly enough, however, malaria has disappeared from large parts of the world without anyone doing much about it at all. The fens of Boston and other northern cities still produce mosquitoes capable of transmitting the parasite, and people carrying the organism still come to these areas from time to time, but it has been many decades since the last case of the disease occurred locally. Observations such as this point to the probability that epidemic malaria is the result of a nicely balanced set of social and economic, as well as biological factors, each one of which has to be present at the appropriate level. We are still completely unable to describe these sufficient conditions with any degree of accuracy, but we know what to do in an epidemic area because we have focused attention on three or four of the most necessary ones.

---

suffered from the disease. The problem with this second explanation is that it did not explain how the parasite came to exist in the bloodstream, until it was found that the mosquito transmitted it. As you can see, the various canons set forth by Mill enable us to identify the relationships that exist among a set of variables. However, they do not help us to name the single factor that causes an effect, just as they did not enable scientists to identify the single factor that causes malaria. This is because the identification of causation can occur only when *no* alternative interpretations for an effect exist other than the one specified. When we have reached this stage, we have essentially identified both the necessary and the sufficient conditions for the occurrence of an event. A **necessary condition** refers to a condition that must be present in order for the effect to occur. (To become an alcoholic, you must consume alcohol.) A **sufficient condition** refers to a condition that will always produce the effect. (Destroying the auditory nerve always results in a loss of hearing.)

Necessary condition A condition that must exist for an effect to occur

Sufficient condition A condition that will always produce the effect under study

A condition must be both necessary and sufficient to qualify as a cause, because in such a situation the effect would never occur unless the condition were present, and whenever the condition was present, the effect would occur. If a condition were only *sufficient*, then the effect could occur in other ways. (There are several ways one can lose one's hearing in addition to destruction of the auditory nerve.) In like manner, a *necessary* condition does not mean that the effect will necessarily occur. (All people who consume alcohol do not become alcoholics, but one must consume alcohol to become an alcoholic.)

To state that we have found the cause for an event means that both the necessary and the sufficient conditions have been found. It means that a complete explanation of the occurrence of the event has been isolated and that, unlike the theory of malaria, the explanation *will never change.*

It is presumptuous to assume that we will ever find the conditions necessary and sufficient for the occurrence of an event, however, since the behavior of organisms is extremely complex. Seldom do we encounter situations or behaviors that cannot be explained in several different ways. Popper (1968) has perhaps been most explicit in his insistence on the necessity of ruling out alternative explanations. According to inductive logic, science must be capable of deciding between the truth and falsity of hypotheses and theories. In other words, if we conduct a scientific experiment testing the hypothesis that depression can be treated with psychotherapy, we should be able to decide if this hypothesis is true or false. Popper rejects the notion of such inductive logic. He maintains that we cannot use the results of one or even several scientific experiments to infer that a given hypothesis or theory is true, or proven. Even if five experiments show the success of treating depressives with psychotherapy, this is not conclusive proof that psychotherapy can successfully treat depressives. The attained relationship could be due to flaws in the experiments or to unknown variables operating simultaneously with the psychotherapy. To Popper, a confirmation of an experiment states only that the hypothesis tested has survived the test. On the other hand, if the experiment fails to confirm a prediction or a theory, the prediction or theory being tested is falsified. Therefore, Popper focuses attention on a **position of falsification** rather than a position of confirmation. For him, a theory or prediction can only achieve the status of "not yet disconfirmed"; it can never be proven. In other words, if an experiment supports the prediction that psychotherapy is beneficial in treating depressives, Popper would not state that the prediction has been confirmed; he would merely state that this prediction has maintained the status of not yet disconfirmed. This status is very precious in science, however, because it means that the theory or prediction has passed the test of rigorous experimentation and thus only states one of the possible true explanations.

Position of falsification
The belief that the best that can be said about a theory or prediction is that it is "not yet falsified"

Deese (1972) provides yet another view of causation. He sees causation as a large network of cause-and-effect relations. Any given cause-and-effect relation that is isolated in a study is only one such relation embedded in a matrix of others. Consider the case of Morison's discussion of malaria, in which he illustrates the covariation between a number of events and malaria, each of which was once considered a specific cause-and-effect relationship. The bad air of the lowlands was found to covary with the incidence of malaria. Later it was found that the presence of a parasite covaried with the appearance of malaria, and even later it was found that a certain species of mosquito caused malaria because it carried the malaria parasite. Note that a number of specific cause-and-effect relationships, in terms of covariation of events, were involved in the history of trying to identify the cause of malaria. It is apparent that none of these specific relationships could be labeled as the cause of malaria, because many of the so-called causative events (such as the mosquitoes) still exist and yet the presumed effect of malaria no longer occurs. For a given effect to occur, as Morison pointed out, a nicely balanced system of interrelated conditions must exist. For malaria to occur, the mosquitoes and parasites must exist in a system of other specific social and economic conditions. Any one condition by itself is not sufficient to produce the effect. Proponents of this view of causation advocate study of the relationship among the levels or amounts of the variables operating within a system rather than study of the covariation between one variable, which can be labeled the cause, and another, which can be labeled the effect. Such a viewpoint sees any given study as representing only a small part of the overall system, and the relationship found in a given study exists only if certain relationships exist among the remainder of the elements of the system.

It is clear that causation is subject to quite a bit of debate. [Brand (1976) provides a detailed discussion of this debate.] Where does this leave the psychologist who is attempting to identify causal relations? The behavior of organisms is extremely complex and multidetermined. It is rarely—if ever—caused by one event. Therefore, not only is it impossible in most instances to name *the* cause of an event, but in reality a single cause for a behavior seldom exists. Given this state of affairs, we must conduct our scientific investigations in a manner that will enable us to identify most of the interacting causes rather than attempting to find the single cause of an event or given behavior.

THE PSYCHOLOGICAL EXPERIMENT

Psychological experiment Objective observation of phenomena that are made to occur in a strictly controlled situation in which one or more factors are varied and the others are kept constant

Zimney (1961, p. 18) defines a **psychological experiment** as "objective observation of phenomena which are made to occur in a strictly controlled situation in which one *or more* factors are varied and the others are kept constant" (italics mine). This definition seems to be a good one because of the components that it includes, each of which will be examined separately. Analysis of this definition, with one minor alteration, should provide a definition of an experiment, an appreciation of the many facets of experimentation, and a general understanding of how experimentation enables causative relationships to be identified.

Objective Observation Impartiality and freedom from bias on the part of the investigator, or objectivity, was previously discussed as a characteristic that the scientist must exhibit. In order to be able to identify causation from the results of the experiment, the experimenter must avoid doing anything that might influence the outcome. Rosenthal (1966) has demonstrated that the experimenter is probably capable of greater biasing effects than one would expect. In spite of this, and recognizing that complete objectivity is probably unattainable, the investigator must strive for freedom from bias.

Science requires that we make empirical observations in order to arrive at answers to the questions that are posed. Observations are necessary because they provide the data base used to attain the answers. To provide correct answers, experimenters must make a concerted effort to avoid mistakes, even though they are only human and therefore are subject to errors in recording and observation. For example, work in impression formation has revealed the biased nature of our impressions of others. Gage and Cronback (1955, p. 420) have stated that social impressions are "dominated far more by what the Judge brings to it than by what he takes in during it." Once scientists realize that they are capable of making mistakes, they can guard against them. Zimney (1961) presents three rules that investigators should follow to minimize recording and observation errors. The first rule is to accept the possibility that mistakes can occur—that we are not perfect, that our perceptions and therefore our responses are influenced by our motives, desires, and other biasing factors. Once we accept this fact, we can then attempt to identify where the mistakes are likely to occur—the second rule. To identify potential mistakes, we must carefully analyze and test each segment of the entire experiment in order to anticipate the potential sources and causes of the errors. Once the situation has been analyzed, then the third rule can be implemented—to take the necessary steps to avoid the errors. Often this involves constructing a more elaborate scenario or just designing equipment and procedures more appropriately. In any event, every effort should be expended to construct the experiment so that accurate observations are recorded.

Phenomenon
A publicly observable
behavior

Of Phenomena That Are Made to Occur Webster's dictionary defines *phenomenon* as "an observable fact or event." In psychological experimentation, **phenomenon** refers to any publicly observable behavior, such as actions, appearances, verbal statements, responses to questionnaires, and physiological recordings. Focusing on such observable behaviors is a must if psychology is to meet the previously discussed characteristics of science. Only by focusing on these phenomena can we satisfy the demands of operational definition and replication of experiments.

Defining a phenomenon as publicly observable behavior would seem to exclude the internal or private processes and states of the individual. In the introductory psychology course, processes such as memory, perception, personality, emotion, and intelligence are discussed. Is it possible to retain these processes if we study only *publicly* observable behavior? Certainly these processes must be retained, since they also play a part in determining an individual's responses. Without getting into the controversy over intervening variables and hypothetical constructs (Marx, 1963, pp. 24–31), let me simply state that such processes are studied diligently by many psychologists. In studying these processes, researchers investigate publicly observable behavior and infer from their observations the existence of internal processes. It is the behavioral manifestation of the inferred processes that is observed. For example, intelligence is inferred from responses to an intelligence test, aggression from verbal or physical attacks on another person.

Not all psychologists accept this position. Notably, B. F. Skinner considered the inferring of internal states as inappropriate and not the subject matter of psychology. He believed that psychology should study only environmental phenomena and ignore anything that cannot be observed or that must be inferred. Psychologists with this opinion believe we should investigate only those environmental sequences, such as stimulus-response reinforcement, that determine behavior, and they have had a great deal of success with this approach. Pribram (1971, p. 253) makes two cogent points regarding this issue. First, the behaving organism is required to define each of the environmental variables. Only the organism can tell us what is reinforcing. Likewise, the response of the organism defines the stimulus and the stimulus defines the response. Second, it is the *internal* processes of the organism that enable the sequencing of events to take place. In order to determine what makes the sequencing occur, we must return to the organism and what is taking place inside that organism.

In the discussion of control as a goal of science, we saw that the psychologist does not have a direct controlling influence on behavior. The psychologist arranges the antecedent conditions that result in the behavior of interest. In an experiment, the experimenter precisely manipulates one or more variables and objectively observes the phenomena *that are made to occur* by this manipulation. This part of the definition of experimentation refers to the fact that the experimenter is manipulating the conditions that cause a certain effect. In this way, experimenters identify the cause-and-effect relationships from experimentation by noting the effect or lack of effect produced by their manipulations.

In a Strictly Controlled Situation   This part of the definition refers to the need to eliminate the influence of variables other than those manipulated by the experimenter (Boring's second meaning of the word *control*). As you have seen, control is one of the most pressing problems facing the experimenter and one to which considerable attention is devoted. Without control, causation could not be identified. Because of the magnitude of this issue, it will be given extended coverage in later chapters.

In Which One or More Factors Are Varied and the Others Are Kept Constant   The ideas expressed in this phase of the definition are epitomized by the *rule of one variable,* which states that all conditions in an experiment must be kept constant except one, which is to be varied along a defined range, and the result of this variation is to be measured on the response variable. The two major ideas expressed in the rule of one variable are constancy and variation. *Constancy* refers to controlling or eliminating the influence of all variables except the one (or ones) of interest. This requirement is necessary to determine the cause of the variation on the response variable. If the constancy requirement is violated, cause for the variation cannot be determined and the experiment is ruined. A learning experiment can easily illustrate the constancy component of the rule. Assume that you are interested in the effect of the length of a list of words on speed of learning. How does increasing the length influence the speed with which one learns that list of words? The length of the list of words could be systematically varied and related to the number of trials needed to learn the list. In such an experiment, some factors that could influence learning speed must be controlled, including the difficulty of the words, the subjects' ability level, the subjects' familiarity with the words, and the subjects' motivation level. Only if these factors are held constant (and therefore do not exert an influence) can you say that the difference in speed of learning is a function of the change in the length of the list of words. The idea of *variation* means that one or more variables must be deliberately and precisely varied by some given amount to determine their effect on behavior. In the learning experiment, the length of the list of words must be changed by an exact, predetermined amount. The questions that frequently arise are *how* and *how much* is the variable to be varied? The answers to these questions will be discussed later in the book.

Advantages of the Experimental Approach

The first and foremost advantage of the experimental approach is the strength with which a causal relationship can be inferred. This inferential strength derives from the degree of control that can be exercised. Control, as stated in Chapter 1, is the most important characteristic of the scientific method, and the experimental approach enables the researcher to effect the greatest degree of control. In an experiment, one is seeking an answer to a specific question. In order to obtain an unambiguous answer, it is necessary to institute control over irrelevant variables by either eliminating their influence or holding their influence constant. Such control can be achieved by bringing the experiment into the laboratory, thereby eliminating noise and other potentially distracting stimuli. Control is also achieved by using such techniques as random assignment and matching, which will be discussed in Chapter 7.

A second advantage of the experimental approach is the ability to precisely manipulate one or more variables of the experimenter's choosing. If a researcher is interested in studying the effects of crowding on a particular behavior, crowding can be manipulated in a very precise and systematic manner by varying the number of people in a constant amount of space. If the researcher is also interested in the effects of sex of the subject and degree of crowding on some subsequent behavior, male and female subjects can be included in both the crowded and noncrowded conditions. In this way, the experimenter can precisely manipulate two variables: sex of subject and degree of crowding. The experimental approach enables one to control precisely the manipulation of variables by specifying the exact conditions of the experiment. The results can then be interpreted unambiguously, because the subjects should be responding primarily to the variables introduced by the experimenter.

A third advantage of the experimental approach is a completely pragmatic one: usefulness. This approach has produced results that have lasted over time, have suggested new studies, and, perhaps most important, have suggested solutions to practical problems.

Disadvantages of the Experimental Approach

The most frequently cited and probably the most severe criticism leveled against the experimental approach is that laboratory findings are obtained in an artificial and sterile atmosphere that precludes any generalization to a real-life situation. The following statement by Bannister (1966, p. 24) epitomizes this point of view:

> In order to behave like scientists we must construct situations in which subjects are totally controlled, manipulated and measured. We must cut our subjects down to size. We must construct situations in which they can behave as little like human beings as possible and we do this in order to allow ourselves to make statements about the nature of their humanity.

Is such a severe criticism of experimentation justified? It seems to me that the case is overstated. Underwood (1959), taking a totally different point of view, does not see artificiality as a problem at all. He states:

> One may view the laboratory as a fast, efficient, convenient way of identifying variables or factors which are likely to be important in real-life situations. Thus, if four or five factors are discovered to influence human learning markedly, and to influence it under a wide range of conditions, it would be reasonable to suspect that these factors would also be important in the classroom. But, one would *not* automatically conclude such; rather, one would make field tests in the classroom situation to deny or confirm the inference concerning the general importance of these variables.[1]

The artificiality issue is a problem only when an individual makes a generalization from an experimental finding without

---

[1] From "Verbal Learning in the Educative Processes" by Benton J. Underwood, Spring 1959, *Harvard Educational Review, 29*, pp. 107–117. Copyright © by President and Fellows of Harvard College.

first determining whether the generalization can be made. Competent psychologists rarely blunder in this fashion because they realize that laboratory experiments are contrived situations.

Additional difficulties of the experimental approach include problems in designing the experiment and the fact that the experiment may be extremely time consuming. It is not unusual for an experimenter to have to go to extreme lengths to set the stage for, motivate, and occasionally deceive the subject. Then, when the experiment is actually conducted, the experimenter and perhaps one or two assistants are often required to spend quite some time with each subject.

A final criticism that has been aimed at the experimental approach is that it is inadequate as a method of scientific inquiry into the study of human behavior. Gadlin and Ingle (1975) believe that a number of anomalies inherent in the experimental approach make it an inappropriate paradigm for studying human behavior. They state that the experimental approach promotes the view that humans are manipulable mechanistic objects because twentieth-century psychology mirrors the mechanistic method and assumptions of nineteenth-century physics. Gadlin and Ingle recommend the search for an alternative methodology that is not fraught with such inadequacies. Under close inspection, however, such criticisms do not hold up. Kruglanski (1976) has refuted each of the criticisms Gadlin and Ingle levied at the experimental approach. For example, the mechanistic manipulable assumption exists only to the extent that the experimenter arranges a set of conditions that may direct the individual's behavior in a given manner. This in no way "suggests that the subject is an empty machine devoid of feelings, thoughts, or a will of his own" (Kruglanski, 1976, p. 656). Again, it appears that we must resort to using the experimental approach to find the answers to our research questions. This does not mean that it is the only approach and that adherence to it will enable us to make great strides in understanding human behavior. At present, however, it seems to be one of the better approaches available to us.

Illustrative Example

To drive home the advantages and disadvantages of the experimental approach, we will take a detailed look at a laboratory experiment conducted by Wellman, Malpas, and Witkler (1981). These investigators tested the hypothesis that phenyl-propanolamine (PPA), the drug used in most over-the-counter diet preparations, causes animals to avoid eating by making them feel bad rather than by suppressing their desire to eat.

To test this hypothesis, researchers initially trained forty albino rats on six consecutive days to consume their daily water intake during a thirty-minute period. On the seventh day, the rats were given a 0.1 percent saccharin solution in-stead of water and then were randomly assigned to receive an injection of either 0.9 percent saline (the placebo, or control, condition) or 10, 20, or 40 mg/kg of PPA. During the next five days, the rats were given access to either water or a 0.1 percent saccharin solution, and the consumption of both water and the saccharin solution was recorded. Wellman et al. then com-puted the ratio of the amount of the saccharin solution that the rats consumed to the amount of water they consumed to determine whether PPA caused the rats to avoid the saccharin and drink primarily the water. Figure 3.1, which presents the results of this study, reveals that the ratio of saccharin solution consumed to water consumed declined as the amount of PPA injected increased.

The advantages of control of extraneous variables and precise manipulation of the variable of interest can easily be demonstrated. This study required control of a number of ex-traneous variables, which was possible because it was a labo-ratory study. The amount of fluid that the rats consumed was controlled by randomly assigning the subjects to the various groups. Additionally, the animals' level of thirst was con-trolled by allowing them to drink only at a certain time of the day. Consequently, all rats were deprived of water for equal periods of time. If this variable had not been controlled, the rats' consumption of water or of the saccharin solution might have been caused by the fact that some of the rats had con-sumed water during the preceding thirty minutes, whereas others had not consumed any water for the last two hours.

It was also necessary to precisely manipulate the dosage of PPA. This was rather easily accomplished in this experi-ment because the animals could be weighed. Once an animal's weight was determined, a dosage corresponding to 10, 20, or 40 mg/kg could be exactly identified and injected into the rat.

This study also demonstrates the primary disadvantage of the laboratory experiment. The laboratory experiment has most frequently been criticized for creating an artificial, sterile

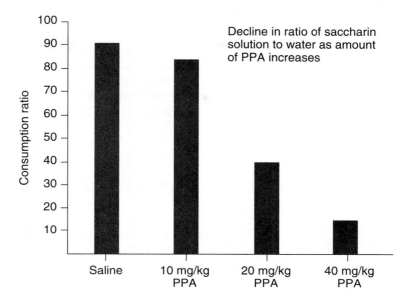

FIGURE 3.1 Ratio of amount of water to amount of saccharin solution consumed with increasing amounts of PPA. (Adapted from "Conditioned Taste Aversion of Unconditioned Suppression of Water Intake Induced by Phenylpropanolamine in Rates" by P. J. Wellman, P. B. Malpas, and K. C. Witkler, 1981, *Physiology and Behavior, 9,* pp. 203–207.)

atmosphere. In this experiment, the rats were individually housed in wire mesh cages with controlled lighting and temperature and were allowed to drink only during a predetermined thirty-minute time period each day. Seldom in real life would organisms confine their fluid consumption to only thirty minutes each day. Critics of laboratory experiments would also attack the Wellman et al. experiment on the basis of the organism used to test the hypothesis of the study. Wellman and his colleagues used rats to test the adverse effects of PPA, but the ultimate goal was to generalize the results to humans.

In most instances we cannot generalize from infrahumans to humans. Instead, we must identify effects on organisms such as rats and then verify the existence of such effects using humans as the research participants. Sulik, Johnston, and Webb (1981) present one of the more dramatic instances in which a direct generalization can be made from infrahumans to humans. Figure 3.2 illustrates the similarity in the pattern of facial malformations that occur in the mouse and in humans as the result of maternal consumption of alcohol, a syndrome called *fetal alcohol syndrome.* This figure reveals that the pattern of facial malformations is strikingly similar, indicating a phenomenon that is common to humans and infrahumans.

EXPERIMENTAL RESEARCH SETTINGS

The experimental approach is used in both laboratory settings and field settings. Although most experimental work has been and will probably continue to be conducted in the laboratory, we are hearing pleas for more field experimentation. Field and laboratory experimentation both use the experimental approach, but they have slightly different attributes which deserve mention.

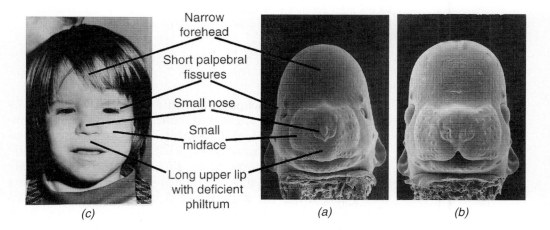

FIGURE 3.2   Malformation of facial features in the mouse and human caused by maternal alcohol consumption: (*a*) mouse embryo with fetal alcohol syndrome; (*b*) normal mouse embryo; (*c*) child with fetal alcohol syndrome. (Reprinted from "Fetal Alcohol Syndrome: Embryogenesis in a Mouse Model" by K. K. Sulik, M. C. Johnston, and M. A. Webb, 1981, *Science, 214,* no. 4523, pp. 936–938 (figure on p. 937), November 20, 1981. Copyright 1981 by the American Association for the Advancement of Science.)

Field Experimentation

Field experiment
An experimental
research study that is
conducted in a real-life
setting

A **field experiment** is an experimental research study that is conducted in a real-life setting. The experimenter actively manipulates variables and carefully controls the influence of as many extraneous variables as the situation will permit. Freedman and Fraser (1966), for example, wanted to find out if people who initially complied with a small request would be more likely to comply with a large request. The basic procedure used was for a researcher to initially ask a group of housewives if they would answer a number of questions about what household products they used. Three days later, the experimenter again contacted these same housewives and asked if they would allow a group of men to come into their homes and spend approximately two hours classifying all of their household products. Another group of housewives was contacted only once, during which time the large request was made. As Figure 3.3 reveals, the housewives who initially complied with the small request were significantly more likely to comply with the large request.

This is an example of a field study because it was conducted in the natural setting of the housewives' homes while they were engaging in daily activities. It also represents an experimental study because variable manipulation was present (small request followed by large request, or just large request) and control was present (the subjects for each group were randomly selected from the telephone directory). Field experiments like this one are not subject to the artificiality problem that exists with laboratory experiments, so field experiments are excellent for studying many problems. Their primary disadvantage is that control of extraneous variables cannot be accomplished as well as it can be with laboratory experiments. In the Freedman and Fraser study, even though subjects were randomly selected from the telephone directory, only those subjects who were actually home could be included in the study. Consequently, a selection bias may have existed. Even though it is more difficult to exercise control in field experiments, such experiments are necessary. We need to get out of the laboratory and get more involved with field experimentation.

Tunnell (1977) has carried such a suggestion a step further. He states that we not only must engage in more field experimentation but also should do so in a manner that makes all variables operational in real-world terms. Consider the

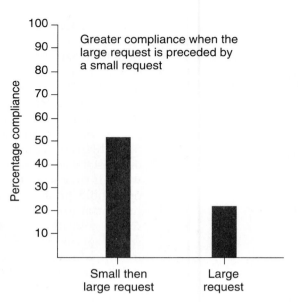

FIGURE 3.3 Percentage of subjects complying with a large request. (Adapted from "Compliance without Pressure: The Foot-in-the-Door Technique" by J. L. Freedman and S. C. Fraser, 1966, *Journal of Personality and Social Psychology, 4,* pp. 195–202.)

study conducted by Ellsworth, Carlsmith, and Henson (1972). They had a confederate pedestrian stare at car drivers who were waiting at a red light. In this way, the researchers hoped to assess the influence of staring on the speed with which the driver left the intersection. As might be expected, staring increases that speed. In this study, Ellsworth et al. included the three dimensions of naturalness identified by Tunnell: natural behavior, natural setting, and natural treatment. The natural behavior investigated was speed of leaving the intersection when the light turned green. The setting was natural because the individual remained in his or her car, as opposed to being brought into a laboratory; the natural treatment was the staring. In reality, the treatment was imposed by a confederate, but it mirrored a behavior that could have occurred naturally. These are the types of behaviors Tunnell says we must strive for when we conduct field experimentation, as opposed to asking subjects to provide self-reports or to recall their own behavior in some prior situation. Asking for such retrospective data only serves to introduce possible bias in the study.

Laboratory Experimentation

The laboratory experiment is the same type of study as the field experiment, but where the field experiment is strong, the laboratory experiment is weak, and where the laboratory experiment is strong, the field experiment is weak. The laboratory experiment epitomizes the ability to control or eliminate the influence of extraneous variables. This is accomplished by bringing the problem into an environment apart from the subjects' normal routines. In this environment, outside influences (such as the presence of others and of noise) can be eliminated. However, the price of this increase in control is the artificiality of the situation created. This issue was covered in detail in the discussion of the disadvantages of the experimental approach. Even though precise results can be obtained from the laboratory, applicability of these results to the real world must always be verified.

Laboratory experiment
An experimental research study that is conducted in the controlled environment of a laboratory

The **laboratory experiment** is a study that is conducted in the laboratory and in which the investigator precisely manipulates one or more variables and controls the influence of all or nearly all of the extraneous variables. For example, Darley and Latané (1968) were interested in trying to explain why people may not help in obvious emergency situations, as in the case of Kitty Genovese. In that situation many people clearly were aware of the situation and realized that it was a true emergency—but no one even bothered to call the police. Darley and Latané speculated that the mere perception that other people are also witnessing the event will decrease the likelihood that any one individual will intervene. In the case of Kitty Genovese, many residents of the apartment house where she lived opened their windows and heard her cries for help, but none of them made any attempt to get involved. To investigate the possible impact that this variable might have on intervention in emergencies, Darley and Latané had to construct a situation in which they precisely manipulated the perception that other people had witnessed an emergency. Additionally, extraneous variables such as knowledge of whether anyone else had actually intervened had to be eliminated.

In order to precisely manipulate this variable and to control for the impact of extraneous variables, Darley and Latané simulated an emergency situation in a laboratory setting. These investigators had subjects, either alone or in the presence of others, complete a variety of questionnaires. While the subjects were completing the questionnaires, smoke started filtering into the room through air vents. Darley and Latané wanted to find out whether it would take longer for someone to report the smoke if more than one person at a time was in the room witnessing the potential emergency. The results of this study revealed that people are less prone to intervene in an emergency when they know that others have also witnessed it.

SUMMARY

The experimental approach is the research method in which one attempts to identify cause-and-effect relationships by conducting an experiment. Causation is a concept that is little understood, and yet it is causative relationships that the experimental method attempts to identify. John Stuart Mill set forth four canons—the methods of agreement, difference, and concomitant variation, and the joint methods of agreement and difference—that he said could be used in identifying causation. However, in order for one to be able to state that *the* cause of a given effect has been found, this condition must qualify as being both necessary and sufficient. Since behavior is multidetermined, it is highly unlikely that one can overrule all possible alternative explanations for behaviors. This is why Popper rejects the confirmationists' position and takes the position of falsification, maintaining that the best status a theory can attain is one of "not yet disconfirmed." Deese believes that causal relations are embedded in a matrix of other causal relations. A given relationship between a cause and an effect will continue to exist only if all the other variables within the matrix or system remain constant.

The psychological experiment achieves the goal of the experimental approach by allowing the researcher to observe, under controlled conditions, the effects of systematically varying one or more variables. The experimental approach has the primary advantage of providing for control of extraneous variables. Other advantages are that it permits the precise manipulation of one or more variables, produces lasting results, suggests new studies, and suggests solutions to practical problems. The experimental approach has the disadvantages of creating an artificial environment and frequently being time consuming and difficult to design.

The experimental approach is used in both field and laboratory settings. In a field setting, the researcher makes use of a real-life situation and thereby avoids criticism for having created an artificial environment. Typically, however, there is not as much control over extraneous variables. In a laboratory setting, the experimenter brings the subjects into the laboratory, where there is maximum control over extraneous variables; however, this usually means creating an artificial environment.

STUDY QUESTIONS

1. What is the primary characteristic of the experimental research approach?
2. Identify and elaborate on the different attempts to explain causation.
3. Why can cause-and-effect statements be made more readily from experimental research than from descriptive research?
4. Define the psychological experiment and explain the significance of each facet of the definition.
5. What are the advantages and disadvantages of the psychological experiment?
6. What are the two settings in which the psychological experiment is conducted, and what are the advantages and disadvantages of each?

KEY TERMS AND CONCEPTS

Experimental research

Causation

Method of agreement

Method of difference

Joint methods of agreement and difference

Method of concomitant variation

Necessary condition

Sufficient condition

Position of falsification

Psychological experiment

Objective observation

Phenomenon

Rule of one variable

Constancy

Field experiment

Laboratory experiment

CHAPTER 4

Problem Identification and Hypothesis Formation

LEARNING OBJECTIVES

1. To learn where to find researchable problems.
2. To learn how to conduct a literature search on a given topic.
3. To learn how to specify a research problem.
4. To learn how to formulate a hypothesis relating to the research question.

Tom, a middle-aged physician, was grief stricken. He had just learned that his daughter had been killed in an automobile accident. His emotional trauma was so great that the only way he could cope with the tragedy was to deny the existence of both his daughter and her death. Over the next several years he mechanically performed his duties as a physician, but he also had a continual infection and a nagging cough. Work became increasingly stressful and more difficult to perform competently. As the girl's birthday approached, three years after her death, the physician again began to grieve. On the eve of her birthday, barely able to breathe, he was admitted to the intensive care unit of the hospital. A lung biopsy revealed that he had a very rare type of pneumonia, and he was given antibiotics, to which he promptly responded. Nevertheless, the infection made him suspicious, because he was bisexual. He researched the literature in the medical library and arrived at a diagnosis of acquired immunodeficiency syndrome (AIDS), which was later confirmed by a laboratory workup.

This case is a true account (Maier & Laudenslager, 1985) of a situation that is increasingly common in the United States. Upon hearing a story like this, most individuals would experience a variety of emotions—sympathy, compassion, perhaps increased apprehension and fear of this killer disease. But if you were a researcher interested in the causes and treatment of AIDS, you would go beyond these emotions and search for factors that might have contributed to AIDS. For example, the physician had experienced a variety of stressful events in his life for the past several years. Did this added stress contribute to or make him more susceptible to contracting AIDS? This is a reasonable hypothesis, as there is a large body of research supporting the theory that medical problems such as ulcers and heart attacks are preceded by emotionally painful events. Recently, data are accumulating that indicate that stress also affects a person's vulnerability to infectious disease. If you knew about this research, you might go beyond asking whether *stress has anything to do with contracting AIDS and ask instead* what *stress might be doing to make a person more vulnerable to infectious diseases like AIDS. These two research questions are currently being researched.*

This illustration may give the impression that legitimate research questions are easy to formulate, and for the veteran researcher this is typically true. Experienced researchers are likely to have more unanswered questions than they have time to investigate. Beginning researchers, however, commonly have difficulty singling out a research problem that they are capable of investigating. In this chapter I will attempt to minimize such difficulty by discussing the origin of researchable problems and the way to convert a problem you have identified into one that can be investigated by the experimental research approach.

INTRODUCTION

Up to this point in the text, I have discussed the general characteristics of the scientific method and the two basic approaches to using this method of inquiry. Using either of these approaches, however, requires that we first have a problem in need of a solution. In the field of psychology, identification of a research idea should be relatively simple because psychology is the scientific study of behavior—including human behavior. Our behavior represents the focus of attention of a great deal of psychological investigation. To convert our observations of behavior into legitimate research questions, we must be inquisitive and ask ourselves why certain types of behavior occur. For example, assume that you hear a person express an extremely resentful, hostile, and prejudiced attitude toward Russians. The next day you see this person interacting with a Russian and note that they are both being very polite and courteous. You have seen a contradiction between the attitude expressed by this individual and her behavior. Two well-founded research questions would be "Why is there a lack of correspondence between attitude and behavior?" and "Under what circumstance do attitudes *not* predict behavior?"

Let us now look at the major sources that can be used to generate research questions.

SOURCES OF RESEARCH IDEAS

Where do ideas or problems originate? Where should we look for a researchable problem? In all fields, there are a number of common sources of problems, such as existing theories and past research. In psychology, we are even more fortunate; we have our own personal experience and everyday events to draw from. The things we see, read about, or hear about can serve as ideas to be turned into research topics. But identifying these ideas as research topics requires an alert and curious scientist. Rather than just passively observing behavior or reading material relating to psychology, we must actively question the reasons for the occurrence of an event or behavior. If you ask "Why?" you will find many researchable topics. A brief glance at the *Psychological Abstract Index* provides an indication of the many areas within psychology that have unsolved problems. Typically, problems originate from one of four sources: everyday life, practical issues, past research, or theory.

Everyday Life

As we proceed through our daily routine, we come into contact with many questions in need of solution. Parents want to know how to handle their children; students want to know how to learn material faster. When we interact with others or see others react, we note many individual differences. If we observe children on a playground, these differences are readily apparent. One child may be very aggressive and another much more reserved, waiting for others to encourage interaction. The responses of a particular person also vary according to the situation. A child who is very aggressive in one situation may be passive in another. Why do these differences exist? What produces these varying responses? Why are some people leaders and others followers? Why do we like some people and not others? Many such researchable questions can be identified from everyone's interactions and personal experiences.

In the late 1960s, Darley and Latané (1968) began a series of investigations that epitomize the use of life's experiences and the events taking place around us as a source of research problems. They were concerned about the fact that bystanders often do not lend assistance in emergency situations. A case in point is the often cited incident discussed in Chapter 3 involving Kitty Genovese, who was stabbed to death in New York City. There were thirty-eight witnesses to the attack, which lasted more than half an hour, and no one even called the police. Many other similar and more recent cases can be recounted. For example, a woman was raped by four men on a barroom pool table while onlookers cheered her attackers on. In St. Louis, a thirteen-year-old girl was raped by two youths as several adults stood around and watched. It took a thirteen-year-old boy to finally summon the police. Darley and Latané asked why. They began to study the conditions that facilitate or inhibit bystander intervention in emergency situations.

Practical Issues

Many experimental problems arise from practical issues that require solution. Private industry faces such problems as low employee morale, absenteeism, turnover, selection, and placement. Work has been and continues to be conducted in these areas. Clinical psychology is in need of a great deal of research to identify more efficient modes of dealing with mental disturbances. Units of the federal and state governments support experimentation designed to solve practical problems, such as finding a cure for cancer. Large expenditures are also being directed toward improving the educational process.

Law enforcement agencies are concerned not only with obtaining accurate eyewitness testimony but also with extracting leads or clues from eyewitnesses. To that end, these agencies are now using hypnosis, under the assumption that hypnosis can extract accurate evidence that otherwise would not be available. The validity of such an assumption was not tested until recently, however. Sanders and Simmons (1983) asked eyewitnesses, some of whom were hypnotized and some of whom were not hypnotized, to identify a thief from a lineup. As Figure 4.1 reveals, hypnotized subjects, contrary to expectations, identified the thief *fewer* times than did the subjects who were not hypnotized. Such evidence suggests that hypnosis is not an effective technique for extracting accurate evidence.

Morin, Charles, and Malyon (1984) have indicated that people with AIDS repeatedly express the perception that their medical needs are being met but that their psychological needs are being neglected. Although these individuals may be getting the best possible medical care, little if any attention has been given to their emotional reaction to being diagnosed as having AIDS. Consequently, research is needed on the psychological impact of this physical disorder and on strategies for helping patients cope with their emotions.

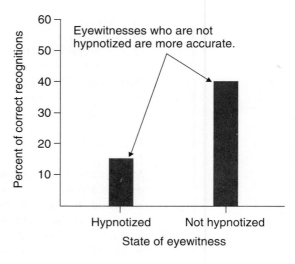

FIGURE 4.1 Accuracy of eyewitness identification as a function of being hypnotized. (Based on data from "Use of Hypnosis to Enhance Eyewitness Accuracy: Does It Work?" by G. S. Sanders and W. L. Simmons, 1983, *Journal of Applied Psychology, 68,* pp. 70–77.)

Past Research

Previously conducted experiments are an excellent source of research ideas. This may sound like a contradiction, since research is designed to answer questions, but one of the interesting features of research is that it tends to generate more questions than it answers. Although each well-designed study does provide additional knowledge, phenomena are multidetermined. In any experiment, only a limited number of variables can be studied. Investigation of these variables may lead to hypotheses about the effects of other variables. The multidimensional nature of phenomena is also frequently the cause of lack of agreement between experimental results. An unidentified variable may be the source of conflict between various studies on a given problem, and experiments must be conducted to uncover this variable and thereby eliminate the apparent contradiction.

To illustrate what has just been said, consider the study conducted by Mellgren, Seybert, and Dyck (1978). They investigated the influence of presenting different orders of schedules of continuous reinforcement, nonreinforcement, and partial reinforcement on resistance to extinction. Previous research had revealed conflicting results when resistance to extinction of subjects who had received continuous reward and then partial reward was compared to that of subjects given only partial reward schedules. Some studies indicated that resistance to extinction decreased, others indicated that it increased, and still others showed that the existence of an increase or decrease in resistance to extinction depended on the stage of extinction. Mellgren et al. attempted to resolve this inconsistency. Results of their study revealed that the greatest resistance to extinction occurred when a large number of nonreinforced trials preceded a partial reinforcement schedule. Although this study showed which schedule produced the greatest resistance to extinction, it left other questions unanswered. For example, it did not provide an explanation of why resistance to extinction increases if a large number of nonreinforced trials precedes partial reinforcement. This led to another study, which attempted to answer this new question. As you can see, each study leads to a subsequent study, and so people can spend their whole lives investigating one particular area. Research is an ongoing process.

Theory

Theory
A group of logically organized and deductively related laws

A **theory,** defined by M. H. Marx (1963, p. 9) as "a group of logically organized (deductively related) laws," is supposed to serve a number of distinct functions. Marx states that theory is both a tool and a goal. The *goal* function is evidenced by the proposition that laws are ordered and integrated by theories; theories summarize and integrate existing knowledge. The *tool* function, evidenced by the proposition that theories guide research, is the function of interest to us. A good theory goes beyond the goal function to suggest new relationships and make new predictions. Thus, it serves as a source of researchable ideas.

Leon Festinger's (1957) theory of cognitive dissonance is an example of a theory that stimulated an extraordinary amount of research in the decade that followed its publication. From this theory, Festinger and Carlsmith (1959) hypothesized and validated the less-than-obvious prediction that, after completing a boring task, subjects who were given $1 to tell a "stooge" that the boring task was interesting and fun actually stated that they enjoyed the task more than did the subjects who were given $20 to do the same thing.

GENDER BIAS IN RESEARCH IDEAS

Although there are many sources of ideas in psychology, it is important that we not overlook significant topics. To do so would lead us to develop a knowledge base that is incomplete. All scientists would probably agree that we need to conduct research on all important topics. However, scientists are human and the questions they ask are shaped not only by the particular topics they think are most significant but also by their gender roles. Because most scientists are males, the research topics they have selected to investigate have been those which males have considered to be most important. This has meant that some significant issues have been neglected or under-researched. Many of these issues are those that revolve about women. For example, feminists argue that menstrual cramps have not been taken seriously by the medical profession (Keller, 1984). Similarly, there is a lot of research focusing on the influence of mothers' working outside the home on their children's psychological welfare. Much less attention has focused on whether fathers' commitment to their work endangers their children's welfare or if the mothers' employment might even benefit their children (Hare-Mustin & Marecek, 1990). It appears that personal biases have influenced the selection of research questions and have led scientists to overlook important aspects of human behavior. Some (e.g., Unger & Crawford, 1992) have suggested that this is because of the preponderance of male scientists. If this is the case, then it is incumbent upon both male and female scientists to constantly assess the most important issues for research. This continual assessment must incorporate listening to the community as well as other scientists. Any gender bias that exists is probably a natural outgrowth of belonging to a particular gender and not a deliberate attempt to avoid particular issues that may be important. To correct such bias and ensure that all topics of importance receive attention it is imperative that the scientific community include scientists of both genders.

IDEAS NOT CAPABLE OF SCIENTIFIC INVESTIGATION

Researchable ideas, as you have just seen, originate from a variety of sources. However, it is important to realize that not all ideas are subject to scientific investigation. One of the criteria that a scientific study must meet is that the research idea must be capable of being confirmed or refuted. There are some ideas that are very important, are debated vigorously, and consume inordinate amounts of time and energy but are not subject to scientific investigation. These ideas typically revolve about issues of morality and religion. Consider, for example, the issue of abortion. This is an issue that has been debated for decades and has polarized the population. A large segment of the population advocates a "pro-choice" position; another large segment advocates a "pro-life" position. Science can investigate the genesis of these positions and mechanisms for changing them but it cannot resolve the issue of which position is the best or correct one.

Similarly, consider religious beliefs and convictions. Science can investigate many facets of religion, such as factors influencing attending church, but it cannot answer questions such as which religion is the best or "true" religion. It also cannot answer religious questions such as whether there is any power in prayer or questions regarding the existence of a supreme being. Such questions lie outside the realm of science.

REVIEW OF THE LITERATURE

After a topic of research has been obtained from one of the sources just mentioned, the next step in the research process is to become familiar with the information available on the topic. For example, assume that you want to conduct research on the impact of environmental stress on AIDS. Before beginning to design such a research project, you should first become familiar with current information on both of these topics. Prior work has been conducted on practically all psychological problems, and the topics of AIDS and environmental stress are no exception.

At this point you might be asking yourself: "Why should I review the literature on my selected topic? Why not just proceed to the laboratory and find an answer to the problem?" There are several good reasons why you should do your home work in the form of a literature review before conducting any experiments. The general purpose of the library search is to gain an understanding of the current state of knowledge about the selected topic. Specifically, a review of the literature will tell you whether the problem you have identified has already been researched. If it has, you should either revise the problem in light of the experimental results or look for another problem. If the topic you have identified has not been investigated, related studies may indicate how you should proceed in attempting to reach an answer to the problem. A literature review should also point out methodological problems specific to the area of study. Are special control groups or special pieces of equipment needed to conduct the research? If so, the literature can give clues as to where to find the equipment or how to identify the particular groups of subjects needed. These are just a few of the salient reasons for conducting a review of the literature.

Assuming you are convinced of the necessity of a literature review, you now need to know how to conduct such a review. Reed and Baxter (1991) and Rosnow and Rosnow (1992) have prepared excellent guides to conducting library research in psychology and to preparing papers that review that research. These guides point out that the two primary sources that should be investigated when reviewing the literature are books dealing with the topic of interest and psychological journals.

Books

Books have been written about most, if not all, areas in psychology. These texts should be examined for material relating to the research topic. The pertinent material may consist of actual information or may point to where information can be obtained. One book that is often very useful is the *Annual Review of Psychology*. Published yearly since 1950, it presents an expert's in-depth discussion of the principal work done during the preceding year on a variety of topics. One of the topics may relate to your own, so it is worthwhile to check this source. Other relevant books and chapters can be identified from a search of *Psychological Abstracts* or PsycLIT (discussed below) as the American Psychological Association is incorporating these references into this data base.

Psychological Journals

Most of the pertinent information about a research topic is usually found in the psychological journals. Frequently, a review that has started with books leads to the journals. Since books are generally the outgrowth of work cited in journals, this progression from books back to journals is a natural one.

How should one proceed in reviewing the work cited in the journals? There are so many psychological journals that it would be impossible to go through each and every one looking for relevant information. This is where *Psychological Abstracts* comes in. **Psychological Abstracts** is a journal consisting of brief abstracts (summaries) of published articles, books, and so forth from sources throughout the world but predominantly within the United States. This journal contains abstracts of over 950 journal and technical reports, monographs, and other scientific documents. Exhibit 4.1 provides a brief outline of how to use *Psychological Abstracts.* By following the steps specified in Exhibit 4.1, you should be able to compile a list of articles pertaining to your study.

Psychological Abstracts
A journal that contains abstracts of books, journal and technical reports, monographs, and other scientific documents

Information Retrieval Systems

In the past, the traditional approach to conducting a literature search was to search the *Psychological Abstracts* manually for relevant books and journals. This approach has become unwieldy and very impractical with the vast amount of information currently available. With the advancements that have been made in computer technology it has become possible to store and access large data sets such as the bibliographic entries published in *Psychological Abstracts.*

In an effort to expedite and improve the accuracy of literature searches a number of comprehensive computerized information storage and retrieval systems have been developed. Within the field of psychology, **PsycINFO,** Psychological Abstracts Information Services, is the primary service for behavioral information. PsycINFO is a family of interrelated information services that provides access to the world's literature in psychology and related behavioral and social sciences.

PsycINFO
Psychological Abstracts Information Services, a family of interrelated computerized information services that provides access to the world's literature in psychology and related behavioral and social sciences

The PsycINFO data base can be accessed through one of several online retrieval services (see Table 4.1) with the use of a computer and the appropriate software. With access to an ordinary telephone line you can communicate with a host computer over a telecommunications network such as Sprint-Net or DialNet. These networks are available in most locations with a local call.

PsycLIT provides the same search flexibility as PsycINFO but frees you of having to connect with a host computer and

---

EXHIBIT 4.1   Procedure for Using *Psychological Abstracts*

Each monthly issue of *Psychological Abstracts* contains summaries listed under sixteen major classification categories. Abstracts are indexed according to subject and author. Each separate issue is indexed, and an expanded and integrated *volume index* is published every six months. This volume index will enable you to use your time most efficiently.

The following procedure is recommended for conducting a literature search on a specific topic:

A. Obtain a copy of the *Thesaurus of Psychological Index Terms* from the library, and identify the terms that relate to your research topic. This reference is an excellent beginning point because the subject index terms used in *Psychological Abstracts* are taken from it.

B. In the subject index section of the volume index of *Psychological Abstracts,* find each term that you identified in the *Thesaurus of Psychological Index Terms,* and read the entries below that term. Each entry will consist of a concise statement of an article's contents, as in the following example:

Subject index   Concise statement
    Ability Level   of the article
    uncertainty about ability as related
    to academic performance in 4th- and
    8th-grade females,
    55731
Number of
the abstract

The number following the description of the article refers to the number of the abstract.

C. From the descriptions, identify those articles that may be related to your research topic, and record the numbers corresponding to the abstracts.

D. For each number you have recorded, find the corresponding abstract in the body of *Psychological Abstracts.*

Abstract number
    55731. Doe, John B. The influence of uncertainty on academic performance. *Journal of Educational Excellence,* 1983, *19,* 476–483. Investigated 4th- and 8th-grade female students' academic performance as a function of their own perceptions of academic ability. Results revealed no difference in 4th-grade students as a function of perceived academic ability. However, 8th-grade students who were unsure of their academic abilities did not perform as well as did 8th-grade students who gave their academic abilities a high evaluation. (15 ref.)—*Journal Abstract.*

From this hypothetical entry, you can see that the abstract not only gives you a brief synopsis of the article but also tells you where you can locate the article and who wrote it. This information should be sufficient to enable you to decide if the article might be of value to you.

E. Record the references of the articles that are related to your research topic so that you can find the journals in which the articles were published.

TABLE 4.1 Online Search Systems

BRS/Search
BRS/After Dark; BRS/COLLEAGUE
  BRS Information Technologies
  Maxwell Online
  8000 Westpark Drive
  McLean, VA 22102
  (800) 955-0906
DIALOG
DIALOG Knowledge Index
  DIALOG Information Services, Inc.
  3460 Hillview Avenue
  Palo Alto, CA 94304
  (800) 334-2564
Data-Star
  D-S Marketing, Inc.
  Suite 110
  485 Devon Park Drive
  Wayne, PA 19087
  (800) 221-7754

---

paying connect charges. PsycLIT is also a computer-based information and retrieval service but it is available for use with personal computers that have CD-ROM capability. This is a service that many libraries currently subscribe to and make available to students. If your library has installed PsycLIT you can conduct a rapid and efficient literature search. PsycLIT provides full bibliographic information and summaries of articles contained in over 1,300 journals; beginning in March of 1992 it included chapter and book summaries as well. This system can be used to access the behavioral literature very rapidly and at no cost to you.

PASAR
The comprehensive information storage and retrieval system operated by the American Psychological Association

If you require only an occasional literature search and do not have access to PsycLIT or PsycINFO, you can make use of PsycINFO's **PASAR** (PsycINFO Assisted Search and Retrieval). PASAR offers comprehensive searches of its available databases at a fee ranging from $50 to $80 per search. A PASAR search can be requested by mail from the American Psychological Association.

There are also other abstracting services and databases that may be useful in obtaining a list of relevant publications. Tables 4.2 and 4.3 summarize these services. If your research topic bridges several different areas or if it is restricted to a

TABLE 4.2   Abstracting and Indexing Services

Abstract or Index Service	Subject Coverage
Psychoanalytic Abstracts	Current research in psychoanalysis
Psychological Abstracts	Psychology and related disciplines
PsycSCAN: Applied Experimental and Engineering Psychology	Human factors, ergonomics, computer applications, environment, safety and accidents, transportation and flight, and working conditions
PsycSCAN: Applied Psychology	Current research on applied psychology
PsycSCAN: Clinical Psychology	Current research on clinical psychology
PsycSCAN: Developmental Psychology	Current research on developmental psychology
PsycSCAN: LD/MR	Articles on learning disabilities, communication disorders, and mental retardation
PsycSCAN: Neuropsychology	Current research on the relationship between the brain and behavior
Science Citation Index	Science and technology
Behavioral Abstracts	Behavioral psychology
Linguistics and Language Behavior Abstracts	Language behavior and linguistics

TABLE 4.3   Databases Incorporating Psychological Publications

Database	Subject Coverage
PsycINFO	Psychology, mental health, biomedicine
PsycALERT	Psychology, mental health, biomedicine
PSYINDEX	Psychology, social sciences
Social Sciences Index	Social science, politics, economics, psychology, humanities
MEDLINE	Medicine, biomedicine, health care
SocioFILE	Sociology and related disciplines

limited area within the field of psychology, you may want to use one or more of these additional abstracting and indexing services.

Additional Information Sources

The regional and national psychological association meetings are an excellent source of *current* information. I emphasize *current* because of the publication lag that exists in journals and books. A research study that appears in a book may be several years old, whereas studies presented at professional meetings are typically much more recent. An additional advantage of securing information at professional meetings is that frequently you can interact with the investigator. Exchanging ideas with the researcher is likely to generate added enthusiasm and many more research ideas.

Many times, the beginning researcher returns from meetings with renewed confidence in his or her developing research skills. Novices often feel that researchers at other institutions are more skilled or more adept. When they attend professional meetings, they find out that others use the same techniques and skills. It is recommended that psychology majors try to attend one of these national or regional meetings. Table 4.4 lists the various regional psychological associations, as well as a variety of other more specialized psychological associations.

TABLE 4.4 Psychological Associations

National	Regional	Selected Others
American Psychological Association American Psychological Society	New England Psychological Association Southwestern Psychological Association Eastern Psychological Association Southeastern Psychological Association Western Psychological Association Midwestern Psychological Assciation Rocky Mountain Psychological Association	Psychonomic Society Association for the Advancement of Behavior Therapy National Academy of Neuropsychologists International Neuropsychological Society

Information can also be gained from direct communication with colleagues. It is not unusual for researchers to call or write each other to inquire about current studies or methodological techniques.

FEASIBILITY OF THE STUDY

After you have completed the literature search, you are ready to decide whether it is feasible for you to conduct the study. Each study varies in its requirements with respect to time, type of subjects, expense, expertise of the experimenter, and ethical sensitivity.

Take, for example, a study by DePaulo, Dull, Greenberg, and Swaim (1989), in which they attempted to determine whether shy individuals would seek help less frequently than would people who were not shy. In conducting this study, the researchers administered a four-item shyness survey to introductory psychology students, and shy and not-shy individuals were selected on the basis of their survey scores. All subjects were then given the impossible task of standing a stick on end when the end was slightly rounded. The number of times shy and not-shy individuals asked for help was recorded. This study was relatively simple to conduct, did not require any special skills on the part of the experimenters or the subjects, was relatively inexpensive, took only a moderate amount of time, and did not violate the subjects' rights.

Contrast this study with one designed to investigate the effectiveness of several forms of therapy in the treatment of outpatient depression (Elkin, Parloff, Hadley & Autry, 1985). About three years of planning and preparation preceded this study, which began in 1982 and was completed in 1986. The study, which cost several million dollars, required access to a sample of clinically depressed individuals and called for trained therapists to diagnose the patients, administer each of the three forms of therapy, and assess the outcome measures. It was also an ethically sensitive study because depression sometimes results in suicide. If the treatment to which a particular depressed patient was randomly assigned was not the most effective one, the resulting delay in administering the most appropriate treatment could increase the probability of suicide. As you can see, this study was very expensive, required a special subject sample, took a tremendous amount of time, required trained and sophisticated researchers, and posed serious ethical issues. Obviously this is a study few researchers and fewer, if any, students could conduct adequately.

These two studies represent opposite ends of the continuum with respect to the issues of time, money, access to subject sample, expertise, and ethics. Although most studies fall somewhere between the two extremes, these examples serve to emphasize the issues that must be considered in selecting a research topic. If the research topic you have selected will take an inordinate amount of time, require funds that you don't have or can't acquire, call for a degree of expertise you don't have, or raise sensitive ethical questions, you should consider altering the project or selecting another topic. If you have considered these issues and find that they are not problematic, then you should proceed with the formulation of your research problem.

FORMULATING THE RESEARCH PROBLEM

You should now be prepared to make a clear and exact statement of the specific problem to be investigated. The literature review has revealed not only what is currently known about the problem but also the ways in which the problem has been attacked in the past. Such information is a tremendous aid in formulating the problem and in indicating how and by what methods the data should be collected. Unfortunately, novices sometimes jump from the selection of a research topic to the data collection stage, leaving the problem unspecified until after data collection. They thus run the risk of not obtaining information on the problem of interest. An exact definition of the problem is very important because it guides the research process.

Defining a Research Problem

Research problem
An interrogative sentence that states the relationship between two variables

What is a **research problem?** Kerlinger (1973, p. 17) defines a problem as "an interrogative sentence or statement that asks: 'What relation exists between two or more variables?' " For example, Milgram (1964a) asked: "Can a group induce a person to deliver punishment of increasing severity to a protesting individual?" This statement conforms to the definition of a problem, since it contains two variables—group pressure and severity of punishment delivered—and asks a question regarding the relation between these variables.

Are all problems that conform to the definition good research problems? Assume that you posed the problem: "How do we know that God influences our behavior?" This question meets the definition of a problem, but it obviously cannot be tested. Kerlinger (1973) presents three criteria that good problems must meet. First, the variables in the problem should express a relation. This criterion, as you can see, was contained in the definition of a problem. The second criterion is that the problem should be stated in question form. The statement of the problem should begin with "What is the effect of . . . ," "Under what conditions do . . . ," "Does the effect of . . . ," or some similar form. Sometimes only the purpose of a study is stated, which does not necessarily communicate the problem to be investigated. The purpose of the Milgram study was to investigate the effect of group pressure on a person's behavior. Asking a question has the benefit of presenting the problem directly, and thereby minimizing interpretation and distortion. The third criterion, and the one that most frequently distinguishes a researchable from a nonresearchable problem, states: "The problem statement should be such as to imply possibilities of empirical testing" (p. 18). Many interesting and important questions fail to meet this criterion and therefore are not amenable to scientific inquiry. Quite a few philosophical and theological questions fall into this category. Milgram's problem, on the other hand, meets all of the criteria. A relation was expressed between the variables, the problem was stated in question form, and it was possible to test the problem empirically. Severity of punishment was measured by the amount of electricity supposedly delivered to the protesting individual, and group pressure was applied by having two confederates suggest increasingly higher shock levels.

Specificity of the Question

Specificity of the research question The preciseness with which the research question is stated

In formulating a problem, **specificity of the research question** is an important consideration. Think of the difficulties facing the experimenter who asks the question "What effect does the environment have on learning ability?" This question meets all the criteria of a problem, and yet it is stated so vaguely that the investigator could not pinpoint what was to be investigated. The concepts of *environment* and *learning ability* are vague (what environmental characteristics? learning of what?). The experimenter must specify what is meant by environment and by learning ability to be able to conduct the experiment. Now contrast this question with the following: "What effect does the amount of exposure to words have on the speed with which they are learned?" This question specifies exactly what the problem is.

The two examples of questions presented above demonstrate the advantages of formulating a specific problem. A specific statement helps to ensure that the experimenters understand the problem. If the problem is stated vaguely, the experimenters probably do not know exactly what they want to study and therefore may design a study that will not solve the problem. A specific problem statement also helps the experimenters make necessary decisions about such factors as subjects, apparatus, instruments, and measures. A vague problem statement helps very little with such decisions. To drive this point home, go back and reread the two questions given in the preceding paragraph and ask yourself: "What subjects should I use? What measures should I use? What apparatus or instruments should I use?"

How specific should one be in formulating a question? The primary purposes of formulating the problem in question form are to ensure that the researcher has a good grasp of the variables to be investigated and to aid the experimenter in designing and carrying out the experiment. If the formulation of the question is pointed enough to serve these purposes, then additional specificity is not needed. To the extent that these purposes are not met, additional specificity and narrowing of the research problem are required. Therefore, the degree of specificity required is dependent on the purpose of the problem statement.

FORMULATING HYPOTHESES

Hypothesis
The best prediction or a
tentative solution to a
problem

After the literature review has been completed and the problem has been stated in question form, you should begin formulating your **hypothesis.** For example, if you were investigating the influence of the number of bystanders on the speed of intervention in emergency situations, you might hypothesize that as the number of bystanders increases, the speed of intervention will decrease. From this sample, you can see that hypotheses represent predictions of the relation that exists among the variables, or tentative solutions to the problem. Formulation of the hypothesis logically follows the statement of the problem, since one could not state a hypothesis without having a problem. This does not mean that the problem is always stated explicitly. In fact, if you survey articles published in journals, you will find that most of the authors do not present a statement of their specific problem. It seems that experienced researchers in a given field have such familiarity with the field that they consider the problems to be self-evident. Their predicted solutions to these problems are not apparent, however, and so these must be stated.

The hypothesis to be tested is often a function of the literature review, although hypotheses are also frequently formulated from theory. As stated earlier, theories guide research, and one of the ways in which they do so is by making predictions of possible relationships among variables. Hypotheses also (but less frequently) come from reasoning based on casual observation of events. In some situations it seems fruitless even to attempt to formulate hypotheses. When one is engaged in exploratory work in a relatively new area, where the important variables and their relationships are not known, hypotheses serve little purpose.

More than one hypothesis can almost always be formulated as the probable solution to the problem. Here again the literature review can be an aid, because a review of prior research can suggest the most probable relationships that may exist among the variables.

Regardless of the source of the hypothesis, it *must* meet one criterion: A hypothesis must be stated so that it is capable of being either refuted or confirmed. In an experiment, it is the hypothesis that is being tested, not the problem. One does not test a question like the one Milgram posed; rather, one tests one or more of the hypotheses that could be derived from this question, such as "group pressure increases the severity of punishment that subjects will administer." A hypothesis that fails to meet the criterion of testability, or is nontestable, removes the problem from the realm of science. Any conclusions reached regarding a nontestable hypothesis do not represent scientific knowledge.

Scientific hypothesis
The predicted relationship among the variables being investigated

Null hypothesis
A statement of no relationship among the variables being investigated

A distinction must be made between the scientific hypothesis and the null hypothesis. The **scientific hypothesis** represents the predicted relationship among the variables being investigated. The **null hypothesis** represents a statement of no relationships among the variables being investigated. For example, Hashtroudi, Parker, DeLisi, and Wyatt (1983) wanted to explore the nature of the memory deficits that occur through the influence of alcohol. One of the research questions these investigators asked was whether the memory deficit induced by alcohol is decreased when intoxicated individuals are forced to generate a meaningful context for a word that is to be recalled. Although not specifically stated, these investigators' scientific hypothesis was that the generation of a meaningful context would reduce the memory deficit produced by the alcohol. The null hypothesis predicted that no difference in recall would be found between intoxicated subjects who generated the meaningful context and those who did not.

Although an experimental study would seem to be directed toward testing the scientific hypothesis, this is not the case. In any study, it is the null hypothesis that is always tested, because the scientific hypothesis does not specify the exact amount or type of influence that is expected. To obtain support for the scientific hypothesis, you must collect evidence that enables you to reject the null hypothesis. (In the Hashtroudi et al. study, the null hypothesis was that no difference would be found between the recall scores of those intoxicated subjects who did and did not use a meaningful context.) Consequently, support for the scientific hypothesis is always obtained indirectly by rejecting the null hypothesis. The exact reason for testing the null hypothesis as opposed to the scientific hypothesis is based on statistical hypothesis-testing theory, which is beyond the scope of this text, but basically the point is that it is necessary to test the null hypothesis in order to obtain evidence that will allow you to reject it so that, indirectly, you can get evidence supportive of the scientific hypothesis.

Why should hypotheses be set up in the first place? Why not just forget about hypotheses and proceed to attempt to answer the question? Hypotheses serve a valuable function. Remember that hypotheses are derived from knowledge obtained from the literature review of other experiments, theories, and so forth. Such prior knowledge serves as the basis for the hypothesis. If the experiment confirms the hypothesis, then, in addition to providing an answer to the question asked, it gives additional support to the literature that suggested the hypothesis. But what if the hypothesis is not confirmed by the experiment? Does this invalidate the prior literature? If the hypothesis is not confirmed, then either the hypothesis is false or some error exists in the conception of the hypothesis. If there is an error in conceptualization, it could be in any of a number of categories. Some of the information obtained from prior experiments may be false, or some relevant information may have been overlooked in the literature review. It is also possible that the experimenter misinterpreted some of the literature. These are a few of the more salient errors that could have taken place. In any event, failure to support a hypothesis may indicate that something is wrong, and it is up to the experimenter to discover what it is. Once the experimenter uncovers what he or she thinks is wrong, a new hypothesis is proposed that can be tested experimentally. The experimenter now has another study to conduct. Such is the continuous process of science. Even if the hypothesis is false, knowledge has been advanced, for now an incorrect hypothesis can be ruled out. Another hypothesis must be formulated and tested in order to reach a solution to the problem.

SUMMARY

In order to conduct research, it is first necessary to identify a problem in need of a solution. Psychological problems arise from several traditional sources: theories, practical issues, and past research. Additionally, in psychology we have our personal experience to draw on for researchable problems, because psychological research is concerned with behavior. Once a researchable problem has been identified, the literature relevant to this problem should be reviewed. A literature review will reveal the current state of knowledge about the selected topic. It will indicate ways of investigating the problem and will point out related methodological problems. The literature review should probably begin with books written on the topic and progress from there to the actual research as reported in journals. In surveying the past research conducted on a topic, the scientist can make use of information retrieval systems, one of which is operated by the American Psychological Association. In addition to using these sources, the researcher can obtain information by attending professional conventions or by calling or writing other individuals conducting research on the given topic.

When the literature review has been completed, the experimenter must determine whether it is feasible for him or her to conduct the study. This means that an assessment must be made of the time, subject population, expertise, and expense requirements, as well as the ethical sensitivity of the study. If this assessment indicates that it is feasible to conduct the study, the experimenter must make a clear and exact statement of the problem to be investigated. This means that the experimenter must formulate an interrogative sentence asking about the relationship between two or more variables. This interrogative sentence must express a relation and be capable of being tested empirically. The question must also be specific enough to assist the experimenter in making decisions about such factors as subjects, apparatus, and general design of the study.

After the question has been stated, the experimenter needs to set down hypotheses. These must be formalized, because they represent the predicted relation that exists among the variables under study. Often, hypotheses are a function of past research. If they are confirmed, the results not only answer the question asked but also provide additional support to the literature that suggested the hypotheses. There is one criterion that any hypothesis must meet: it must be stated so that it is capable of being either refuted or confirmed. Always remember that it is actually the null hypothesis, and not the scientific hypothesis, that is being tested in a study.

STUDY QUESTIONS

1. What are the different sources of research ideas?
2. What purpose is served by a review of the literature, and what sources are available to assist in the literature review?
3. What are the characteristics of a good research problem, and why is it important to start with a clear definition of the research problem?

4. What is a hypothesis, what function does it serve, and what criteria must it meet?
5. Why is a research study directed toward testing the null—as opposed to the scientific—hypothesis?

KEY TERMS AND CONCEPTS

Theory
Gender bias
Psychological Abstracts
PsycINFO
PASAR
Research problem

Specificity of the research question
Hypothesis
Scientific hypothesis
Null hypothesis

CHAPTER 5

Ethics

LEARNING OBJECTIVES

1. To gain an understanding of the need to consider ethical issues when designing and conducting research.
2. To gain an understanding of the guidelines that must be followed in conducting research with humans.
3. To understand the implication of adhering to the guidelines for conducting research with humans.
4. To gain an understanding of the guidelines that must be followed in conducting research with animals.

In November 1986 it was revealed that members of the White House staff were involved in a complicated covert operation in which weapons were sold to Iran in return for release of Americans held captive in Beirut. Additionally, the profit from the arms sale was being diverted illegally to purchase weapons for the contras in Nicaragua. Two of the most significant questions asked were whether President Reagan knew of the operation and whether he had given it his approval. Initially the president and the White House staff stated firmly that President Reagan was completely ignorant of the operation. As the investigation into the Iran arms deal continued, however, it became increasingly clear that the president was not totally uninformed.

For many reporters, this incident brought back memories of another national trauma—Watergate, which happened over a decade earlier. In 1972 Richard M. Nixon was reelected president of the United States, partly as a result of the popularity he had gained from his foreign relations trips to Peking and Moscow. This popularity, however, was short lived. Newspaper reporters had revealed an attempted burglary and wiretapping of the Democratic National Committee headquarters at the Watergate complex during the election campaign; the burglars turned out to be men hired by some of the president's closest advisers. Some excellent investigative reporting uncovered a larger picture of political corruption, which was ultimately traced to the White House. As a result of this continuing investigation, the Judiciary Committee of the House of Representatives eventually recommended to the House that President Nixon be impeached. This led to Nixon's resignation in 1974 as president of the United States.

In an attempt to explain the actions of President Nixon's administration, the press focused on the paranoid style of the administration, the recruitment of an amoral staff, and even the notion that the Nixon administration had been injected with "moral penicillin." Members of the Nixon administration, on the other hand, explained their actions as prompted by the potentially violent plans of the radical left, which they claimed necessitated such a course of action. In other words, the press attempted to attribute the actions to underlying personality or dispositional characteristics of members of the Nixon administration, whereas members of the administration attributed their ac-

tions to the characteristics of the situation in which they were placed.

West, Gunn, and Chernicky (1975) conducted a study in which they attempted to determine if a situation could be constructed in which individuals would engage in illegal and corrupt acts. This study was conducted to determine whether outsiders (for example, the reporters investigating the Watergate scandal) would attribute the illegal acts to personality or dispositional characteristics of the actors, whereas the actors (for example, members of the Nixon administration) would attribute their actions to surrounding circumstances. In conducting this experiment, West et al. constructed an elaborate scheme in which a private investigator individually approached undergraduate majors in a criminology class and asked them to meet him at a local restaurant at a designated time. At this meeting each research participant was told that the Internal Revenue Service wanted to hire someone (that is, the subject) to burglarize a local advertising firm for the purpose of microfilming an allegedly illegal set of accounting records used to defraud the U.S. government of several million tax dollars each year. The research participants were randomly assigned to one of four experimental conditions: they were told either that they would be given immunity if apprehended, that there would be no immunity if they were apprehended, that they would be given $2,000 for their participation in the crime, or that the crime was being committed just to see whether the plans would work. The results of this study supported the observations made regarding the various explanations of the Watergate break-in. The research participants stated that they agreed to participate in the crime because of the circumstances surrounding it, such as guaranteed immunity or the promised payment of $2,000, whereas observers tended to attribute their actions to dispositional or personality characteristics.

Although the West et al. study legitimately examined the cause of the divergent explanations of the Watergate break-in, it also generated a variety of ethical concerns. Through an elaborate set of deceptions, many of the research participants were enticed to agree to participate in an illegal and possibly immoral activity. Although the activity was never carried out and the research participants were informed of the deception after making a decision to participate and completing a variety of other measures of dependent variables, participation in the experiment had the potential to alter the subject. Subjects may have experienced a loss of self-esteem or feelings of guilt, embarrassment, or anxiety after realizing that they had agreed to participate in a burglary, which turned out to be a psychological experiment. Such issues must be considered before conducting any experiment.

Research with humans is not the only type of research that is subject to ethical scrutiny. Animal research is also scrutinized and periodically criticized as being unethical. For example, it has been pointed out that animals are deprived of food, placed in a stressful environment, shocked, and surgically altered—all in the name of science. Is it appropriate to subject animals to such treatment? This is a question that always confronts the scientist. Thus, psychological research has the potential for endangering the physical and psychological well-being of the research participants. If all scientific investigations consisted of innocuous studies, there would be little need to consider the welfare of the research participant or the ethical issues surrounding the study. Unfortunately, as the foregoing examples illustrate, many psychological studies have been conducted that arouse ethical concerns. There is a need for a code of ethics to provide guidelines for researchers to follow in conducting psychological research. This chapter will cover the issues surrounding the ethics of psychological research.

INTRODUCTION

Once you have constructed your research problem and formulated the hypothesis, you are ready to begin to develop the research design. The design will specify how you will collect data that will enable you to test your hypothesis and arrive at some answer to the research question. However, at the same time that you are designing the research study, you must pay attention to ethical issues involved in research.

In their pursuit of knowledge relating to the behavior of organisms, psychologists conduct surveys, manipulate the type of experience that individuals receive, or vary the stimuli presented to individuals and then observe the subjects' reactions to these stimuli. Such manipulations and observations are necessary in order to identify the influence of various experiences or stimuli. At the same time, scientists recognize that individuals have the right to privacy and to protest surveillance of their behavior carried out without their consent. People also have the right to know if their behavior is being manipulated and, if so, why. The scientific community is confronted with the problem of trying to satisfy public demand for solutions to problems such as cancer, arthritis, alcoholism, child abuse, and penal reform without infringing on people's rights. For a psychologist trained in research techniques, a decision *not* to do research is also a matter of ethical concern.

In order to advance knowledge and to find answers to questions, it is often necessary to impinge on well-recognized rights of individuals. Consideration of ethical issues is, therefore, integral to the development of a research proposal and to the conduct of research (Sieber & Stanley, 1988). It is very difficult to investigate topics such as child abuse, for example, without violating the right to privacy, because it is necessary to obtain information about the child abuser and/or the child being abused. Such factors create an ethical dilemma: whether to conduct the research and violate certain rights of individuals for the purpose of gaining knowledge or to sacrifice a gain in knowledge for the purpose of preserving human rights. Ethical principles are vital to the research enterprise because they assist the scientist in preventing abuses that may otherwise occur and delineate the responsibilities of the investigator.

RESEARCH ETHICS: WHAT ARE THEY?

Research ethics
A set of guidelines to assist the experimenter in making difficult research decisions

When most people think of ethics, they think of moralistic sermons and endless philosophical debates. However, **research ethics** should not be a set of moralistic dictates imposed on the research community by a group of self-righteous busybodies. Rather, they should be a set of principles that will assist the community of experimenters in deciding which goals are most important in reconciling conflicting values (Diener & Crandall, 1978).

Within the social and behavioral sciences, ethical concerns can be divided into three different areas (Diener & Crandall, 1978): (1) the relationship between society and science, (2) professional issues, and (3) treatment of subjects.

The ethical issue concerning the relationship between society and science revolves about the extent to which societal concerns and cultural values should direct the course of scientific investigation. Traditionally, science has been conceived of as trying to uncover the laws of nature. It is assumed that the scientist examines the phenomenon being investigated in an objective and unbiased manner. However, the literature dealing with experimenter effects reveals that the scientist can never be totally objective. Similarly, the society surrounding the scientist dictates to a great extent which issues will be investigated. The federal government spends millions of dollars each year on research, and it sets priorities for how the money is to be spent. To increase the probability of obtaining research funds, investigators orient their research proposals toward these same priorities, which means that the federal government at least partially dictates the type of research conducted. AIDS research provides an excellent illustration. Prior to 1980, AIDS (acquired immunodeficiency syndrome) was virtually unheard of. Few federal dollars were committed to investigating this disorder. But when AIDS turned up within the U.S. population and its lethal characteristic was identified, it rapidly became a national concern. Millions of dollars were immediately earmarked for research to investigate causes and possible cures. Many researchers reoriented their interests and investigations to the AIDS problem because of the availability of research funds.

Societal and cultural values also enter into science to the extent that the phenomenon a scientist chooses to investigate is often determined by that scientist's own culturally based interests (for example, a female psychologist might study sex discrimination in the work force, or a black psychologist might study racial attitudes). The scientific enterprise is not value-free; rather, society's values as well as the scientist's own can creep into the research process in subtle and unnoticed ways.

The category of professional issues includes the expanding problem of scientific misconduct. The primary issue and the one that has received the most attention is fraudulent activity by scientists. A scientist is trained to ask questions, to be skeptical, and to use the scientific method in searching for truth. This search for truth is completely antithetical to engaging in any type of deception. The most serious crime in the scientific profession is to cheat or present fraudulent results. Although fraudulent activity is condemned on all fronts, in the past decade there has been a disturbing increase in the number of reports of scientists who forge or falsify data, manipulate results to support a theory, or selectively report data, as illustrated in Exhibit 5.1. Between 1950 and 1979 there were only fourteen documented cases of serious scientific misconduct, whereas there were twenty-six cases between 1980 and 1987 (Woolf, 1988)—and these data may reflect an underreporting of the actual incidence.

The cost of such fraudulent activity is enormous, both to the profession and to the scientist. Not only is the whole scientific enterprise discredited, but the professional career of the individual is destroyed. There is no justification for faking or altering scientific data.

Although fraudulent activity is obviously the most serious form of scientific misconduct, there is a broader range of less serious, though still unacceptable, practices that are beginning to receive attention (Hilgartner, 1990). These include such practices as listing as "honorary authors" on a publication people who have contributed nothing substantive to the paper or assigning subjects to conditions nonrandomly when the research design calls for random assignment. In addition to such issues the Committee on Standards in Research within the American Psychological Association is studying fraud and misconduct in relation to research publication, use and archiving of videotaped data, research with vulnerable populations, and financial conflicts of interests (Grisso et al., 1991).

The two issues of concern with respect to research publication are partial publication and dual publication. **Partial publication** refers to collecting data for one study and then publishing several articles based on this one large set of data rather than publishing all the findings and data in one article. The APA *Publication Manual* (APA, 1983) states that such piecemeal publication is undesirable and is viewed as duplicate publication. However, the author of such partial publication typically derives benefits from it. Promotion and tenure as well as salary raises are partially contingent on the number of articles published, so it is to the benefit of the author to publish as many articles as possible in good journals—which promotes piecemeal publication.

Dual publication refers to publishing the same data and results in more than one journal or other publication. The APA

Partial publication
The publication of several articles based on one large set of data, a practice viewed as unacceptable by the scientific community

Dual publication
Publication of the same data and results in more than one journal or other publication simultaneously, a practice explicitly forbidden in the scientific community

EXHIBIT 5.1 Two Cases of Reportedly Fraudulent Research

Although most known cases of fraudulent research have occurred in the field of medicine, several very significant instances have recently been identified in the field of psychology. Two of the most infamous cases are described in this exhibit.

Cyril Burt, the first British psychologist to be knighted, received considerable acclaim in both Great Britain and the United States for his research on intelligence and its genetic basis. A biographical sketch published upon his death depicted a man with unflagging enthusiasm for research, analysis, and criticism. Shortly after his death, however, questions about the authenticity of his research began to appear. Ambiguities and oddities were identified in his research papers. A close examination of his data revealed that correlation coefficients did not change across samples or across sample sizes, suggesting that he may have fabricated data. Attempts to locate one of Burt's important collaborators were unsuccessful. Dorfman (1978) conducted an in-depth analysis of Burt's data and showed beyond a reasonable doubt that Burt fabricated his data on the relationship between intelligence and social class.

More recently, the National Institute of Mental Health (NIMH) conducted an investigation of alleged research fraud by one of its grantees, Steven E. Breuning. Breuning received his doctorate from the Illinois Institute of Technology in 1977 and several years later obtained a position at the Coldwater Regional Center in Michigan. At Coldwater, Breuning was invited to collaborate on an NIMH-funded study of the use of neuroleptics on institutionalized mentally retarded people. In January 1981 he was appointed director of the John Merck program at Pittsburgh's Western Psychiatric Institute and Clinic, where he continued to report on the results of the Coldwater research and even obtained his own NIMH grant to study the effects of stimulant medication on retarded subjects. During this time Breuning gained considerable prominence and was considered one of the field's leading researchers. In 1983, however, questions were raised about the validity of Breuning's work. The individual who had initially taken Breuning on as an investigator started questioning a paper in which Breuning reported results having impossibly high reliability. This prompted a further review of Breuning's published work, and contacts were made with personnel at Coldwater, where the research had supposedly been conducted. Coldwater's director of psychology had never heard of the study and was not aware that Breuning had conducted any research while at Coldwater. NIMH was informed of the allegations in December of 1983. Following a three-year investigation, an NIMH team concluded that Breuning "knowingly, willfully, and repeatedly engaged in misleading and deceptive practices in reporting his research." He reportedly had not carried out the research that was described, and only a few of the experimental subjects had ever been studied. It was concluded that Breuning had engaged in serious scientific misconduct (Holden, 1987).

Publication Manual (APA, 1983) explicitly forbids dual publication. However, different outlets such as scientific journals and books and articles for the popular press reach different audiences and serve different purposes. The ethical issue under consideration is whether presenting the same data to these different audiences represents an ethical violation.

An ethical issue arises with regard to the use and archiving of videotapes because a videotape represents a permanent record of behavior. The person or persons who are videotaped give their informed consent to participate in the study prompting the videotaping. However, the tapes may contain data that provide an answer to the research question posed by another investigator. The ethical question is whether these videotapes can be used for this second study without the informed consent of the participants.

Research with vulnerable populations gives rise to a host of potential ethical problems. Arranging transportation to a clinic or laboratory, for example, may represent an intolerable burden for impoverished individuals. Consideration must also be given to the concerns of the parents or guardians of vulnerable persons such as the profoundly mentally retarded or persons with serious mental illness. What procedures must be taken to assure their rights to privacy, and what constitutes a valid informed consent to participate in a study if an individual, such as a profoundly mentally retarded person, is not capable of providing the traditional written consent?

The problem of financial conflict of interest in science arises from the fact that a scientific breakthrough may have commercial applications that allow the scientist to reap financial rewards from the research. This possible outcome may entice researchers to engage in unethical activities, as did a National Institutes of Health AIDS researcher who apparently set up a company that did business with his laboratory. He did not disclose this relationship because he helped the company to win federal contracts (Culliton, 1990).

The increased frequency of and interest in scientific misconduct has naturally stimulated discussion about its cause and the type of action that needs to be taken to reduce the frequency of misconduct. Hilgartner (1990) has pointed out that there are three kinds of responsibility with regard to scientific misconduct: causal, moral, and political responsibility.

Causal responsibility focuses on the etiology, or origin, of the problem. Knight (1984) believes that there are both per-

Causal responsibility The source of a problem of fraudulent research or scientifc misconduct; the source may be either a personal or a nonpersonal factor

sonal and nonpersonal factors that contribute to scientific misconduct. Nonpersonal factors include such things as the pressure to publish and the competition for research funding. Most science is conducted within the confines of research institutions. These institutions evaluate scientists on the basis of the grants they receive and the articles they publish. Receiving a promotion or even keeping one's position may be contingent upon the number of articles published and grants obtained. The constant pressure to publish often takes on more importance than the desire to further scientific knowledge. Such pressure is frequently reported by scientists who engage in fraudulent activities. Other nonpersonal factors include inadequate supervision of trainees, inadequate procedures for keeping records or retaining data, and the diffusion of responsibility for jointly authored studies. Consequently, the nonpersonal factors include mechanisms that operate at several levels ranging from the laboratory to the training of graduate students to the mechanism for funding research.

Moral responsibility
The responsibility borne by the person who engages in fraudulent research or scientific misconduct

Moral responsibility focuses on determining who is to be blamed for the problem. The person who engaged in the deviant behavior reaps most of the blame. This person is often viewed as having a psychological problem or illness, which represents a personal factor (Knight, 1984). The fraudulent activity is seen as the researcher's reaction to the extreme stress resulting from participation in a highly competitive academic research environment. To the extent that such an explanation absolves the scientist of moral responsibility by reducing the deviant behavior to a personality defect, it may be nothing more than a form of rationalization. Stress can bring about profound behavioral changes, however, and cause people to do things they otherwise would not do.

Although the individual engaged in the scientific misconduct always retains most of the blame, some (e.g., Hilgartner, 1990) believe that the blame should be shared by a number of others. A co-investigator who fails to carefully check data or an investigator who neglects his or her laboratory should assume part of the blame.

Political responsibility
The responsibility borne by the person who must do something about a problem of fraudulent research or scientific misconduct

Political responsibility focuses on who is responsible for doing something about the problem. Traditionally it has been assumed that controls within the scientific enterprise can operate as deterrents. First, the results of scientific experiments are open to inspection by other scientists, who are trained to critique one another's work and to search for defects and inappropriate methodologies. Each study must be reviewed by other scientists for appropriateness before it is published in a reputable journal. Even after a study passes this peer review process, the ultimate deterrent to fraud is replication. Any scientific result must be capable of being repeated by other scientists to be viewed as valid. Just because a study cannot be replicated does not mean that it was fraudulently conducted or reported, however. There is a difference between disconfirming the results of a prior study and failing to replicate. Some successful experimentation involves subtle elements that are not explicitly described in the published report, and without knowledge of these subtle elements it may be impossible to successfully replicate the study. Distinguishing between disconfirmation and fraud can take years, which is why fraudulent activity sometimes continues for what seems to be an inordinate amount of time. The crucial point is that if the research is at all important, any deceit will eventually be detected because replication will be attempted by other investigators.

Although the control mechanisms of peer review and replication had seemed adequate for controlling episodes of scientific misconduct heretofore, the recent increase in such deviant activity has resulted in a number of proposals suggesting changes in the way scientific research is conducted. There are four general orientations that have received attention (Hilgartner, 1990). The first approach takes a "law enforcement" perspective, which focuses on detection, deterrence, and punishment. It emphasizes the litigation of allegations, swift and severe punishment, and implementation of a witness protection program to prevent retaliation against whistleblowers. Consistent with this law enforcement approach, the U.S. Public Health Service (1989) established the Office of Scientific Integrity to deal with investigations of misconduct in science.

A second approach takes an oversight perspective. This involves implementing procedures that would provide additional assurance of quality control in science by intensifying the routine scrutiny of research results, data, and laboratory practices. This approach boils down to implementing procedures concerning the recording and retention of data.

A third approach takes an educational orientation that emphasizes the training of researchers. This training would place more emphasis on the articulation of research ethics in addition to the methods of research.

A final perspective stresses a reward system and seeks to change the rules regarding the way academic appointments, promotions, and grants are allocated so as to reduce the motivation to cheat. One proposal is to base promotion on the quality of publications and less on their number, which would reduce the motivation for partial publication of studies.

Proposals such as these have generated a lot of discussion and controversy. Implementing such suggestions not only would change the way much research is conducted but might intensify the bureaucratic regulation of science. They also raise the possibility of stifling creativity by creating undue barriers to scientific research.

The treatment of research subjects is the most fundamental issue confronted by scientists. The conduct of research with humans can potentially create a great deal of physical and psychological harm, as the examples in Exhibit 5.2 illustrate. Experiments designed to investigate important psychological issues may subject participants to humiliation, physical pain, and embarrassment. In planning an experiment, a scientist is obligated to consider the ethics of conducting the necessary research. Unfortunately, some studies cannot be designed in such a way that the possibility of physical and psychological harm is eliminated. Hence the researcher often faces the dilemma of having to determine whether the research study should be conducted at all. Since it is so important, we will consider this issue in some detail.

ETHICAL DILEMMAS

Ethical dilemma
The investigator's conflict in weighing the potential cost to the subject against the potential gain to be accrued from the research project

The scientific enterprise in which the research psychologist engages creates a special set of dilemmas. On the one hand, the research psychologist is trained in the scientific method and feels an obligation to conduct research; on the other hand, doing so may necessitate subjecting research participants to stress, failure, pain, aggression, or deception. Thus, there arises the **ethical dilemma** of having to determine if the potential gain in knowledge from the research study outweighs the cost to the subject. In weighing the pros and cons of such a question, the researcher must give primary consideration to the welfare of the subject. Unfortunately, there is no formula or rule that can help investigators. The decision must be based on a subjective judgment, which should not be made entirely by

EXHIBIT 5.2   Psychological Experiments That Have the Potential for Creating Physical or Psychological Harm

A number of experiments that have been conducted within the field of psychology have the potential for creating either physical or psychological harm. Three such experiments are described below.

Humphreys (1970) was interested in learning about the motives of individuals who commit fellatio in public restrooms. To accomplish this purpose, Humphreys offered to serve as "watchqueen"—the individual who keeps watch and gives a warning when a police car or a stranger approaches. In this capacity, Humphreys observed hundreds of impersonal sexual acts. He even gained the confidence of some of the men whom he observed and persuaded them to tell him about the rest of their lives and their motives. In other cases he recorded the automobile license numbers of the men and then obtained their addresses from the department of public safety. At a later time he appeared at their homes, claiming to be a health service interviewer, and then interviewed them about their marital status, jobs, and so on. Clearly Humphreys deceived the men he observed and questioned, creating the potential for anger, embarrassment, and even more extreme reactions from the subjects.

Berkun, Bialek, Kern, and Yagi (1962) reported on a variety of studies conducted by the military to investigate the impact of stress. In one experiment, army recruits were placed in a DC-3 that apparently became disabled and was preparing to crash-land. Before the impending crash, the recruits were asked to complete a questionnaire that assessed their opinions regarding the disposition of their earthly possessions in case of death, their knowledge of emergency landing procedures, and so forth. After all the questionnaires had been completed, the responses were jettisoned in a metal container so as not to be destroyed, and then the plane landed safely. Only then did the recruits realize that the whole incident was an experiment.

Middlemist, Knowles, and Matter (1976) investigated the impact of an invasion of personal space on physical responses. In collecting data on this topic, the authors used hidden periscopes to study the effect of closeness of others on the speed and rate of urination by men in a public lavatory. To manipulate the closeness variable, confederates would rush in and begin to urinate in the urinal next to those subjects assigned to the crowded condition.

the researcher or his or her colleagues, since such individuals might be so involved in the study that they might tend to exaggerate its scientific merit and potential contribution. Investigators must seek the recommendations of others, such as scientists in related fields, students, or lay individuals. In making the final judgment, however, the investigator must always remember that no amount of advice or counsel can alter the fact that the final ethical responsibility lies with the researcher conducting the study.

DEVELOPMENT OF THE APA CODE OF ETHICS

Nazi scientists during World War II conducted some grossly inhumane experiments that were universally condemned as being unethical. For example, they immersed people in ice water to determine how long it would take them to freeze to death. However, these experiments were conducted by individuals living in what is thought to have been a demented society, and it seems to be assumed that such studies could not be performed in our culture. Prior to the decade of the 1960s, comments about the ethics of research were few and far between. Edgar Vinacke's (1954) statement about the ethics of deceiving subjects represents one of the few expressed.

In the mid-1960s ethical issues became a dominant concern, as it grew increasingly clear that science did not invariably operate to benefit others and did not always conduct its experiments in a manner that ensured the safety of participants. In the medical field, Pappworth (1967) cited numerous examples of research that violated the ethical rights of human subjects. One issue of *Daedalus* in 1969 was devoted to the ethics of human experimentation, particularly as it related to medical research. The Tuskegee experiment (Jones, 1981), described in Exhibit 5.3, probably epitomizes the type of unethical experimentation that was conducted within the medical field. There was an equal concern about the violation of the rights of human subjects in psychological research. Kelman (1967, 1968, 1972) has been by far the most outspoken on this issue, although others, such as Seeman (1969) and Beckman and Bishop (1970), have also contributed. More recently, entire books have been devoted to this issue (for example, Klockars and O'Connor, 1979). This widespread concern has led to the development of a code of ethics to be used by psychologists for guidance in the conduct of research using human subjects.

Before presenting the code of ethics that has been adopted by the American Psychological Association (APA), I want to describe the procedure that was followed in its development to give you an appreciation of the work and thought that went into it.

EXHIBIT 5.3   The Tuskegee Syphilis Experiment

In July 1972 the Associated Press released a story that revealed that the U.S. Public Health Service (PHS) had for forty years been conducting a study of the effects of untreated syphilis on black men in Macon County, Alabama. The study consisted of conducting a variety of medical tests (including an examination) on 399 black men who were in the late stages of the disease and on 200 controls. Although a formal description of the experiment could never be found (apparently one never existed), a set of procedures evolved in which physicians employed by the PHS administered a variety of blood tests and routine autopsies to learn more about the serious complications that resulted from the final stages of the disease.

This study had nothing to do with the treatment of syphilis; no drugs or alternative therapies were tested. It was a study aimed strictly at compiling data on the effects of the disease. The various components of the study, and not the attempt to learn more about syphilis, made it an extremely unethical experiment. The subjects in the study were mostly poor and illiterate, and the PHS offered incentives to participate, including free physical examinations, free rides to and from the clinic, hot meals, free treatment for other ailments, and a $50 burial stipend. The participants were not told the purpose of the study or what they were or were not being treated for. Even more damning is the fact that the subjects were monitored by a PHS nurse, who informed local physicians that those individuals were taking part in the study and that they were not to be treated for syphilis. Participants who were offered treatment by other physicians were advised that they would be dropped from the study if they took the treatment.

As you can see, the participants were not aware of the purpose of the study or the danger it posed to them, and no attempt was ever made to explain the situation to them. In fact, subjects were enticed with a variety of inducements and were followed to ensure that they did not receive treatment from other physicians. This study seems to have included just about every possible violation of our present standard of ethics for research with humans.

(From Jones, 1981.)

BACKGROUND OF THE DEVELOPMENT OF THE APA CODE OF ETHICS

The Committee on Ethical Standards in Psychological Research was appointed by the board of directors of the American Psychological Association to revise the 1953 code of ethics for research using human subjects. This committee patterned its work after that developed by the previous ethics committee, incorporating two distinctive features of the previous committee's method. First, the members of the profession were asked to supply ethical problems to serve as the raw material for the formation of the ethical principles. Nine thousand members of the American Psychological Association and the entire membership of selected groups such as the Division of Developmental Psychology were sent a questionnaire requesting a description of research studies that posed ethical problems. From these massive surveys, 5,000 research descriptions were obtained that made up the raw material for the committee. The committee members also interviewed thirty-five individuals who either had demonstrated a high level of concern for ethical issues or had had a great deal of exposure to a variety of research projects.

With this wealth of information in hand, the committee began the process of drafting the proposed principles. Once the initial draft had been completed, the second distinctive feature of the 1953 ethics committee's method was introduced. This involved distributing the proposed principles throughout the profession so that they could be reviewed and criticized by its members. Those attending city, state, regional, and national meetings and conventions, as well as psychology departments and individual psychologists interested in research ethics, discussed these principles. The proposed principles also appeared in the APA *Monitor*, the monthly newspaper of the association, to ensure that all members had the opportunity to review them. Many reactions were received, as well as reactions to the reactions. The commentaries by Alumbaugh (1972), Baumrind (1971, 1972), Kerlinger (1972), May (1972), and Pellegrini (1972) are all thought-provoking examples of the reactions. The committee then prepared a new draft of the principles and subjected them to the same type of review process that the older draft had undergone. The revised ethical principles received general acceptance from the APA membership. In view of this, the committee recommended adoption of ten principles. In 1973 these ten principles were adopted by the American Psychological Association and distributed to its membership.

The committee also recommended a mandatory review of the principles at five-year intervals. Consistent with that recommendation, in 1978 the APA established the Committee for the Protection of Human Participants in Research to review and make recommendations regarding the official position of the association on the use of human participants in research. This committee made several changes in the principles, including addition of a more detailed explication of the ethical issues and elaboration of the sections on deception, informed consent, and field research.

The revisions were then submitted to the Board of Scientific Affairs, the Council of Graduate Departments of Psychology, and all members of the first ethics committee. A notice of the availability of the revisions appeared in the APA *Monitor*, and comments were solicited from all those who requested a copy of the revisions. On the basis of these comments, a second draft was prepared, and its availability was announced in the *Monitor*. It was sent for review to many of the committees, departments, associations, and individuals who had commented on the first draft. Reactions to the second draft indicated that it represented a better balance of opinions. Consequently, the committee recommended approval of the revised set of ethical tenets in August 1982. It is this set of revised principles that follows.

ETHICAL PRINCIPLES

Any psychologist conducting research must ensure that the dignity and welfare of the research participants are maintained and that the investigation is carried out in accordance with federal and state regulations and with the standards set forth by the American Psychological Association. These standards, consisting of ten principles, were published by the American Psychological Association in *Ethical Principles in the Conduct of Research with Human Participants*. Each of the ten principles will be discussed briefly to clarify the principle, to focus on potential questions about application of the rule, and to identify the factors that should be given consideration in answering these questions.

Principle A

In planning a study, the investigator has the responsibility to make a careful evaluation of its ethical acceptability. To the extent that the weighing of scientific and human values suggests a compromise of any principle, the investigator incurs a correspondingly serious obligation to seek ethical advice and to observe stringent safeguards to protect the rights of human participants.[1]

The first principle states that it is the investigator's responsibility to evaluate his or her study in light of each of the following ethical principles. If, in the investigator's opinion, any aspect of the study suggests a compromise of the principles, the investigator incurs the obligation to seek the advice and counsel of others to ensure that the rights and welfare of the research participants are maintained and that the study maximizes the potential knowledge that can be gained. In seeking advice, the investigator can turn to scientific review groups or to colleagues, but such individuals are likely to be more concerned with the scientific merit of the study than with the welfare of the research participants. A better choice would be to turn to groups that were established for the purpose of safeguarding the rights and welfare of research participants, such as departmental ethics committees or institutional review boards. In any case, the investigator should keep in mind that others' roles are always advisory—responsibility for the ethical acceptability of the study ultimately resides with the investigator.

Principle B

> Considering whether a participant in a planned study will be a "subject at risk" or "subject at minimal risk," according to recognized standards, is of primary ethical concern to the investigator.

The second principle states that a basic ethical concern of the investigator is to determine the degree of risk imposed by the study. Virtually any research study will impose some demands on the research participant, such as the requirement that the subject devote the length of time it takes to complete the study, and these demands could be perceived as costs to the subject. Usually, however, such costs are minimal, and there is minimal risk to the subject. Exceptions are experiments that use deception, subject the participants to stressful conditions,

---

[1]Excerpts throughout this section are from *Ethical Principles in the Conduct of Research with Human Participants*, 1982. Washington, D.C.: American Psychological Association. Copyright 1982 by the American Psychological Association. Reprinted by permission of the publisher.

or require the subjects to take drugs; these certainly pose a threat to the welfare and dignity of the participant. It is the ethical responsibility of the investigator to distinguish between these two conditions and, when the study poses a threat to the subject, to seek the counsel of others to make sure that the subjects are protected.

The decision-making process involves weighing the costs and risks to the participants against the potential benefits that may result from the study. It is not appropriate to conclude that research cannot be conducted if it imposes some risk to the participant. Each study has to be considered separately, and a subjective judgment made about the potential gains and costs. A cost–benefit analysis must be conducted on any study that imposes risk to the participants. If the potential gain exceeds the potential cost, advisory groups typically recommend that the investigator conduct the study. Where the risks are relatively great, the investigator has the obligation to consider alternative approaches that would impose smaller risks.

Principle C

> The investigator always retains the responsibility for insuring ethical practice in research. The investigator is also responsible for the ethical treatment of research participants by collaborators, assistants, students, and employees, all of whom, however, incur similar obligations.

In Principle C, it is stated that the final responsibility for the decision to conduct a study falls on the investigator, regardless of the advice received from others, as does the responsibility for maintaining ethical practice throughout the research study. This issue becomes particularly significant when there is more than one person involved in conducting the study, as there is a tendency to let the other person take responsibility for ethical conduct. Principle C says that this should never happen. In no case does the addition of individuals dilute one's ethical responsibility; instead, it merely multiplies it. Each coprincipal investigator is responsible for the ethical conduct of the study, and where students or assistants are involved, it is the principal investigator's responsibility to train them to be ethically responsible and supervise them to ensure that they act in an ethical manner. This does not relieve research assistants of all responsibility, however. They too should be sensitive to the ethical conduct of the study, and, if the assistant feels a moral reluctance to conduct the study, then the investigator or supervisor must not pressure the assistant to complete it.

Principle D

> Except in minimal-risk research, the investigator establishes a clear and fair agreement with research participants, prior to their participation, that clarifies the obligations and responsibilities of each. The investigator has the obligation to honor all promises and commitments included in that agreement. The investigator informs the participants of all aspects of the research that might reasonably be expected to influence willingness to participate and explains all other aspects of the research about which the participants inquire. Failure to make full disclosure prior to obtaining informed consent requires additional safeguards to protect the welfare and dignity of the research participants. Research with children or with participants who have impairments that would limit understanding and/or communication requires special safeguarding procedures.

Principle D states that, except in minimal-risk situations, the investigator has the ethical responsibility to inform the research participants of their obligations and responsibilities, to inform the participants about all aspects of the research that might influence willingness to participate, and to answer any other questions regarding the project. No one can argue with such a practice, since it tells the participants exactly what is expected of them and of the investigator. With this knowledge, the prospective subjects can evaluate the research and come to a free decision regarding whether or not to participate. The problem is that this ideal cannot always be attained. Some potential subjects (for example, children) do not have the competence to make the necessary decision and to provide informed consent. In such cases, the legal as well as ethical practice is to obtain the informed consent from a person whose primary interest is the subject's welfare, generally the parent or legal guardian. Even when such consent has been obtained, the subject should also be asked to give permission before participating.

Principle E

> Methodological requirements of a study may make the use of concealment or deception necessary. Before conducting such a

study, the investigator has a special responsibility to (1) determine whether the use of such techniques is justified by the study's prospective scientific, educational, or applied value; (2) determine whether alternative procedures are available that do not use concealment or deception; and (3) ensure that the participants are provided with sufficient explanation as soon as possible.

Principle E is actually an extension of Principle D, which states that the investigator should fully inform the participant about all aspects of the study. In many instances, it is not possible to obtain informed consent, either because the information is too technical for the individual to evaluate, the subject is incapable of making a responsible judgment, or, as is most commonly the case, valid data could not be gathered if participants were fully informed about the study. In disguised field experiments, the "people in the street" unknowingly become research participants. In such instances, the ethical issue revolves around the fact that the subjects are not informed about the nature of the experiment or even that they are participating in a study. Sometimes participants are misinformed or deceived about the purpose of an experiment or implications of their behavior. Some people believe that any research involving concealment or deception is unethical and should not be conducted. Others believe in the necessity of such research, but at the same time acknowledge that it may result in an ethical dilemma for the investigator. Principle E directs the investigator to determine if alternative procedures are available that do not employ deception or concealment. If not, then the researcher must determine if the concealment or deception is justified in terms of the scientific, educational, or applied value of the study. Once the study has been completed, the participant must be given an explanation of the whole research procedure as soon as possible. Principle E provides a guide to be followed when, in the judgment of the investigator, fully informed consent cannot be obtained.

Principle E does not reduce the ethical responsibility of the investigator; rather, it provides criteria for the investigator and his or her advisors to use in making a judgment regarding the ethical dilemma. The American Psychological Association (1982, p. 41) gives four conditions that may make the use of deception more acceptable. These are as follows:

    a. The research objective is of great importance and cannot be achieved without the use of deception;

b. on being fully informed later (Principle E), participants are expected to find the procedures reasonable and to suffer no loss of confidence in the integrity of the investigator or of others involved;

c. research participants are allowed to withdraw from the study at any time (Principle F), and are free to withdraw their data when the concealment or misrepresentation is revealed (Principle H), and

d. investigators take full responsibility for detecting and removing stressful after-effects of the experience (Principle I).

Principle F

The investigator respects the individual's freedom to decline to participate in or to withdraw from the research at any time. The obligation to protect this freedom requires careful thought and consideration when the investigator is in a position of authority or influence over the participant. Such positions of authority include, but are not limited to, situations in which research participation is required as part of employment or in which the participant is a student, client, or employee of the investigator.

Principle F asserts that it is ethically unacceptable to coerce a subject to participate in research. One of our human rights is freedom of choice, and this extends to participation in psychological research. One runs into two related problems in attempting to implement this principle. The first difficulty, which Kelman (1972) has eloquently discussed, is the power relationship that often exists between the researcher and the subject. Typically, the subject is in a less powerful position, which places him or her at a perceived disadvantage in feeling free to decline participation. In implementing Principle F, an investigator might ask the group of potential subjects for volunteers to participate in the research and tell them that they do not have to volunteer. But if these potential subjects are children or the investigator's students or patients, they may not feel free *not* to volunteer without encountering some penalty. Investigators must be aware of the power they hold relative to their subjects and consider this when trying to implement Principle F.

The second difficulty in trying to fulfill Principle F is that the psychologist has an obligation to conduct research and to advance a segment of behavioral science. To accomplish this, one needs an available supply of subjects. For those studying humans, there is a need for a supply of warm bodies, which means that somehow some people need to be motivated to volunteer for psychological research studies. This has led many psychology departments to form a subject pool by requiring all introductory psychology students to participate in one or more psychology experiments. A number of individuals feel that Principle F is violated by such a requirement. The APA ethics committee (American Psychological Association, 1982, pp. 47–48) has suggested a set of procedures a department can follow in order to maintain Principle F in spirit while still providing a pool of subjects to serve as research participants. These suggestions are as follows:

a. Students are informed about the research requirement before they enroll in the course, typically by an announcement in an official listing of courses. In addition, during the first class meeting, the instructor provides a detailed description of the requirement, frequently in written form, covering the following points: the amount of participation required; the available alternatives to actual research participation; in a general way, the kinds of studies among which the student can choose; the right of the student to drop out of a given research project at any time without penalty; any penalties to be imposed for failure to complete the requirement or for nonappearance after agreeing to take part; the benefits to the student to be gained from participation; the obligation of the researcher to provide the student with an explanation of the research; the obligation of the researcher to treat the participant with respect and dignity; the procedures to be followed if the student is mistreated in any way; and an explanation of the scientific purposes of the research carried on in the departmental laboratories.

b. Prior approval of research proposals, sometimes by a single faculty member but more often by a departmental committee or an Institutional Review Board, is recommended. The following considerations are appropriate for the review: Will dangerous or potentially harmful procedures be employed? If so, what precautions have been taken to protect the participants from the possible damaging effects of the procedure? Will inordinate demands be made upon the participants' time? Will the research involve deception or withholding information? If so, what plans have been developed for subsequently informing the participants? What plans have been made for providing the participants with an explanation of the study? In general, what will the participants gain?

c. Alternative opportunities for research participation are provided. This provision lets students choose the type of research experience and (often of more consequence to students) the time and place where they will participate.

Providing options commensurate in time and effort that do not require service as a research participant is necessary. The student may observe ongoing research and prepare a report based upon this experience or submit a short paper based upon the reading of research reports. Care should be taken to ensure that selecting such options has no punitive consequences.

d. Before beginning participation, the student receives a description of the procedures to be employed and is reminded of the option to drop out later without penalty if so desired. At this point consent is sought, and the student may be asked to document consent in writing. In any event, participation in any teacher's own research project should be optional for all students.

e. Steps are taken to ensure that the student is treated with respect and courtesy.

f. Participants receive some kind of reward for their participation. At a minimum this reward involves as full an explanation of the purposes of the research as is possible. In addition, some departments may reward research participation with better grades, although many critics would question the educational propriety of this practice. The assignment of a grade of "incomplete" as a sanction against nonfulfillment is common, although some critics regard this as too coercive. Where this sanction is used, procedures exist for allowing the student to fulfill the requirement later on.

g. There is a mechanism by which students may report any mistreatment. Usually, the mechanism involves reporting questionable conduct on the part of a researcher to the instructor, the departmental ethics committee, or the chair of the department.

h. The recruiting procedure is under constant review. Assessments of student attitudes toward the requirement are obtained at the end of each course having such a requirement each time the course is offered. These data, together with evaluations of the workability of the procedures by the investigators, provide the basis for modifying the procedures in subsequent years.

Principle G

The investigator protects the participant from physical and mental discomfort, harm, and danger that may arise from research procedures. If risks of such consequences exist, the investigator informs the participant of that fact. Research procedures likely to cause serious or lasting harm to a participant are not used unless the failure to use these procedures might expose the participant to risk of greater harm or unless the research has great potential benefit and fully informed and voluntary consent is obtained from each participant. The participant should be informed of procedures for contacting the investigator within a reasonable time period following participation should stress, potential harm, or related questions or concerns arise.

Principle G begins by stating the ideal—that research participants should be protected from any physical or mental discomfort, harm, or danger. Fortunately, most psychological research is relatively innocuous and exposes the participants to little suffering or danger. However, a review of the psychological literature indicates that some studies have indeed subjected research participants to psychological and physical risks. For example, studies have induced physical discomfort and risk through administration of drugs and electric shock. Subjects have also suffered by being exposed to failure, anxiety, frustration, and scenes of extreme human suffering. When exploring such potentially risky issues, a psychologist has the obligation first to investigate alternative means of studying the problem, such as using animals or studying individuals who are in naturally occurring but unavoidable stressful situations (for example, an individual waiting to undergo an operation). When such alternatives cannot be found, the investigator is faced with an ethical dilemma. If the investigator determines that the benefits outweigh the costs, he or she has the obligation to inform the participant of the risks. If the research procedure is likely to cause serious or lasting harm, it is not used unless the study has *great* potential benefit or unless failure to use the procedure would expose the participant to even greater risk.

Fully informing the research participant of all risks that may be involved and still attaining valid data is difficult in some studies. If one were investigating failure, for valid data to be generated it would be necessary for the participants to believe that they had actually failed. Where deception can be justified, the investigator incurs an additional responsibility— to ensure that the participants experience the least possible psychological damage. In all instances, the participant should be informed of the procedures for contacting the investigator if any stress, harm, or questions arise.

Principle H

> After the data are collected, the investigator provides the participant with information about the nature of the study and attempts to remove any misconceptions that may have arisen. Where scientific or humane values justify delaying or withhold-

ing this information, the investigator incurs a special responsibility to monitor the research and to ensure that there are no damaging consequences for the participant.

Several times in this text it has been stated that good research design sometimes necessitates deceiving or withholding information from the research participants. Where such a situation exists, Principle H applies; the investigator has a responsibility to debrief the subject or provide the full details of the study, including why the deception was necessary or why information was withheld. The investigator must also ensure that there are no damaging consequences for the participant.

A number of difficulties can be encountered in trying to satisfy Principle H. The debriefing may anger or disillusion the participants because they learn that they were deceived. Aronson and Carlsmith (1968) discuss this issue at length and propose a debriefing procedure (outlined in Chapter 12) that should minimize this possibility. The fact that the debriefing may not be believed is a problem where the study involves a double deception (two consecutive deception procedures in the same study). The subject has been deceived twice and may think that this is another trick. In cases where children are the research participants, the primary objective of debriefing should be to ensure that they leave with no undesirable side effects because of their relative inability to understand complex explanations.

The timing of the postinterview can also pose a problem in some studies. The investigator may feel that the debriefing should occur en masse after completion of data collection, or, in a multiple-session study, the research design may call for delaying debriefing until the end of the second or third session, even though deception occurred in the first session. In such an instance, the investigator must consider the possible detrimental effects that may accrue from allowing the person to leave the laboratory either misinformed or uninformed. In cases where such detrimental effects may exist, the researcher should try to find alternative designs or perhaps abandon the study. At the very least, the investigator should seek the counsel of others regarding the proper procedure to follow.

Principle I

Where research procedures result in undesirable consequences for the individual participant, the investigator has the responsibility to detect and remove or correct these consequences, including long-term effects.

Although most psychological research is quite innocuous, occasionally stress or some other undesirable consequence is either deliberately or inadvertently imposed on the participants. Principle I states that, regardless of the source of the stress or other undesirable consequence, the investigator is responsible for detecting and removing it. This means that the investigator must be alert to stressful reactions that may occur during the conduct of the experiment and take corrective action immediately. Under some conditions, a long-term follow-up may be necessary to ensure that detrimental effects do not persist over time.

Some studies require the use of a control group to test for the influence of a therapeutic procedure, drug, or educational experience. The typical approach is to withhold the treatment procedure or experience from the control group. In such instances, beneficial alternative procedures should be considered (for example, giving the control group a different treatment of known benefit).

Principle I also requires the investigator to provide control groups with access to beneficial treatments if these treatments prove to be efficacious. If debriefing the research participants about the various components of the study may be detrimental to them, however, the investigator becomes involved in a special conflict. On the one hand, the researcher is obligated to inform the participants about the study; on the other hand, the investigator must avoid harming the participants. The resolution of this conflict seems to be that the participants should be debriefed only if the harmful information might be uncovered by the subjects at a later date and only if the investigator conducting the debriefing is qualified to handle any resultant distress the participants may experience.

Principle J

Information obtained about the research participant during the course of an investigation is confidential unless otherwise agreed upon in advance. When the possibility exists that others may obtain access to such information, this possibility, together with plans for protecting confidentiality, is explained to the participant as part of the procedure for obtaining informed consent.

Every person has a right to privacy that cannot be violated without his or her permission. In a research study, the participants provide informed consent that allows the investigator to observe and record their behavior. Consequently, the investigator has a great deal of personal information that could be passed on to others. Principle J asserts that any data obtained about a research participant must be kept confidential. There are a number of situations in which the experimenter may be pressured to release information. Parents, friends, a teacher, a therapist, a school, a clinic, or an industrial organization may request information. The investigator, however, is ethically bound to maintain confidentiality unless consent is given by the research participant to release the information.

Some other situations pose an even greater threat to Principle J. The law does not provide protection for the confidentiality of research data, so the courts can demand that the investigator release the data. Also, if the experimenter uncovers information suggesting that the participants might harm themselves or others, he or she has an obligation to disclose this information to others. Loss of confidentiality can occur when the data obtained from the participants are published or placed in a data bank. According to Principle J, the investigator has a responsibility to inform the research participants of the possibility that the data may not remain confidential and also of the steps that will be taken to ensure confidentiality. For example, the investigator could promise to store and code the information in a way that makes identification impossible.

POSSIBLE CONSEQUENCES OF ADOPTING THE APA CODE OF ETHICS

The ten principles summarized above have been adopted by the association as its official position (American Psychological Association, 1982) and, therefore, are to be used by psychologists conducting human research.

What are the potential consequences of adopting this particular ethical posture? West and Gunn (1978) found that research proposals have received increased scrutiny since the adoption of the original code of ethics. In some cases, independent variable manipulations that previously would have been used and considered appropriate are having to be altered. For example, anger manipulations traditionally used in aggression research are being changed so that a less severe anger manipulation is used (Baron, 1976). The ethical guidelines have created a shift toward the use of milder and less deceptive manipulations. The implications of this, according to West and Gunn (1978), are twofold. First, the number of subjects used in a given experiment will have to be increased in order to detect a difference between treatment conditions. Second, there will be an accelerated trend toward nonmotivational theories because the experimental manipulations generated by psychologists will be so mild that they probably will not arouse motives that will support a motivational interpretation.

Gergen (1973a, p. 912) summarized his position as follows:

> I have argued that from a research standpoint, there is little to merit the promulgation of the proposed ethical principles. Not only have we failed to demonstrate that our present procedures are detrimental to human subjects, but there is good reason to suspect that the principles would be detrimental to the profession and to the enhancement of knowledge should they be adopted. It has further been maintained that great danger lies behind our attempts at moralizing. What is needed is factual advice about the possible harmful consequences of various research strategies. Such advice could be embodied in a series of advisory statements for researchers. While these statements would primarily be conjectural at this point in time, we should ultimately be able to replace conjecture with fact.

Gergen's position is that we should determine the effects of each principle before incorporating it into a set of ethical standards for everyone to use. As Gergen (1973a, p. 907) stated:

> If subjects remain unaffected by variations along these dimensions, then the establishment of the principles becomes highly questionable. If subjects are generally unconcerned about what is to happen to them, if they find experimental deceptions rather intriguing, if they do not generally care about the rationale of the research, and if their attitudes about life and themselves remain untouched regardless of whether the ethical principles are experimentally realized, then establishing and reinforcing the principles simply pose unnecessary hardships for the scientist. The life of the research psychologist is difficult enough without harnessing him with research restrictions that have little real-world consequence.

When the original ethical principles (American Psychological Association, 1973) were adopted, little research had been conducted on their influence. Today we find ourselves in a slightly different position. Results of a number of studies seem to have been incorporated into the current revision of the ethical principles (American Psychological Association, 1982). For example, whereas the initial set of principles said that participants must be informed of all features of the study that might affect their willingness to participate (Principle D), the current edition recognizes that concealment or deception within a study may be necessary (Principle E). This alteration may have been prompted by research such as that conducted by Resnick and Schwartz (1973). These investigators attempted to determine the impact of following the informed consent principle to its logical extreme in a simple but widely used verbal conditioning task developed by Taffel (1955). The control, or noninformed, group was given typical instructions, which gave them a rationale for the study and informed them of the task that they were to complete. The experimental, or informed, group received *complete* instructions regarding the true reason for conducting the experiment and the *exact* nature of the Taffel procedure. Figure 5.1 depicts the results of the data obtained from the fourteen subjects in each treatment condition. The uninformed subjects performed in the expected manner, demonstrating verbal conditioning. The informed group, however, revealed a reversal in the conditioning rate. Such data show that maintaining maximum ethical conditions alters the knowledge that we accumulate. This altered information might represent inaccurate information, which would create a *lack* of external validity.

In addition to finding a drastic difference in response on the part of the informed subjects, Resnick and Schwartz also found that informing subjects of the entire experiment apparently destroys any incentive to participate in the study. Uninformed subjects were enthusiastic and appeared at the scheduled time, but informed subjects were generally uncooperative and "often haughty, insisting that they had only one time slot to spare which we could either take or leave" (p. 137). It actually took Resnick and Schwartz five weeks to collect the data on the fourteen informed subjects for such reasons! These researchers suggested that completely informing subjects makes some subjects very suspicious and may cause them to stay away. For most participants, however, a full disclosure of the research causes them to lose interest, which "suggests that people enjoy an element of risk and nondisclosure and become bored rapidly with the prospect of participating in something of which they already have full knowledge" (Resnick & Schwartz, 1973, p. 137). This view is supported by a survey of

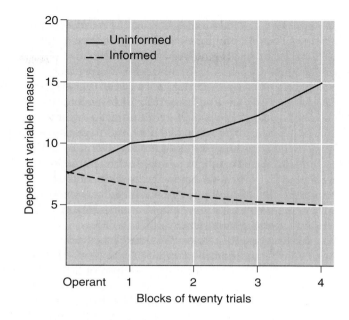

FIGURE 5.1 Verbal conditioning data obtained by Resnick and Schwartz. (Adapted from "Ethical Standards as an Independent Variable in Psychological Research" by J. H. Resnick and T. Schwartz, 1973, *American Psychologist, 28*, p. 136. Copyright 1973 by the American Psychological Association. Reprinted by permission of the author.)

student opinion conducted by Epstein, Suedfeld, and Silverstein (1973). They found that more than 70 percent of the college students they surveyed did not expect to be told the purpose of the experiment and that deception, though not desirable, was not an inappropriate feature of the research setting.

Let us now look at the three issues raised by the code of ethics: informed consent, deception, and the possibility of coercion in relation to participants' freedom to decline to participate in or to withdraw from the study at any time (Principles D, E, and F).

Informed Consent

Informed consent refers to the principle that the research participant should be fully informed about all aspects of the study.

Informed consent
Consent to participate in a scientific study, given by a potential subject who has received an explanation of the procedure to be followed and a description of any possible adverse results from participating

This means that the research participant must be given an explanation of the procedure to be followed and any possible adverse reactions that could accrue from participation. With this information the research participant can make an informed decision and either decline to participate in the study or give his or her informed consent.

Gaining a subject's informed consent is considered to be vital because of the sacredness of the principle that individuals have a fundamental right to determine what is done to their minds and bodies. Once a person is provided with all available information, it is assumed that he or she can make a free decision as to whether to participate and in this manner subjects can avoid experimental procedures they would consider objectionable.

Although informed consent is required as part of a research protocol to protect the subject from participating in objectionable research, Loftus and Fries (1979) have pointed out that providing informed consent may harm the subject. This harm can arise from our suggestible nature. Suggestion can have positive effects, as has been noted in medical research for some time. Some individuals derive benefits from consuming sugar pills (placebos) if they believe the suggestion that they are actually taking a drug. However, negative side effects can also arise from human suggestibility. For example, Loftus and Fries (1979) report on the results of a drug study in which subjects had provided informed consent. In this study some of the subjects who were given an injection of the placebo reported experiencing dizziness, nausea, vomiting, and even depression. One subject reported that the effects were so strong that they caused him to have an automobile accident.

Such evidence suggests that obtaining a subject's informed consent should not automatically be considered positive and that the possibility of a subject's experiencing harm as a direct result of the consent procedure should be considered. This does not mean that all features of the consent procedure should be eliminated if a decision is made that the subject may be harmed by some aspect of it. Rather, the component that produces the harm (e.g., a statement suggesting that a drug may have specific and improbable side effects) must be isolated and provided only to those who specifically request it. The features that protect subjects, such as statements describing the procedures to be followed and the general level of risk, must be retained.

Deception

A number of investigators have attempted to determine the extent to which deception is used in psychological studies. Table 5.1 summarizes the results of such surveys. It is obvious from these surveys that the use of deception steadily increased between 1948 and the 1970s. This increase occurred primarily within the fields of personality and social psychology, although deception is not uncommon in other kinds of studies. For example, the CBS TV news program "60 Minutes" recently staged an elaborate deception in its investigation of the use of polygraph tests by private employers (Saxe, 1991). Staff from "60 Minutes" randomly selected four polygraph examiners from the telephone directory and hired them to examine four

TABLE 5.1   Use of Deception in Psychological Research

Author(s)	Journals	Year	Percentage of Deception Studies
Seeman (1969)	Journal of Abnormal and Social Psychology	1948	14.3
Seeman (1969)	Journal of Personality	1948	23.8
Seeman (1969)	Journal of Consulting Psychology	1948	2.9
Seeman (1969)	Journal of Experimental Psychology	1948	14.6
Menges (1973)	Journal of Abnormal and Social Psychology	1961	16.3
Seeman (1969)	Journal of Abnormal and Social Psychology	1963	36.8
Seeman (1969)	Journal of Personality	1963	43.9
Seeman (1969)	Journal of Consulting Psychology	1963	9.3
Seeman (1969)	Journal of Experimental Psychology	1963	10.8
Menges (1973)	Journal of Personality and Social Psychology	1971	47.2
Menges (1973)	Journal of Abnormal and Social Psychology	1971	21.5
Menges (1973)	Journal of Educational Psychology	1971	8.3
Menges (1973)	Journal of Consulting Psychology	1971	6.3
Menges (1973)	Journal of Experimental Psychology	1971	3.1
Carlson (1971)	Journal of Personality and Journal of Personality and Social Psychology	1968	57.0
Levenson, Gray, and Ingram (1976)	Journal of Personality	1973	42.0
Levenson, Gray, and Ingram (1976)	Journal of Personality and Social Psychology	1973	62.0
Adair, Dushenko, and Lindsay (1985)	Journal of Personality and Social Psychology	1979	58.5

people who supposedly worked for the CBS-owned magazine *Popular Photography*. The polygraphers were instructed to examine the four employee suspects in an attempt to determine if one of them had stolen $500.00 worth of camera equipment. All four polygraphers were told that all suspects had access to the stolen camera, and each polygrapher was told that a different person was probably the guilty party. Unbeknownst to the polygraphers, there had been no theft of property and the four suspects were confederates who were to be paid $50 if they could convince the polygraphers of their innocence. Additionally, the office in which the polygraph examinations were to be conducted was modified to enable surreptitious filming. Much to the delight of the "60 Minutes" team, all four polygraphers fingered as the guilty party the individual whom they had been told was the suspect, although none of the confederates stole anything or confessed to a theft.

Given that deception is here to stay and that alternatives to deception such as role playing (Kelman, 1967) are inadequate substitutes (Miller, 1972), we need to take a look at two questions.

First, is it possible to devise a compromise between the need for deception and the ethical principle of informed consent? This compromise could take the form of informing all potential subjects in, say, the departmental subject pool that some of the experiments they are to engage in will include some form of deception. Campbell (1969, p. 370) has suggested the following as a possible procedure:

> Announce to all members of the subject pool at the beginning of the term, "In about half of the experiments you will be participating in this semester, it will be necessary for the validity of the experiment for the experimenter to deceive you in whole or in part as to his exact purpose. Nor will we be able to inform you as to which experiments these were or as to what their real purpose was, until after all the data for the experiment have been collected. We give you our guarantee that no possible danger or invasion of privacy will be involved, and that your responses will be held in complete anonymity and privacy. We ask you at this time to sign the required permission form, agreeing to participate in experiments under these conditions." This would merely be making explicit what is now generally understood, and probably would not worsen the problem of awareness and suspicion that now exists.[2]

---

[2]From "Prospective: Artifact and Control" by D. Campbell, in *Artifact in Behavioral Research*, edited by R. Rosenthal and R. L. Rosnow, 1969. New York: Academic Press. Reprinted by permission of the author and the publisher.

Again, we must ask if this procedure is a legitimate alternative that would still enable us to obtain unbiased data. Holmes and Bennett (1974) provided data on this question. One of their groups of subjects, the informed group, was told that deception was involved in some psychological experiments. The researchers found that giving these subjects such information in no way affected their performance; they performed in the same manner as did the deceived subjects. Consequently, it seems that telling subjects that they may be deceived without telling them the exact nature of the deception does not alter results.

The second question is, What effect may deception have on the subject? More than two decades ago Kelman (1967) predicted that the persistent use of deception would cause research subjects to become distrustful of psychologists and undermine psychologists' relations with subjects. Fortunately, Sharpe, Adair, and Roese (1992) revealed that such a dire prediction was unfounded and that current research participants are as accepting of arguments justifying the use of deception as they were twenty years ago.

It also seems to be generally accepted that deception has a significant potential for wronging and harming subjects. Sieber's (1982) stance perhaps epitomizes this view. She identified seven types of deception as well as the types of potential harm or ethical objections that accompany each. As Table 5.2 shows, Sieber listed six ethical objections or sources of potential harm that can be generated by deception and associated at least one of these with each type of deception. For example, a study that deceived subjects and did not obtain their informed consent would involve an invasion of privacy, denial of self-determination, concealment, and lying. Based on her listing of types of deception, Sieber (1983a) provided a taxonomy of factors to be used in assessing the risk/benefit ratio of deception research. Expositions such as those presented by Sieber give the impression that deception is not only ethically objectionable but potentially harmful as well. However, these writings seem to be based on the authors' opinions, which may or may not conform to reality. The only way to determine if such expositions represent moral philosophizing or actual ethical objections is to investigate the perceptions of participants who have been subjected to deception.

Christensen (1988) summarized the results of studies investigating the reactions of subjects to deception experiments. The literature consistently revealed that research participants

TABLE 5.2 Potential Harms from or Ethical Objections to Kinds of Deception in Research

	Potential Harm or Ethical Objection					
Kind of Deception	Invades Privacy	No Consent	No Self-Determination	No Debriefing	Lies	Researcher Conceals
Deception not by the researcher						
Self-deception	4	3	3	3	2	3
Third-person deception	4	4	4	3	2	4
Deception by the researcher						
Informed consent	1	1	1	1	1	4
Consent to deception	3	1	1	1	4	4
Waives informing	3	4	1	2	4	4
False informing	4	4	4	2	4	4
No informing	4	4	4	3	4	4

Source: From "Deception in Social Research: I. Kinds of Deception and the Wrongs They May Involve" by J. E. Sieber, 1982, *IRB: A Review of Human Subjects Research, 4*(9), pp. 1–5. Reprinted by permission of The Hastings Center. © Institute of Society, Ethics, and the Life Sciences, 360 Broadway, Hastings-on-Hudson, NY 10706.

Key: 4 = accompanies, 3 = likely to accompany, 2 = unlikely to accompany, 1 = does not accompany

do not perceive that they were harmed and do not seem to mind having been misled. For example, Pihl, Zacchia, and Zeichner (1981) conducted a follow-up investigation of subjects who had participated in a series of studies that included deception and potential physical and mental stress. Subjects who had been in a study of the effects of alcohol on aggression were telephoned by a caller who asked them to recall all components of the experiment and then asked if they would complete a questionnaire that would be mailed to them. This questionnaire asked if anything about the experiment bothered them (and, if so, to what degree and for how long), if they felt free to discontinue the experiment at any time, and if they would be willing to participate in a similar study in the future. Results showed that, of the subjects contacted, only 19 percent reported being bothered by any aspect of the experiment, and 4 percent said they were bothered by the deception. Most of the factors that bothered the subjects were aspects of the experiment other than deception or the fact that the experiment required them to deliver shocks. The components that upset the subjects were mostly rather trivial (one subject felt that using a cloth holder for a drinking glass was unsanitary). The greatest distress surrounded the type of alcohol consumed, the dose, and the speed with which it had to be consumed. One subject reported being bothered for several days because "laboratory and not commercial alcohol was consumed" (Pihl et al., 1981, p. 930). Interestingly, this subject was in a placebo group that did not even consume alcohol. It is also interesting to note that the distress surrounding the deception and averse stimuli variables lasted less time than did the distress surrounding other seemingly trivial variables such as boredom. As is illustrated in Figure 5.2, the duration of distress over the deception or shock was only an hour or less, whereas the dissatisfaction with the alcohol lasted an average of twenty hours.

Smith and Richardson (1983) have produced results that apparently confirm the findings of Pihl et al. Smith and Richardson questioned 215 male students and 249 female students at the end of the academic quarter, following the students' participation in the minimum number of required

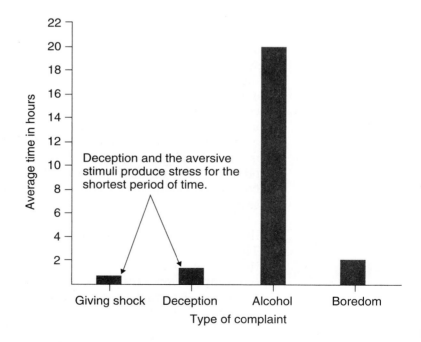

FIGURE 5.2 Average length of distress for four categories of complaints. (Based on data from "Follow-up Analysis of the Use of Deception and Aversive Contingencies in Psychological Experiments" by R. O. Pihl, C. Zacchia, and A. Zeichner, 1981, *Psychological Reports, 48,* pp. 927–930.)

psychology experiments. The subjects were questioned on a variety of issues, including the potential harmful effects that deception may have on research participants. This study showed, contrary to statements made by Sieber (1982), that the subjects who had participated in deception experiments reported enjoying the experiment more, felt that they had received more educational benefit from the experiment, and perceived their participation in the research as being more satisfactory than did other subjects. Not only did Smith and Richardson not find support for the notion that deception is harmful—they provided data suggesting that deception may be advantageous.

Although research participants consistently report that they do not mind having been misled and were not harmed by deception experiments, a case could be made for the view that the detrimental effects of deception depend on the type of study being conducted, and the study conducted by Smith and Richardson averaged the responses of subjects who had participated in many different types of studies. Such an argument might be valid if Pihl et al. had also averaged across many deception studies, but they did not. Also, Gerdes (1979) had subjects participate in several different experiments in which the type and the degree of deception were varied. This study did not show any differences in subjects' reactions to the deception, but it did indicate that subjects did not mind being misled and were not annoyed by having details of the experiment withheld. Christensen (1988) pointed out that deception is viewed as less acceptable ethically if the study investigates private behaviors such as sexual experiences or if the experimental procedure has significant potential to harm the research participant.

Thus, the evidence shows that deception is not necessarily the harmful component many individuals assume it to be. This does not mean that the potential harmful effects of deception can be forgotten, though. One of the primary modes used to eliminate any harmful effects of deception is debriefing, or a postexperimental interview with the subject, in which the components of the experiment are explained. All the studies that investigated the impact of deception incorporated a debriefing procedure, and if such a procedure does in fact eliminate any harmful effects of deception, this may explain the positive findings of these experiments.

In his reply to Baumrind's (1964) criticism of his earlier (Milgram, 1963) study, which investigated the extent to which subjects would obey instructions to inflict extreme pain on confederates, Milgram (1964b) reported that, after extensive debriefing, only 1.3 percent of his subjects reported any negative feelings about their experiences in the experiment. Such evidence indicates that the debriefing was effective in eliminating the extreme anguish that these subjects apparently experienced.

Ring, Wallston, and Corey (1970) lent considerable support to Milgram's 1964 findings. In their quasi-replication of Milgram's 1963 experiment, they found that only 4 percent of the subjects who had been debriefed indicated they regretted having participated in the experiment, and only 4 percent believed that the experiment should not be permitted to continue. On the other hand, about 50 percent of the subjects who had not been debriefed responded in this manner. Berscheid, Baron, Dermer, and Libman (1973) found similar ameliorative effects of debriefing on consent-related responses. Holmes (1973) and Holmes and Bennett (1974) took an even more convincing approach and demonstrated that debriefing reduced the arousal generated in a stress-producing experiment (expected electric shock) to the prearousal level, as assessed by both physiological and self-report measures.

Smith and Richardson (1983) asserted that their deceived subjects received better debriefings than did their nondeceived subjects and that this more effective debriefing may have been the factor that caused the deceived subjects to have more positive responses than did the nondeceived subjects.

Dehoaxing
Debriefing the participants about any deception that was used in the experiment

Desensitizing
Eliminating any undesirable influence that the experiment may have had on the subject

This suggests that debriefing is quite effective in eliminating the stress produced by the experimental treatment condition. However, Holmes (1976a, 1976b) has appropriately pointed out that there are two goals of debriefing and both must be met for debriefing to be maximally effective. These two goals are dehoaxing and desensitizing. **Dehoaxing** refers to debriefing the subjects about any deception that the experimenter may have used. In the dehoaxing process, the problem is one of convincing the subject that the fraudulent information given was, in fact, fraudulent. **Desensitizing** refers to debriefing the subjects about their behavior. If the experiment has made subjects aware that they have some undesirable features (for example, that they could and would inflict harm on others), then the debriefing procedure should attempt to help the subjects deal with this new information. This is typically done by suggesting that the undesirable behavior was caused by some situational variable rather than by some dispositional characteristic of the subject. Another tactic used by experimenters is to point out that the subjects' behavior was not abnormal or extreme. The big question is whether or not such tactics are effective in desensitizing or dehoaxing the subjects. In Holmes's (1976a, 1976b) review of the literature relating to these two techniques, he concluded that they were effective; however, this means only that effective debriefing is *possible.* These results hold only if the debriefing is carried out properly. A sloppy or improperly prepared debriefing session may very well have a different effect. Additionally, the beneficial impact of debriefing can be experienced only if the experimental procedure includes a debriefing session. Adair, Dushenko, and Lindsay (1985), in their survey of the literature, found that only 66 percent of all the deception studies reported in the *Journal of Personality and Social Psychology* in 1979 included debriefing. This suggests that researchers should be more diligent about debriefing their subjects.

One more point needs to be made about debriefing. Perhaps debriefing should not be universally applied in all experiments. Aronson and Carlsmith (1968) and Campbell (1969) have discussed the potentially painful effects that debriefing can have if the subject learns of his or her own gullibility, cruelty, or bias. It is for this reason that Campbell (1969) suggested that debriefing be eliminated when the experimental treatment condition falls within the subject's range of ordinary experiences. This recommendation has also been supported by survey data collected by Rugg (1975).

The results of the studies presented so far have indicated that deception seems to have a more negative effect on researchers than it does on the participants. Sullivan and Deiker (1973), Rugg (1975), and Collins, Kuhn, and King (1979) have all investigated this question and found that, of the individuals surveyed, the participants were much more lenient in their perceptions of the ethical issues of human research than were the researchers. Rugg (1975) extended this survey to include additional significant groups, such as ethics committee members and law professors, and found these individuals without exception to be much stricter in their interpretation of the ethics of human research than were the participants (college students). It appears that individuals conducting human research constitute a strict self-regulating force and are sensitive to the rights and welfare of their research participants. Wilson and Donnerstein (1976) do, however, make the point that the participants' attitudes toward a given research procedure should be considered, since they found that some procedures elicited a negative reaction from a substantial minority of the individuals surveyed.

Coercion and Freedom to Decline Participation

Principle F of the code of ethics is quite explicit in ensuring freedom from coercion to participate in psychological experiments. Use of "subject pools" consisting of all students enrolled in certain courses is widespread. Concern over the possibly coercive nature of the relationship between professors and students led the American Psychological Association (1982) to issue its set of eight guidelines to follow in the use of subject pools. However, given the evidence regarding the apparent positive influence of deception experiments, one must wonder whether the research participants who are exposed to the potentially coercive measures really perceive the measures to be that negative and whether this coercion harms the participants in any way. Leak (1981) investigated subjects' reactions to their research experience after being induced to participate in research by the offer of extra credit points. Leak found that the students were divided on their perception of the coercive nature of the means used for attaining participation. About half of the students thought the means were coercive and half did not; however, all students overwhelmingly viewed the research participation as positive. Overall, the subjects viewed the research experience as being worthwhile, as contributing to their knowledge and interest in psychology, and as having scientific merit. Also, somewhat surprisingly, the students did not resent or object to being offered the extra credit for participation, even though about 50 percent of them considered it to be coercive. Leak's results, then, indicate that most students do not view the subject pool negatively. Britton (1979) reached the same conclusion. He found that students evaluated their experience as research participants very favorably and that only 4 percent acknowledged experiencing any discomfort from the experience. Such evidence implies not that concern about the coercive influence surrounding the subject pool should be diminished, but rather that the current procedures to minimize coercion must be operating effectively.

In addition to the issue of coercion, Principle F discusses the necessity of ensuring that individuals always feel free to decline to participate in or to withdraw from the research at any time. This principle seems quite reasonable and relatively innocuous. Gardner (1978), however, has asserted that such a perception, though ethically required, can influence the outcome of some studies. The subtle influence of telling subjects that they were free to discontinue participation was discovered quite accidentally. Gardner had been experimenting on the detrimental impact of environmental noise. Prior to the incorporation of a statement informing potential subjects that they could decline to participate without penalty, he always found that environmental noise produced a negative aftereffect. After he incorporated this statement, however, he could not produce the effect. In order to verify that a statement regarding freedom to withdraw was the factor causing the elimination of the negative aftereffect of environmental noise, Gardner replicated the experiment, telling subjects in one group that they could decline to participate any time without penalty and not making this statement to subjects in another group. As Figure 5.3

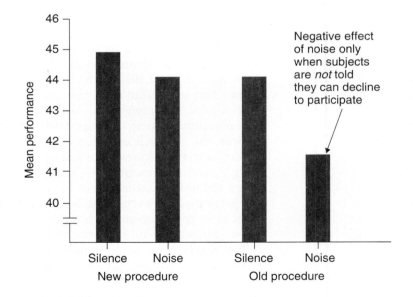

FIGURE 5.3 Accuracy of performance of subjects during silence or environmental noise conditions after being instructed or not instructed that they can decline to participate. (Based on data from "Effects of Federal Human Subjects Regulations on Data Obtained in Environmental Stressor Research" by G. T. Gardner, 1978, *Journal of Personality and Social Psychology, 36*, pp. 628–634.)

illustrates, the environmental noise caused a decline in performance under the old procedures but not under the new procedures. This study indicates the very subtle effects that ethical principles can have and suggests that such effects should be considered when prior results are not replicated and the only difference in procedure is the incorporation of the ethical principles.

ETHICS OF ANIMAL RESEARCH

Considerable attention has been devoted to the ethics of human research. In about 7 percent of the studies they conduct, however, psychologists use animals as their research subjects in order to gain control over many potentially contaminating factors (Gallup and Suarez, 1985) or to investigate the influence of a variable that might be judged too dangerous to test on humans.

During the 1980s, groups such as People for the Ethical Treatment of Animals, the Animal Rights Coalition, and the Society Against Vivisection protested the use of animals in research. Many of these activists believe that the rights accorded humans should be extended to nonhuman animals (Dewsbury, 1990). We should not require anything of animals that we would not require of humans. This position is articulated most effectively by one of the codirectors of People for the Ethical Treatment of Animals, who stated that "There is no rational basis for separating out the human animal. A rat is a pig is a dog is a boy. They're all mammals" (McCabe, 1986).

The topic on which animal rights groups have been most vocal is the use of animals in research. They have argued that use of animals in research is immoral, is of no value in treating human disorders, and is unacceptably cruel (Greenough, 1991). Although these arguments have been demonstrated to be untrue in many respects (e.g., Greenough, 1991; Lansdell, 1988), this has not deterred the animal rights groups. These groups have targeted psychological research as being particularly abusive. J. McArdle, director of laboratory animal welfare for the Humane Society of the United States, articulated the position of these groups toward psychological research most effectively. He stated that psychological research exposes animals to pain, stress, heat, cold, electric shock, starvation, and mutilation in an attempt to understand the complex functions of the human mind and to create animal models of human maladies. This pain and suffering might be considered worthwhile if animal research produced its intended results, but the Humane Society and many other animal rights groups take the position that animal experimentation in psychology does not meet its intended goal and that "among all scientific disciplines it is the ideal candidate for complete elimination" because it will not lead to a major scientific discovery (McArdle, 1984).

It is true that animals used in psychological research have been shocked, deprived of food and water, and subjected to stress and various surgical operations. To state that psychological experimentation on animals did not or will not lead to important scientific discoveries, however, is not only inaccurate but shortsighted. Psychological research on animals has benefited both humans and animals. For example, whereas traditionally animals that damaged flocks or crops were poisoned or shot, the results of psychological experimentation on animal behavior have provided us with more humane alternatives. It has been shown that a scarecrow consisting of a model owl mounted on a weathervane and grasping a model crow in its talons is 80 percent more effective in reducing crop loss than is a traditional scarecrow. Psychological experimentation on animals has also led to discoveries in classical and instrumental conditioning, which in turn have been important in the treatment of disorders such as enuresis and anorexia nervosa (Miller, 1985).

The fact that benefits from psychological research on animals are relatively easy to identify has not deterred the animal rights groups or made them less militant. There are some reports (Cowley et al., 1988) that the number of animal protection groups is increasing. Their activities range from letter writing and distortion of the truth to harassment, burglary, and violence. The director of the Washington, D.C., office of the National Anti-Vivisection Society has stated that seventy million animals are used each year in research (Pardes, West, & Pincus, 1991). The actual figure is between seventeen and twenty-two million, 90 percent of which are rats. Promoting violence and causing deliberate harm to others are also tactics that have been used. The Animal Liberation Front is listed by the FBI as one of the country's ten most dangerous organizations (Pardes et al., 1991). An animal activist was recently convicted of attempted murder, possession of explosives, and bomb manufacturing after planting a remote-controlled pipe bomb near the parking space of the chairman of the U.S. Surgical Corporation. Trans Species Unlimited, a Pennsylvania-based animal rights group, picketed the laboratory of a noted drug researcher, made constant phone calls, and wrote letters condemning her experiments, causing her to forfeit a $530,000 three-year grant.

All the attention attracted by the animal rights groups suggests that they have widespread support among the general population. Gallup and Beckstead (1988) found this not to be the case, however, at least within a college population. College students are concerned about the pain and suffering that research animals may have to endure, but they also appreciate and support the need to use animals in research.

The allegations of the animal rights groups have not gone unnoticed, and alternative procedures have been investigated. Gallup and Suarez (1985) reviewed all the alternatives suggested to date and concluded that these proposed alternatives, such as the use of naturalistic observation, embryos, or even plants, are complements to the use of animals but not a substitute for it. At present there seems to be no adequate substitute for the use of animals as research subjects. This does not mean that researchers have a license to subject animals to any procedure that can be justified in the name of research. Rather, it places an additional burden on the researcher to provide humane care and treatment of the animals and to minimize the pain and suffering they may experience. This may be accomplished in part by using pets as subjects and conducting experiments in the owner's home. Home testing has the advantage of not requiring the animal to be confined to a laboratory, which may produce a calmer and more tractable animal and one that does not display the fear and hyperexploration associated with a novel setting; in addition this approach is cost effective (Devenport and Devenport, 1990). However, only relatively noninvasive studies can be conducted, which means that home testing can be used in only a percentage of psychological studies.

Regardless of the environment in which a psychological study is conducted, the researcher using animals as research participants faces an ethical dilemma similar to that faced by the researcher using humans as research participants. The potential gain in knowledge that may accrue from the research must be weighed against the cost to the subject. This subjective decision must be made by the investigator after receiving recommendations from scientists in related fields as well as from lay individuals. To assist the researcher in this process,

the American Psychological Association has established a set
of guidelines for the care and use of animals. In 1992 these
guidelines were revised and presented to the association members for their approval. This revision, which was approved in
February of 1993, includes the following points.

General Issues

The acquisition, care, housing, use, and disposition of animals
should be in compliance with the appropriate federal, state,
local, and institutional laws and regulations and with international conventions to which the United States is a party. APA
authors must state in writing that they have complied with
the ethical standards when submitting a research article for
publication. Violations by an APA member should be reported
to the APA Ethics Committee, and any questions regarding the
guidelines should be addressed to the APA Committee on Animal Research and Ethics (CARE).

I. Personnel All personnel involved in animal research
should be familiar with the guidelines. Any procedure used by
the research personnel must conform with federal regulations
regarding personnel, supervision, record-keeping, and veterinary care. Both psychologists and their research assistants
must be informed about the behavioral characteristics of their
research animals so that unusual behaviors that could fore-
warn of health problems can be identified. Psychologists
should ensure that anyone working for them when conducting animal research receives instruction in the care, maintenance, and handling of the species being studied. The
responsibilities and activities of anyone dealing with animals should be consistent with their competencies, training, and experience regardless of whether the setting is the
laboratory or the field.

II. Care and Housing of Animals The "psychological well-
being" of animals is a topic that is currently being debated.
This is a complex issue because the procedures that may
promote the psychological well-being of one species may not
be appropriate for another. For this reason the APA does not
stipulate any specific guidelines but rather states that psychologists familiar with a given species should take measures,
such as enriching the environment, to enhance the psycho-

logical well-being of the species. For example, the famous Yerkes laboratory and New York University's LEMSIP (Laboratory for Experimental Medicine and Surgery in Primates) have constructed wire mesh tunnels between the animals' cages to promote social contact.

In addition to providing for the animals' psychological well-being, the facilities housing animals should conform to current USDA (1990, 1991) regulations and guidelines and are to be inspected twice a year (USDA, 1989). Any research procedures used on animals are to be reviewed by the institutional animal care and use committee (IACUC) to ensure that they are appropriate and humane. This committee essentially supervises the psychologist who has the responsibility for providing the research animals with humane care and healthful conditions during their stay at the research facility.

III. Acquisition of Animals Animals used in laboratory experimentation should be lawfully purchased from a qualified supplier or bred in the psychologist's facility. When animals are purchased from a qualified supplier, they should be transported in a manner that provides adequate food, water, ventilation, and space and that imposes no unnecessary stress on the animals. If animals must be taken from the wild, they must be trapped in a humane manner. Endangered species should be used only with full attention to required permits and ethical concerns.

IV. Justification of the Research Research using animals should be undertaken only when there is a clear scientific purpose and a reasonable expectation that the research will increase our knowledge of the processes underlying behavior, increase our understanding of the species under study, or result in benefits to the health or welfare of humans or other animals. Any study conducted should have sufficient potential importance to justify the use of animals and any procedure that would produce pain in humans should be assumed to also produce pain in animals.

The species chosen for use in a study should be the one best suited to answer the research question. However, before a research project is initiated, alternatives or procedures that would minimize the number of animals used should be considered. Regardless of the type of species or number of animals used, the research may not be conducted until the protocol has been reviewed by the IACUC. After the study is initiated, the psychologist must continuously monitor the research and the animals' welfare.

V. Experimental Procedures   The design and conduct of the study should involve humane consideration for the animals' well-being. In addition to the procedures governed by guideline IV, "Justification of the Research," the researcher should adhere to the following points.

1. Studies, such as observational and other noninvasive procedures, that involve no aversive stimulation and create no overt signs of distress are acceptable.
2. Alternative procedures that minimize discomfort to the animal should be used when available. When the aim of the research requires use of aversive conditions, the minimal level of aversive stimulation should be used. Psychologists engaged in such studies are encouraged to test the painful stimuli on themselves.
3. It is generally acceptable to anesthetize an animal prior to a painful procedure if the animal is then euthanized before it can regain consciousness.
4. Subjecting an animal to more than momentary or slight pain which is not relieved by medication or some other procedure should be undertaken only when the goals of the research cannot be met by any other method.
5. Any experimental procedure requiring exposure to prolonged aversive conditions, such as tissue damage, exposure to extreme environments, or experimentally induced prey killing, requires greater justification and surveillance. Animals that are experiencing unalleviated distress and are not essential to the research should be euthanized immediately.
6. Procedures using restraint must conform to federal guidelines and regulations.
7. It is unacceptable to use a paralytic drug or muscle relaxants during surgery without a general anesthetic.
8. Surgical procedures should be closely supervised by a person competent in the procedure, and aseptic techniques that minimize risk of infection must be used on warm-blooded animals. Animals should remain under anesthesia until a procedure is ended

unless there is good justification for doing otherwise. Animals should be given post-operative monitoring and care to minimize discomfort and prevent infection or other consequences of the procedure. No surgical procedure can be performed unless it is required by the research or it is for the well-being of the animal. Multiple surgeries on the same animal must receive special approval from the IACUC.

9. Alternatives to euthanasia should be considered when an animal is no longer required for a research study. Any alternative taken should be compatible with the goals of the research and the welfare of the animal. This action should not expose the animal to multiple surgeries.

10. Laboratory-reared animals should not be released because, in most cases, they cannot survive or their survival may disrupt the natural ecology. Returning wild-caught animals to the field also carries risks both to the animal and to the ecosystem.

11. Euthanasia, when it must occur, should be accomplished in the most humane manner and in a way that ensures immediate death and is in accordance with the American Veterinary Medical Association panel on euthanasia. Disposal of the animals should be consistent with all relevant legislation and with health, environmental, and aesthetic concerns and should be approved by the IACUC.

VI. Field Research Field research, because of its potential for damaging sensitive ecosystems and communities, must receive IACUC approval, although observational research may be exempt. Psychologists conducting field research should disturb their populations as little as possible and make every effort to minimize potential harmful effects on the population under investigation. Research conducted in inhabited areas must be done so that the privacy and property of any human inhabitants are respected. The study of endangered species requires particular justification and must receive IACUC approval.

VII. Educational Use of Animals Discussion of the ethics and value of animal research in all courses is encouraged. Although animals may be used for educational purposes after review of the planned use by the appropriate institutional committee, some procedures that may be appropriate for research purposes may not be justified for educational purposes. Classroom demonstrations using live animals can be valuable instructional aids—as can videotapes, films, and other alternatives. The anticipated instructional gain should direct the type of demonstration.

SUMMARY

A great deal of psychological research requires the use of humans as subjects. These individuals have certain rights, such as the right to privacy, that must be violated if researchers are to attempt to arrive at answers to many significant questions. This naturally imposes a dilemma on the researcher as to whether to conduct the research and violate the rights of the research participant, or abandon the research project. Increased attention to ethical concerns has resulted in the development of a set of ethical principles published by the American Psychological Association. This code of ethics consists of ten principles to be followed by research psychologists in conducting their research.

A great deal of time, effort, and thought went into the APA code of ethics. Since the adoption of the 1973 set of principles, some studies have been conducted to assess the impact these principles may have. The 1982 revised ethical principles incorporated the results of these studies. There are still issues that need to be discussed, however, including the effects of deception, the pressure placed on students to participate, and the freedom of subjects to withdraw.

A number of individuals have suggested alternatives to deception, such as role playing, but research studies have shown that such alternatives are poor substitutes. Therefore deception remains a part of numerous psychological studies, and its potential effects must be considered. It is generally assumed that deception creates stress and that this stress or invasion of privacy is ethically objectionable and perhaps harmful to the subjects. Yet research indicates that participants do not view deception as detrimental and that those who have been involved in deceptive studies view their research experience as more valuable than do those who have not. This phenomenon may be due to the increased attention given in deception studies to debriefing, which seems to be effective in eliminating the negative effects of deception as well as any stress that may have occurred.

Investigators also are quite concerned about coercing students to become research participants. Experiments investigating the perceptions of research participants drawn from a subject pool reveal that they generally view their research experience quite positively.

Although there does not seem to be a negative effect resulting from deception or from the use of subject pools, it has been demonstrated that informing subjects that they are free to withdraw at any time without penalty can influence the outcome of some experiments.

More recently, increased concern for the ethical treatment of animals in research has surfaced. Spearheading this trend are a variety of animal rights groups who claim that psychologists have treated their research animals in an inhumane manner. Although a survey of the literature demonstrates that the claims of animal rights groups are unfounded, their efforts have resulted in the development of institutional animal care and use committees and a set of guidelines adopted by the American Psychological Association for use by psychologists working with animals. These guidelines address different issues, ranging from where the animals are housed to how the animals are disposed of. Psychologists using animals for research or educational purposes should be familiar with these guidelines and adhere to them.

STUDY QUESTIONS

1. Define research ethics. In what areas are they of primary concern?
2. What is meant by an ethical dilemma?
3. What led to the development of the APA code of ethics?
4. Identify and explain each of the ten principles in the code of ethics relating to human research participants.
5. What alterations might accompany the adoption of the code of ethics?
6. Discuss the ethical issues surrounding the use of deception.

7. Discuss the ethical issue surrounding the use of subject pools.
8. Why is there increasing concern about animal rights, and what issues should be considered in using animals in research studies?

KEY TERMS AND CONCEPTS

Research ethics
Partial publication
Dual publication
Causal responsibility
Moral responsibility
Political responsibility

Ethical dilemma
Informed consent
Deception
Dehoaxing
Desensitizing

CHAPTER 6

Variables Used in Experimentation

LEARNING OBJECTIVES

1. To develop an understanding of independent and dependent variables.
2. To learn how to formulate independent and dependent variables.
3. To be able to identify the appropriate number of independent and dependent variables to use in a study.
4. To develop an appreciation of the factors that must be considered when specifying independent and dependent variables.

Although most people think of the brain as a single structure, the brain is actually divided into two halves, linked together by several different bundles of nerve fibers. The largest of these bundles is called the corpus callosum, and for many years its function baffled scientists. In several studies, investigators cut the corpus callosum of experimental animals, usually monkeys, eliminating the only possible connection between the two halves of the brain. Amazingly, this operation, called split-brain *surgery, seemed to have little consequence for the monkeys. Split-brain monkeys, for example, seemed to behave no differently than they did before the surgery. Parallel data also existed for humans, leading some to suggest that the only function of this band of fibers is to hold the two halves of the brain together to keep them from sagging. Split-brain surgery has been used as a treatment for severe epilepsy in humans, and again, this operation seems to leave the patient unchanged. Typically, following recovery from the surgery, the patient's epilepsy improves, with no change in personality, intelligence, or behavior.*

Although this has been the case for most of the split-brain patients, there have been a few bizarre occurrences. One patient described an instance in which his right and left hands were trying to do opposite things. One morning, when he got out of bed and tried to put on his pants, he found that while his right hand was trying to pull his pants up, his left hand was trying to pull them down. Another time, this same patient got angry at his wife. His left hand reached out to grab her, but at the same time his right hand grabbed the left in an attempt to stop it. Such cases, although they are the exception and not the rule, suggest that split-brain surgery can have some behavioral consequences.

If you were interested in the behavioral effects of split-brain surgery, as Sperry (1968) was, you would design a psychological experiment to identify the significant effects. Designing an experiment requires that you make a variety of decisions. You must choose the variables that are to be investigated in the experiment, the variables that may introduce a confounding influence, and the techniques that must be employed in order to eliminate such a confounding influence. Only after making such decisions can you describe the final design of the experiment. This chapter will focus attention on the specification and formulation of the independent variable and the dependent variable—the two variables that must be included in any experiment.

INTRODUCTION

One of the first decisions that must be made after the research problem and the hypothesis have been specified is which variable or variables are to serve as the independent variable and which variable or variables are to serve as the dependent variable. By **variable** I mean any characteristic of an organism, environment, or experimental situation that can vary from one organism to another, from one environment to another, or from one experimental situation to another. Therefore, independent and dependent variables can be any of the numerous characteristics or phenomena that can take on different values, such as IQ, speed of response, number of trials required to learn something, or amount of a particular drug consumed. The researcher's task is to select one or more of these variables as the independent variable and another of these variables as the dependent variable. The **independent variable** is the variable that the experimenter changes within a defined range; it is the variable in whose effect the experimenter is interested. The **dependent variable,** on the other hand, is the variable that measures the influence of the independent variable. For example, if you were studying the effectiveness of several teaching techniques, the task of the dependent variable would be to assess effectiveness. Consequently, the dependent variable is linked to the independent variable.

In any given study, there are many possible independent and dependent variables that could be used. How do we identify the ones that are to be included? The independent and dependent variables for a study are specified by the research problem. You will recall from Chapter 4 that the research problem asks a question about the probable relationship between two variables. For example, one of the research questions Flaherty and Checke (1982) asked was whether the concentration of a solution of sugar and water was important in determining the extent to which rats would decrease their consumption of a solution of saccharin and water. Such a research question specifies which variable must be independent and which must be dependent. Because Flaherty and Checke varied the sugar–water concentration consumed, it had to be the independent variable. The influence of the sugar–water concentration was assessed by observing the degree to which it reduced the consumption of the saccharin–water solution. The magnitude of this decrease represented the response of the organisms (albino rats, in this instance) and

Variable
Any characteristic or phenomenon that can vary across organisms, situations, or environments

Independent variable
One of the antecedent conditions manipulated by the experimenter

Dependent variable
The response of the organism; the variable that measures the influence of the independent variable

measured the influence of the sugar–water concentration. Therefore, the suppression measure represented the dependent variable.

Although the research problem may specify both the independent and the dependent variables, it is not always a simple task to design an experiment that uses these independent and dependent variables. For example, assume that a research problem involves the investigation of aggression in rats. If aggression is specified to be the independent variable, you must identify ways to vary aggression. If it is specified to be the dependent variable, you have to identify ways of measuring aggression. As you can see, many decisions and a great deal of thought may be involved in the development of these variables. This chapter will discuss the types of variables and the factors that must be considered when constructing the independent and the dependent variables for an experiment.

TYPES OF VARIABLES

Within the field of psychology many different areas are investigated. Studies have been conducted on topics such as attraction, learning, types of abnormal behavior, and child development. Within each topic area there are a number of variables that can be studied. For example, the topic of child development focuses on variables such as stranger anxiety, attractiveness, ethnic background, and language development. Ethnic background and language development are not, however, the same type of variable. To understand the nature of variables and how variables are used in psychological research, it is important to understand the distinctions made among types of variables. We will consider two approaches often used in psychological research to distinguish between variables.

Discrete variable
A variable that comes in whole units or categories

Continuous variable
A variable that forms a continuum and that can be represented by fractional and whole units

One way to distinguish between variables is to categorize them as either discrete or continuous. **Discrete variables** are variables that come in whole units or categories. A family has a specific number of children. A person is either sick or well, male or female. A stimulus can be either present or absent. **Continuous variables,** on the other hand, are variables that form a continuum and can be represented by both whole and fractional units. Attitudes toward President Clinton, for example, vary from extremely positive to extremely negative. The latency period prior to making a response can be measured in minutes, seconds, or milliseconds. Consequently, a continuous variable can be measured with varying degrees of precision. The precision with which it is measured is limited only by the precision of the measuring instrument. For example, if latency to responding were measured by a clock that recorded only hours, minutes, and seconds, then the most precise measurement we could obtain would be in seconds. Another more precise instrument could measure latency in milliseconds.

Qualitative variable
A variable that varies in kind

Quantitative variable
A variable that varies in amount

Variables can also be distinguished by categorizing them as either quantitative or qualitative. **Qualitative variables** are variables that vary in kind. Psychiatric patients, for example, are given one of several diagnoses, such as bulimia or anorexia. Individuals can be labeled attractive or nonattractive. **Quantitative variables** are variables that vary in amount. Loudness is measured in decibels; latency to responding is measured in seconds, minutes, or some other time measurement.

These are just two of the dimensions on which variables can be categorized. Others could be identified. Any one variable may fit into several different categories on different dimensions. For example, using the variable of hair color, one might classify people as blond, brunette, or redhead. Such a classification is representative of a discrete variable, because a person can be placed into only one of these categories. However, the variable of hair color also represents a qualitative variable, because the three hair colors vary in kind.

THE INDEPENDENT VARIABLE

The independent variable has been defined as the variable manipulated by the experimenter. It is of interest to the investigator because it is the variable hypothesized to be one of the causes of the presumed effect. To obtain evidence of this predicted causal relationship, the investigator manipulates this variable independently of the others. In the Flaherty and Checke (1982) experiment, sucrose concentration was the independent variable manipulated by the experimenter. Marks-Kaufman and Lipeles (1982) investigated the influence of chronic self-administration of morphine on the food rats ate. In this study, rats either were allowed or were not allowed to administer the morphine to themselves, so the ability to self-administer morphine was the independent variable. In an experiment that examines the influence of rate of presentation of words on speed of learning, the independent variable is speed of presentation. Variation in the rate of presentation from one to three seconds provides an independent manipulation that, along with the control of other factors such as ability, enables one to identify the effect of rate of presentation on learning speed.

These examples demonstrate the ease with which one can pick out the independent variable from a study. They also illustrate the requirements necessary for a variable to qualify as an independent variable. In all of the foregoing examples, the independent variable involved variation—variation in rate of presentation of words, sucrose concentration, or self-administration of morphine. This variation was not random but was under the direct control of the experimenter. In all cases, the experimenter created the conditions that provided the type of variation desired. Here we have the two requirements necessary for a variable to qualify as an independent variable: variation and control of the variation. We shall look at each requirement separately and also discuss other issues related to the independent variable.

Variation

To qualify as an independent variable, a variable must be manipulable. The variable must be presented in at least two forms. There are several ways in which the desired variation in the independent variable can be achieved. We will take a look at each of these.

Presence versus Absence The presence-versus-absence technique for achieving variation is exactly what the name implies: one group of subjects receives the treatment condition and the other group does not. The two groups are then compared to see if the group that received the treatment condition differs from the group that did not. A drug study like the one illustrated in Figure 6.1 illustrates this type of variation. One group of subjects is given a drug, and a second group is given a placebo. The two groups of subjects are then compared on some measure, such as reaction time, to determine if the drug group had significantly different reaction times than did the

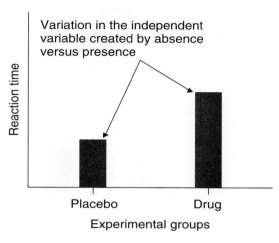

Variation in the independent variable created by absence versus presence

Reaction time

Placebo   Drug

Experimental groups

FIGURE 6.1 Illustration of presence-versus-absence variation in the independent variable.

placebo group. If it did, then the difference is attributed to the drug.

The work that Sperry and others conducted on the behavioral effects of split-brain surgery also represents a good example of the presence-versus-absence technique for manipulating the independent variable. The group of subjects with the split-brain surgery represents the *presence* condition, and the group of subjects that did not receive this surgery represents the *absence* condition.

Amount of a Variable A second basic technique for achieving variation in the independent variable is to administer different amounts of the variable to each of several groups. For example, Ryan and Isaacson (1983) varied the amount of the drug ACTH administered to rats to determine the minimum dose of ACTH required to induce excessive grooming. Rats were injected with either 0, 20, 50, 80, or 1,000 nanograms of ACTH in the area of the brain known as the nucleus accumbens. One to two minutes following injection of ACTH, the rats' grooming behavior was recorded. As can be seen from Figure 6.2, the results reveal that even a dose of ACTH as low as 20 nanograms induced excessive grooming.

This study shows not only variation in the amount of a variable but also presence–absence variation, because the first condition consisted of injecting zero nanograms of ACTH, or an absence of ACTH. Actually, one microliter of a saline solution was injected in order to create a placebo condition. The

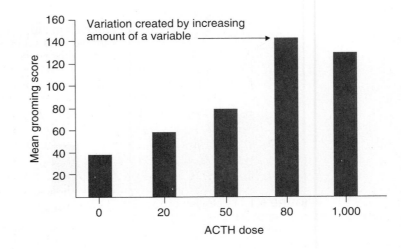

FIGURE 6.2 Illustration of variation of the independent variable by amount of a variable. (Based on data from "Intraaccumbens Injections of ACTH Induce Excessive Grooming in Rats" by J. P. Ryan and R. L. Isaacson, 1983, *Physiological Psychology, 11*, pp. 54–58.)

technique of varying the amount of a variable can be combined with the presence–absence technique. This combination of techniques is frequently necessary so that the experimenter can tell not only whether the independent variable has an effect but also what influence varying amounts of the independent variable may have. Using a combination of the amount and the presence-versus-absence techniques, Ryan and Isaacson could tell not only that ACTH induced grooming but also that grooming behavior was affected by different amounts of ACTH. However, you should not assume that all studies varying the amount of the independent variable also use a presence–absence technique—in some studies this is not possible. For example, if you were investigating the influence of exposure durations on recognition of different types of words, all subjects would have to be exposed to the words for some period of time. It would not make sense to ask people to identify a word to which they had never been exposed. Consequently, you would only vary the amount of time of exposure to the various words and would not include an absence condition.

One question that comes up with regard to establishing variation concerns the number of levels of variation to induce. An exact answer cannot be given other than that there must be at least two levels of variation and that these two must differ from each other. The research problem, past research, and the experience of the investigator should provide some indication as to the number of levels of variation that should be incorporated in a given experiment.

Also, the type of inference that is to be drawn from the results of the study will suggest the number of levels of variation that need to be included. If, for example, the objective of a particular drug study is to determine if a drug produces a given effect, you would probably use only two levels of variation. One group of subjects would receive a large dose of the drug, and another group would receive the placebo. But if you were concerned with identifying the specific drug dosage that produced a given effect, you would probably have many levels of variation, ranging in small increments from none to a massive dose.

Type of a Variable A third means of generating variation in the independent variable is to vary the type of variable under investigation. Assume that you were interested in determining whether or not a person's reactions to others were affected by the label these others were given. Such a study could be conducted by having a school psychologist and a teacher discuss the teacher's pupils at the beginning of a school year. In this discussion, the school psychologist could let the teacher know the type of student he or she would be facing. As illustrated in Figure 6.3, some of the students would be labeled as

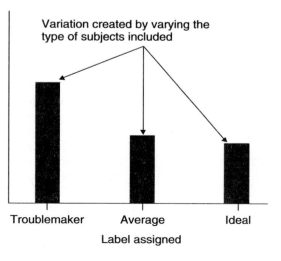

FIGURE 6.3 Illustration of variation in the independent variable created by type of a variable.

troublemakers, some would be portrayed as average, run-of-the-mill students who may occasionally create a disturbance, and a third group would be labeled ideal students who never give any trouble. In actuality, a matched group of students would have been randomly assigned to the three groups. Some time after the school term had begun, the teacher would be required to assess the students in terms of problem behavior. The teacher might be asked to rate each of the students in terms of the degree to which they were considered to be problem children, or perhaps to rank the students in order from those who never gave any trouble to those who were constantly a problem. If the assessments initially provided by the school psychologist were confirmed by the teacher's ratings or rank ordering, then support would be given to the hypothesis that giving a child a certain type of label tends to generate the type of behavior labeled. In this hypothetical example, variation was generated in the type of behavioral label given to each child.

Establishing Variation in the Independent Variable

You have just seen that there are three basic techniques for creating variation in the independent variable. These techniques seem straightforward and relatively simple. Remember, however, that the variation created must be under the control of the experimenter. It is much easier to establish this controlled variation for some variables than for others. In drug studies, it is relatively easy to establish controlled variation, because different doses of a drug can be measured quite accurately. If a presence-versus-absence form of variation were to be used, a placebo could be administered to one group and a specific amount of the drug to the other group. In this way, exact control would be maintained over the independent variable—the amount of the drug administered.

It is not always so easy to establish controlled variation, though. Assume that you want to investigate the influence of anxiety or fear on the desire of individuals to be together. You have decided to use the presence-versus-absence technique for varying the level of anxiety or fear. How do you create a controlled variation of this independent variable? Do you tell the subjects that you are going to hurt them in the hope that this anticipation will create anxiety or fear? Or do you try to create anxiety or fear in some other manner? Clearly, it seldom suffices to state that you are going to achieve variation by a certain technique. You must identify exactly how you plan to establish the variation—by presence or absence, amount, or type.

Next, we will take a look at two concrete ways in which variation can be achieved and the difficulties each technique produces.

Experimental manipulation
The controlled adjustment of the independent variable

Experimental Manipulation The term **experimental manipulation** of an independent variable refers to a situation in which the experimenter administers one specific controlled amount of a variable to one group of individuals and a different specific controlled amount of the same variable to a second group of individuals. For example, one of the research questions Shuell (1981) wanted to investigate was, What is the influence of type of practice on long-term retention? Shuell identified two types of practice: distributed and massed. Shuell operationally defined massed practice as six learning trials administered on the same day and distributed practice as six learning trials distributed over three days. He had one group of subjects learn a list of words using massed practice and another group learn a list of words using distributed practice. The experimenter not only defined the independent variable of massed or distributed practice operationally but also had total control over the administration of the practice schedule. Thus the researcher, in a controlled manner, experimentally manipulated the independent variable.

The two basic ways of experimentally manipulating the independent variable are instructional manipulation and event manipulation.

Manipulation of Instructions. One of the techniques available for creating variation in the independent variable is **instructional manipulation.** One group of subjects receives one set of instructions, and another group receives another set of instructions. Barlow, Sakheim, and Beck (1983), for example, investigated the hypothesis that increases in anxiety level would increase sexual arousal. In order to create a variation in anxiety, experimenters told one group of subjects that there was a 60 percent chance they would receive an electric shock when a light came on. Subjects in a second group were told that there was a 60 percent chance they would receive an electric shock if their level of arousal was less than the average

Instructional manipulation
Varying the independent variable by giving different sets of instructions to the subjects

of all of the research subjects. The last group was the control group, who were told that the light had no meaning. In reality, no subject received an electric shock. The instructional set was administered only to generate the anticipation of possibly receiving shock. Since shock is unpleasant and anxiety provoking to most individuals, it is probably safe to assume that this instructional manipulation generated anxiety in those who were told they had a 60 percent chance of getting shocked. Those subjects who were not given such instructions would not become anxious in anticipation of shock. Thus, instructional manipulation can enable the researcher to establish experimental manipulation.

Manipulation of variables through instruction is not without its dangers, two of which can be readily identified. First, there is the risk that some subjects will be inattentive when the instructions are given. These subjects will miss part or all of the instructions and therefore will not be operating according to the appropriate manipulation, thereby introducing error into the results. The second danger is the possibility that subject-to-subject variation exists in the interpretation of the instructions. Some subjects may interpret the instructions to mean one thing, and others interpret them in a different way. In this case, an unintentional variation is introduced that represents error, or, actually, an uncontrolled variable. The danger of misinterpretation can be minimized if instructions are kept simple, given emphatically, and related to the activity at hand. Probably no more than one variable should be manipulated through instructions. Manipulation of more than one variable will often result in instructions that are too complex and too long, rendering the manipulations ineffective by virtue of increasing inattentiveness, misinterpretation, and forgetting.

Manipulation of Events. A second means of establishing variation in the independent variable is **event manipulation.** In order to investigate the behavioral effects of surgically cutting the corpus callosum, it is necessary either to identify humans who have undergone this surgery for medical reasons or to actually perform this surgery on animals and then subject them to a variety of behavioral tests. Drug research varies events such as drug dosages, and learning experiments vary events such as meaningfulness of the material presented to subjects. Most human experiments and almost all animal experiments use event manipulation to achieve variation. Communication skills have been developed in chimpanzees (Fouts, 1973), enabling researchers to use instruction with these infra-

Event manipulation
Affecting the independent variable by altering the events that subjects experience

humans. When a choice exists between using instructions and events to create the variation, the best choice in most cases is to use events, because events are more realistic and thus have more impact on the subject.

Consider the experiment conducted by Aronson and Linder (1965). They wanted to determine if liking for another is partially determined by the behavior exhibited by that other. To investigate this problem, they designed an experiment in which a confederate and a subject interacted on seven different occasions. After each of the seven sessions, the subject overheard the confederate's evaluation of her. These evaluations were either all positive, all negative, initially negative and then positive, or initially positive and then negative. After overhearing all evaluations, subjects recorded their impressions of the confederate. In this experiment, the event manipulated was the overhearing of an evaluation of the subject's performance. This manipulation could also have taken place through instructions—the experimenter could have told the subjects how they performed—but the event manipulation was more meaningful and realistic to the subjects. The advantage of this increased meaningfulness in experimentation is that the problems of inattentiveness and misinterpretation are minimized. Additionally, the response emitted by the subject in a realistic situation is probably a better representation of how he or she would respond outside the experimental environment because the realism probably removes much of the artificiality created by the experiment.

In animal research, the issue of realism seldom exists, for two reasons. First, events are used to manipulate the independent variable. Second, the conditions that are used to motivate the animal to respond seem to create the realism. Researchers motivate animals to respond by depriving them of food or water or by administering electric shock. Such conditions seem to be real and meaningful to these subjects, even though this is something about which an animal researcher rarely speculates.

That is not to say that all manipulations of events get the subjects involved or are as meaningful as real life. In fact, many experiments have little impact on the individual and therefore tend to have little realism. Many person–perception experiments have, for example, manipulated exposure by requiring judges to predict responses of others after viewing different amounts of a filmed interview of them. The filmed interview removes the realism from the experiment. Requiring the judge and the others to spend either ten minutes or half an hour together before the judge makes any predictions would increase the realism. Why is this not done? The answer is related to the problem of control. Filming allows control over variables that would not be controlled in the free interaction situation. In the free interaction situation, many differences would exist between the judges and the others, such as what the judge and the other talked about and how each responded. Consequently, judges would have different information on which to base their judgments, and thus the study would contain many uncontrolled variables. In the filmed interview, each judge receives the same information, so there is control over both the type and the amount of information received.

As realism and meaningfulness of social experiences are increased in an attempt to increase impact, control is decreased. As control over variables is increased, the impact of the experiment is decreased. This is a real dilemma for the scientist trying to investigate social interaction experimentally. On the one hand, the scientist wants control over variables. On the other hand, he or she wants to eliminate the sterile atmosphere of the experiment, because such an atmosphere may fail to involve the subject and therefore may have no significant influence on behavior. There is no simple solution to this problem of maximizing realism and control. However, the researcher's first concern should be to make sure that the desired experimental effect actually occurs. If this effect does not occur, it makes no sense to worry about other aspects of the experiment.

Individual difference manipulation
Varying the independent variable by selecting subjects that differ in the amount or type of a measured internal state

Individual Difference Manipulation **Individual difference manipulation** refers to the situation in which the independent variable is varied by selecting subjects that differ in terms of some internal state (for example, self-esteem or anxiety level). The assumption underlying such a manipulation is that each individual possesses a certain amount of a variety of variables, commonly labeled *personality variables*. One efficient means of achieving a manipulation would be to select subjects having different levels of a given variable (such as anxiety) and then look for effects of the difference. As illustrated in Figure 6.4, the typical procedure is to administer to a large sample of people an instrument measuring the internal state of interest. From the test results, two smaller groups of subjects are selected: one group that has scored high on the variable of interest and one group that has scored low. These two groups are required to perform a task, and then the two groups are com-

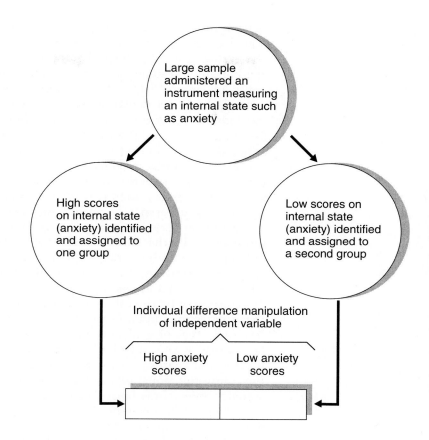

FIGURE 6.4 Variation created by the manipulation of individual differences.

pared to determine if a difference exists in the task performance. If a difference does exist, it is typically attributed to the differences in the internal states. Ritchie and Phares (1969) were interested in the relationship among internal–external control, communicator status, and attitude change. To examine this relationship, researchers first had to classify subjects as internals or externals, so they administered the Locus of Control Scale to 152 female subjects, from whom 42 internals and 42 externals were chosen. The degree of attitude change of these two groups of subjects was compared for different levels of communicator status. Results indicated that a high-prestige communicator produces more change in externals

than in internals. This result was attributed to the externals'
higher expectancy of reinforcement.

What is wrong with such an experiment? It seems to be
appropriately designed and statistically analyzable. The difficulty lies in the fact that the states—internal and external
control—were not experimentally manipulated. Subjects were
not randomly assigned to conditions but were nonrandomly
assigned, or selected, on the basis of their test scores. Thus, it
is possible that another variable, correlated with the internal
and external states, produced the differences in performance.
If, in a learning experiment, motivation was highly correlated
with tested anxiety, then it would be impossible to determine
whether the difference in performance between a high-anxiety
and a low-anxiety group was due to the anxiety factor or to
motivation. In the Ritchie and Phares study, if some other
variable was highly correlated with the subjects' locus of control, then, similarly, one could not determine if the observed
effect was caused by one's perceived locus of control (internal
or external) or by the other correlated variable. An experiment
that tries to manipulate variables in this way is therefore an ex
post facto type of study, even though it may be conducted in a
laboratory, with control over other variables. In fairness to
Ritchie and Phares, it must be stated that they did manipulate
the communicator status variable, so the study they conducted
actually did include both an ex post facto component and an
experimental component. I should also add that individual
difference manipulations are not inherently undesirable. A
great deal of high-quality research has been generated through
the use of such manipulations. However, you should remain
aware of their limitations when using them.

Constructing the Independent Variable

In addition to deciding how variation in the independent variable is to be achieved, the researcher must also decide how the
independent variable is to be constructed or operationally defined. In an experiment, we are trying to determine if the
independent variable actually produces the hypothesized effect. In order to test the hypothesis and achieve the necessary
variation, we must translate the independent variable into
concrete operational terms; in other words, we must specify
the exact empirical operations that define the independent
variable. If the independent variable represents different learn-

ing techniques, then we must operationally define each one of the learning techniques. The ease with which the independent variable is translated into operational terms varies greatly among research problems. For example, if the independent variable consists of a drug, then there is no difficulty in operationally defining the independent variable because a specific empirical referent exists in the form of the drug. All that is needed is to obtain the drug and then to administer different dosages to different groups of subjects. Similarly, if the independent variable consists of either the length of time various words are exposed to subjects or the area of a rat's brain that is destroyed, it is simple to translate the independent variables into concrete operations. The only decision that must be made is the length of time the words should be exposed to the subjects or the area of the brain that must be destroyed.

Difficulty in translating the independent variable into concrete operational terms arises when the independent variable consists of some abstract construct such as attitude, frustration, anxiety, learning, or emotional disturbance. The problem stems from the fact that there is no one single definition or empirical referent for such constructs. For example, learning in albino rats could be operationally defined as speed of acquisition, number of trials to extinction, or latency of response. Aggression in some instances may be defined as including "intent to harm," whereas in other instances "intent to harm" may be irrelevant. The scientist's task is to identify the specific empirical referents that correspond to the meaning denoted by the way the construct is used as the independent variable in the research question. For example, if we were investigating the aggressiveness with which women pursue their jobs, the component of "intent to harm" would probably not be included in the operational definition of aggressiveness. However, if we were investigating the influence of children's aggressive responses on their popularity, then the empirical referent of "intent to harm" probably would be included.

For some types of research, operationally defining a conceptual variable that does not have specific empirical referents is not a major problem, because standard agreed-on techniques exist. A study of the effect of schedules of reinforcement on strength of a response requires the construction of the different reinforcement patterns. The schedules of reinforcement identified by Skinner are standard and accepted in the field (the interested student is referred to Ferster and Skinner, 1957). These could immediately be incorporated into the study.

In other areas of research, such as social psychology, difficulty is frequently encountered in constructing conditions that represent a realization of the independent variable specified in the problem. The reason is that relatively few standard techniques exist for manipulating the conceptual variable in such areas. Few of the manipulations of such variables as conformity, commitment, and aggression are identical. Aronson and Carlsmith (1968, p. 40) state that this lack of development of specific techniques is a function of the fact that the variables with which the social psychologist works must be adapted to the particular population with which he or she is working. A standard technique would not work with all populations. Therefore, the researchers must use ingenuity in accomplishing the task, capitalizing on previous work by borrowing ideas and innovations and incorporating them. The problem is that many researchers simply use prior ideas and innovations without creating a better translation of the abstract concept. They "cling to settings and techniques that have been used before" (Ellsworth, 1977). Every translation has had its own quirks and characteristic sources of error, and, as Stevens (1939) has pointed out, each operational definition or translation of an abstract concept represents merely a partial representation of that concept. Therefore, one should not automatically assume that prior translations are the best or even the most appropriate. In fact, in any translation the investigator has to compromise, sacrificing some methodological advantages for others. Given the problem of translating abstract concepts into operational terms, you should first determine how the topic or phenomenon has been studied in the past. Once that has been determined, you should then decide whether or not any of these translations are appropriate for your study. In making this decision, consider the overall research that has been conducted on the topic. It may be that most prior studies have focused only on a narrow but typical translation of an abstract concept. If this is the case, then it would be appropriate to identify another translation that would help approximate a more complete representation of the concept. For example, fear has typically been studied in the laboratory. However, the range of fear that can be generated in this setting is limited by the ethical restrictions on imposing stimuli that may create extreme fear. Given this situation, it may be more appropriate to search for a setting in which extreme fear is created naturally.

In other areas of research, the problem is not so much *how* the conceptual variable will be translated into specific experimental operations but *which* of the many available techniques will be used. Plutchik (1974) lists eight different techniques (including approach–avoidance conflict and physiological measures) that Miller (1957) identified for either producing or studying fear in animals. If you were to study the conceptual variable of fear in animals, which one should you use? What specific operations will be used to represent this conceptual variable of fear? To answer this question, you have to determine which techniques most adequately represent the variable. This issue will be discussed later in the chapter.

Why do several different techniques exist for constructing a single variable such as learning, emotion, or fear? Variables such as fear refer to a general state or condition of the body. Therefore, there is probably no single index that is *the* way to produce such a concept. Some ways are probably better than others, but no one index can provide complete understanding. To obtain a better understanding of the concept, you should investigate several indexes, even though they may initially seem to give contradictory results. As our knowledge of the idea increases, the initial results that appear to be contradictory will probably be integrated.

Construct Validity of the Independent Variable

After deciding on the experimental operations that you are going to use as your manifestation of the conceptual variable and the manner in which this variable will be varied, it is a good idea to look back at these operations and ask the following questions: Do the experimental operations represent the conceptual variable that I had in mind? Will the different levels of variation of the independent variable make subjects behave differently? In what ways should they behave differently? These questions are asked in an attempt to ensure that construct validity (Cook and Campbell, 1979) has been attained. By **construct validity,** I mean the extent to which the abstract construct or conceptual variable of interest can be inferred from the operational definition of that construct. For example, Davis and Memmott (1983) used the rate with which rats pressed a lever to infer that rats can indeed count. Essentially,

Construct validity
The extent to which the abstract construct can be inferred from the operational definition of that construct

Davis and Memmott believed that rats could count (the abstract construct) if their speed of lever pressing changed when three shocks had been administered (the empirical operation indicative of counting). If the change in response following the third shock does indeed indicate counting ability, then construct validity has been attained; but if it does not indicate counting ability, construct validity has not been attained. The problem of construct validity exists because sometimes we must deal with imperfect translations of our abstract construct, and it is possible that the experimental operations do not represent the construct of interest to the experimenter.

Figure 6.5 illustrates the circularity involved in construct validity as well as the possible imperfect translations that can exist, as embodied in the procedure used by Schachter and Singer (1962) in their study of euphoria. In order to create

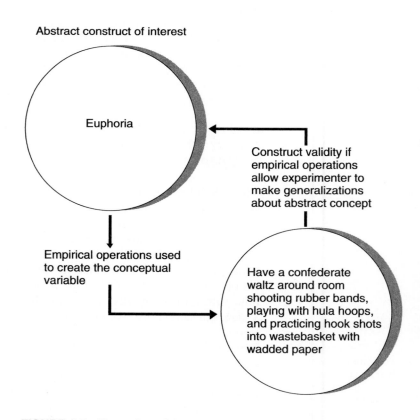

FIGURE 6.5 Illustration of the requirements for construct validity.

euphoria in subjects, Schachter and Singer had a confederate waltz around the room shooting rubber bands, playing with a hula hoop, and practicing hook shots in a wastebasket with wadded paper. The question that these researchers then had to answer was whether these imperfect operations actually created euphoria in their subjects. Unfortunately, whether such empirical operations have construct validity is often a matter of judgment on the part of the experimenter.

Cook and Campbell (1979) provide several suggestions for achieving construct validity. According to them, the first step in achieving construct validity is to provide a clear definition of the abstract construct. Such preexperimental explication of the construct seems to be a logical first step, since the empirical operations used to represent this construct must arise from its definition. Only by providing a clear definition of euphoria could Schachter and Singer proceed to specify the empirical operations that would generate euphoria.

After the abstract construct has been defined and translated into specific experimental operations, several types of data can be collected to ensure that construct validity has been attained. First, data can be collected to demonstrate that the empirical representation of the independent variable produces the outcomes expected. If a rat were hungry, we would expect it to engage in such behavior as consuming more food, tolerating more quinine in the food, increasing its rate of bar pressing to obtain food, and producing more stomach contractions. Miller (1957) actually conducted such a study. Although these four measures did not produce identical results, depriving the rats of food did result in the expected behavior, giving additional evidence that food deprivation appropriately represents hunger.

Second, data can be obtained to show that the empirical representation of the independent variable does not vary with measures of related but different conceptual variables. For example, an operational definition of *communicator expertise* should be related to the level of knowledge possessed by the communicator but not to the communicator's power, attractiveness, position, or trustworthiness. Therefore, evidence for construct validity can be obtained by collecting both convergent (the expected outcomes are produced) and divergent (outcomes are unrelated to the conceptual variables) data.

Unfortunately, the procedures just suggested are seldom used in psychological experimentation. Most of the time, an experimenter operationally defines the independent variable, and it is up to the reader to judge whether or not an empirical realization of the conceptual variable was actually produced. All too often, no data are given to provide validating evidence that an adequate translation of the conceptual variable was actually attained.

One reason the techniques discussed above are rarely used is that they are time consuming. It is much easier to select the operational definition and then proceed to investigate the problem area. Another reason is that the techniques are often difficult to carry out. In order to satisfy the divergence criteria, it is necessary not only to identify the related-but-different variables but also to define these variables operationally. For example, to obtain divergent evidence regarding communicator expertise, one would first have to identify the related-but-different variables, such as communicator trustworthiness. Then one would have to operationally define both conceptual variables—trustworthiness and expertise—and collect data on both variables. Therefore, obtaining such evidence typically involves conducting another experiment.

The procedure just discussed for providing assurance of construct validity can be used with both human and infrahuman subjects. When dealing with human subjects, one can use several additional techniques to provide some evidence of whether the conceptual variable is generating the observed results. These additional techniques generally fall under the heading of *manipulation checks* because they represent checks on the manipulation of the independent variable.

The first approach is to conduct interviews with the subject after the independent variable has been introduced, to determine whether the desired construct was actually attained. Asch (1956), in a study on conformity, wanted to create pressure on each subject to say something that was obviously not correct in order to conform. In such an experiment, one needs to determine whether that pressure has indeed been created. One means of doing so is to generate group pressure on several subjects and then to interview them to find out if the pressure was actually created. Naturally, there are difficulties in conducting these interviews. Subjects often identify the response that is wanted and then intentionally provide this response. If the postinterviews are conducted correctly, however, this error is minimized. The appropriate means of conducting debriefing sessions or postinterviews will be discussed in a later chapter in this book. Among the benefits that can accrue from this interview, the greatest is that the interview can point out possible weaknesses in the operational definition of the independent variable and indicate what needs to be done to eliminate these weaknesses. It is important that these interviews follow immediately after the administration of the treatment conditions rather than after the response data are collected, for the response may influence the subjects' introspective reports. Research on scapegoating supports this statement; it has shown that a variable such as aroused hostility may be diminished by giving the subjects an opportunity to displace it. Since these interviews follow the administration of the treatment conditions, they must, as you might suspect, be conducted on subjects who are part of a pilot study conducted for the purpose of investigating the influence of the experimental operations. Otherwise the interview may affect the response data.

A more difficult but superior technique involves getting a behavioral indicator that the experimental operations are arousing the desired effect. If you want to arouse anxiety in subjects, get some indication, such as a galvanic skin response (GSR), that anxiety is actually being aroused. Zimbardo, Cohen, Weisenberg, Dworkin, and Firestone (1966) measured subjects' GSR to monitor anxiety through the course of their experiments. Aronson and Carlsmith (1963) attempted to create mild and severe threat conditions in children. Intuitively, they seem to have manipulated threat. They could have verified the manipulation of the threat conditions by pretesting the children with toys of varying degrees of attractiveness and with various threat conditions until they had a condition in which more children disregarded the mild threat than the severe threat in order to play with the forbidden toy. In this way, they would have had a behavioral indication that threat was actually varied.

Pretesting can also give some indication of the level of intensity of the manipulation. If the hypothesis under investigation involves three levels of intensity of threat, then three levels of threat have to be created. How does one know that three levels are created and how intense they are? Either of the foregoing methods could give some indication. A verbal report or response to a questionnaire may indicate level of intensity. In a study that attempted to generate three levels of threat, level of intensity of the severity of threat could be determined behaviorally by noting the percentage of children that disregarded the threat. If in one condition 10 percent of the children disregarded the threat, in the second condition 50 percent of

the children disregarded the threat, and in the third condition
80 percent of the children disregarded the threat, one would
have some evidence of the intensity of each of the threat
conditions.

The Number of Independent Variables

How many independent variables should be used in an experiment? In looking through the literature in any area of psychology, you will find some studies that used only one independent
variable and others that used two or more. What criterion
dictated the number used in each study? Unfortunately, no
rule can be stated to answer this question. We do know that
behavior is multidetermined, and inclusion of more than one
independent variable is often desirable because of the added

Interaction
The differential effect
that one independent
variable has for each
level of one or more
additional independent
variables

information it will give in the form of an interaction. **Interaction** refers to the differential effect that one independent variable has for each level of one or more additional independent
variables. For example, Mellgren, Nation, and Wrather (1975)
investigated the relationship between magnitude of negative
reinforcement and schedule of reinforcement for producing
resistance to extinction. As predicted, they found that the
effect of magnitude of reinforcement was dependent on the
schedule of reinforcement used. The subjects (albino rats) that
were performing under a partial reinforcement schedule revealed greater resistance to extinction when administered a
large negative reinforcement; rats performing under a continuous schedule of reinforcement displayed greater resistance to
extinction when administered a small negative reinforcement.
In other words, the effect of reinforcement magnitude is dependent on the schedule of reinforcement. If the second variable of reinforcement schedule had not been included, the
study would probably have revealed, as demonstrated by other
studies, that no difference existed between magnitudes of
negative reinforcement in terms of the ability to affect resistance to extinction. Inclusion of this variable showed that
magnitude of negative reinforcement did affect resistance to
extinction, but its effect depended on the schedule of reinforcement. Experiments like the one conducted by Mellgren et al.
reveal the advantage and even the necessity of varying more
than one independent variable.

Although theoretically and statistically there is no limit
to the number of variables that can be varied, realistically

there is. From the subject's point of view, as the number of variables increases, there are more things to be done, such as participating in more events or taking more tests. The subject is apt to become bored, irritated, or resentful and thereby introduce a confounding variable into the experiment. From the experimenter's point of view, as the number of variables increases, the difficulty in making sense out of the data increases along with the difficulty in setting up the experiment. Aronson and Carlsmith (1968, p. 51) give the following rule of thumb. They say that the experiment "should only be as complex as is necessary for the important relationships to emerge in a clear manner."[1] In other words, do not use the "why not" approach, in which a variable is included in the experiment because there is no real reason not to include it. Only include those variables that seem to be necessary to reveal the important relationships.

THE DEPENDENT VARIABLE

The dependent variable has been defined as the behavioral variable designed to measure the effect of the variation of the independent variable. This definition, like the definition of the independent variable, seems straightforward and simple enough. Also, like the independent variable, the dependent variable is relatively easy to identify in a given study. Aronson and Mills wanted to investigate the influence of severity of initiation on liking for a group. Liking for a group represented the dependent variable. Ritchie and Phares (1969) investigated attitude change as a function of communicator status and locus of control. Attitude change represented their dependent variable. However, many decisions must be made to secure the most appropriate measure of the effect of the variation in the independent variable.

A psychological experiment is conducted to answer a question (What is the effect of . . . ?) and to test the corresponding hypothesis (A certain change in x will result in a certain

---

[1]Reprinted by special permission from "Experimentation in Social Psychology" by E. Aronson and J. M. Carlsmith, in *The Handbook of Social Psychology*, 2nd edition, volume 2, edited by G. Lindzey and E. Aronson, 1968, p. 51. Reading, Mass.: Addison-Wesley.

change in y). In order to answer the question and test the hypothesis, the researcher varies the independent variable to determine whether it produces the desired or hypothesized effect. The experimenter's concern is to make sure that he or she actually obtains an indication of the effect produced by the variation in the independent variable. To accomplish this task, the experimenter must select a dependent variable that will be sensitive to, or able to pick up, the influence exerted by the independent variable. Often, researchers believe that an effect was produced because they think they saw behavioral change exhibited, yet their study indicates that the independent variable produces no effect. Such a case may indicate distorted perception, or it may mean that the dependent variable was not sensitive to the effects produced by the independent variable.

The initial observations of the impact of split-brain surgery on humans are a good example of failing to observe the appropriate dependent variables. As indicated at the beginning of this chapter, it was assumed at first that split-brain surgery left the person unchanged because there seemed to be little alteration in personality or behavior. This conclusion was reached because the dependent variables assessed consisted primarily of observations of individuals performing their daily activities. This assessment revealed that the individuals could continue their normal daily activities with little, if any, apparent change. But when Roger Sperry and others changed the test environment and assessed these subjects' responses on a variety of specific dependent variables, they demonstrated that split-brain surgery has a rather significant impact. For example, Zaidel and Sperry (1974) focused on the dependent variable of short-term memory and demonstrated that split-brain subjects have a deficiency in this area.

It is the task of the dependent variable to determine whether the independent variable did or did not produce an effect. If an effect was produced, the dependent variable must indicate whether the effect was a facilitating one or an inhibiting one and must reveal the magnitude of this effect. If the dependent variable can accomplish these tasks, the experimenter has identified and used a good, sensitive dependent variable. The first decision for the experimenter, then, is what specific measure to use to assess the effect of the independent variable. In making this decision it is important to consider the gender of the experimental subjects. This is an important issue because the gender of the subjects can influence their response to the dependent variable measure. Consider, for example, a study that investigated the influence of siblings on empathy. In this study empathy would represent the dependent variable and the presence of siblings would represent the independent variable. Assume further that the operational definition you selected for the dependent variable was a self-report in response to the question "When your best friend is feeling sad, do you feel sad?" A gender bias might arise from the use of such an assessment of the dependent variable, because males may be more hesitant than females to report empathic feelings (Matlin, 1993). In such an instance the dependent variable of empathy would not represent an appropriate assessment of the independent variable (the presence of siblings) for males. Such a biased assessment of the dependent variable would result in data that would lead one to conclude incorrectly that females are more empathic than males, when the truth may be that there are no differences. To demonstrate such a difference a gender-neutral measure would have to be used—perhaps assessment of people's facial expressions when they looked at a sad movie (Matlin, 1993).

Once the dependent variable has been selected, the experimenter still confronts several problems. The experimenter must somehow ensure that the subject is taking the measurement seriously and is doing his or her best. The experimenter must also make sure that the subject is responding in a truthful manner rather than "cooperating" with the experimenter by responding in a manner that he or she feels will be most helpful to the experimenter. The last two problems are most crucial in human experiments.

The Response to Be Used as the Dependent Variable

What response should be selected as the dependent variable? We just saw that the foremost criterion for the dependent variable is sensitivity to the effect of the independent variable. To my knowledge, there is no specific rule that will tell you how to select a dependent variable. Psychologists use as dependent variables a wide variety of responses, ranging from questionnaire responses to verbal reports, overt behavior, and physiological responses. The task is to select the response that is the most sensitive to the effect produced by the independent variable. In most experiments, there are several different measures that could be used, and the experimenter must choose among them. For example, attitudes can be measured by a response to a questionnaire, by a physiological response, or by observation of a subject's response.

The difficulty in identifying the most appropriate dependent variable seems to stem from the fact that psychologists study processes, attitudes, or outcomes of the human and infrahuman organism. When an independent variable is introduced, our task is to determine the effect of the independent variable on phenomena such as learning, attitudes, or intelligence. Are these processes, attributes, or outcomes facilitated, inhibited, or affected in some other way? The problem is that the processes, attributes, or outcomes are not directly observable. Since direct observation is not possible, some result of the construct under study that can be observed must be selected for observation to allow inference back to the construct.

Consider learning as an example. It is impossible to study the learning process directly. But if a student sits down and studies certain material for an hour and then can answer questions he or she previously could not, we say that learning has taken place. In this case, learning is inferred from an increase in performance. In such a way, we can acquire information about a phenomenon. The decision that the scientist faces is selecting the aspect or type of response that will provide the best representation of change in the construct as a result of the variation in the independent variable. Previous experimentation can help one make such a decision. Prior research has been conducted on most phenomena, and many dependent variables have been used in these studies. Results of these studies should provide clues to the responses that would be most sensitive.

Aronson and Carlsmith (1968, p. 54) address the problem of selecting the dependent variable with research conducted on humans. They discuss some of the advantages and disadvantages of various techniques that can be used to measure the dependent variable in social psychological research. One very significant point they make is that the more commitment demanded of the subject by the dependent variable, the greater the degree of confidence we can have in the results of our experiment. Why is this so? First, making a commitment to a course of action reduces the probability of faking on the part of the subject because it helps to ensure that subjects take the dependent variable measure seriously. If we wanted to find out which person in a group is most liked by a particular individual, we could have that individual rate each of the group members on a liking rating scale. Or we could have him or her choose a member of the group as a roommate for the next year, with the contingency that the person picked will in fact be the roommate. In this case, if the request were credible, the subject would be motivated to respond truthfully, because he or she would have to live with the decision. The second advantage of requiring the subject to make a commitment is that it often increases one's confidence that the dependent variable of interest is really being measured. Recording the frequency with which fights are initiated is a better index of aggression than having a subject verbally state that he or she is angry or evaluate the degree of anger on a rating scale.

Behavior that involves a commitment is probably the best type of dependent variable to use. But because of cost, time, or some other constraint, sometimes it is not feasible to use such a dependent variable. In that case, a questionnaire or a verbal report must be used instead. Although questionnaires and verbal reports yield a great deal of useful data, there is an increased likelihood of error with these measures, because the subjects either may not take the measure seriously or may "cooperate" with the experimenter by producing the results they think the experimenter desires.

Reliability and Validity of the Dependent Variable

Reliability
The extent to which the same results are obtained when responses are measured at different times

Validity
The extent to which you are measuring what you want to measure

In deciding on the dependent variable to use as an index of the influence of the independent variable, one must consider the issues of reliability and validity. **Reliability** refers to consistency or stability; **validity** refers to whether you are measuring what you want to measure. Ideally, the dependent variable will be both reliable and valid. However, this is an aspect of research about which experimental psychologists have been extremely lax, whereas those involved in the field of test construction have been justifiably rigorous. In constructing a test, first one should establish the reliability of that test and then proceed to attempt to get evidence of its validity. Psychologists conducting experiments seem to assume that reliability and validity exist intrinsically if the dependent variable has been operationally defined. Such an assumption should not be made, because if the dependent variable is not reliable and valid, the experiment is worthless.

Reliability In most experiments, reliability of the dependent variable can be established by determining the consistency with which responses are made on the dependent variable. If organisms consistently respond in the same way on the dependent variable, then that variable is reliable. Most experiments, however, are conducted as single-occasion events, which means that the dependent variable is observed only once. In other words, the subject comes to the experiment one time, and the dependent variable is only measured once. Consequently, the reliability of the dependent variable cannot be assessed. According to Epstein (1981), if reliability could be measured, low reliability (usually less than .30) would exist for a variety of measures including self-ratings, other ratings, behavioral responses, personality test results, and physiological responses. However, as shown in Figure 6.6, Epstein (1979) found that the stability of each of these responses increased as they were averaged over several days. In other words, the reliability of the dependent variable responses increased as the responses were aggregated over time. The lowest reliabil-

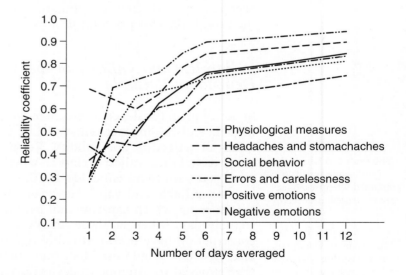

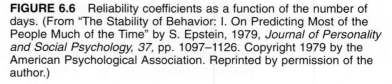

FIGURE 6.6 Reliability coefficients as a function of the number of days. (From "The Stability of Behavior: I. On Predicting Most of the People Much of the Time" by S. Epstein, 1979, *Journal of Personality and Social Psychology, 37,* pp. 1097–1126. Copyright 1979 by the American Psychological Association. Reprinted by permission of the author.)

ity existed when reliability was assessed by correlating the responses obtained on one day with those obtained on the second day. When reliability was assessed by correlating the average response obtained over one twelve-day period with the average response obtained over another twelve-day period, excellent reliability was found to exist.

Such evidence strongly suggests that most dependent variables used in the traditional single-occasion experiment are unreliable. This in turn means that the results obtained from our single-session studies do not produce stable findings and are therefore suspect. Where does this leave us? The results of Epstein's study indicate that, for our findings to be reliable, we must repeat each experiment a number of times and then use the average response as our dependent variable. Many studies cannot be repeated on the same group of subjects, however, because the repetition would alter the subjects' responses. So in many instances we are locked into the single-occasion experiment. Although Epstein's study strongly suggests that a single study, by itself, should not be given much credence, replication of the results of single-occasion studies provides a good measure of the reliability of dependent measures because replication of an effect cannot be attained if the dependent variable measure is unreliable. Thus, the results from any one study should not be considered seriously except as those results contribute to a population of studies; it is the results from such a population of studies that give us reliable information.

Validity Establishing the validity of the dependent variable involves obtaining evidence to support the hypothesis that the dependent variable actually measures the construct we want it to measure. For example, does latency of response really represent degree of learning? Does selection of a person as a roommate for the next year really represent the degree to which the subject likes the person? As you can see, establishing the validity of the dependent variable is a difficult task. The individuals concerned with test construction have wrestled with this problem and have devised a number of techniques for obtaining evidence of validity. Experimenters often do not even consider the validity question because, I believe, it has been obscured by operationism. For example, *intelligence* has been operationally defined as what intelligence tests measure. No one can argue with this operational definition. What you can argue with is whether intelligence tests do really measure intelligence. Operational definitions are necessary for scientific communication, but they say nothing about validity. Yet many researchers seem to write off the validity question once they have satisfied the criterion of operational definition.

How do we obtain an index of validity of our operationally defined dependent variable? Again, there is no simple solution. The discussion of commitment alluded to the validity question. As the commitment required of the dependent variable increases, we can increase our faith in the results of the experiment because commitment reduces faking and seems to indicate a more valid representation of the dependent variable that we wanted to measure. Thus, as an index of aggression, fighting seems to be more valid than a rating of anger. Commitment, therefore, seems to be related to validity.

To get additional indicators of validity, we must resort to an assessment of the construct validity of the dependent variable. This process is actually very similar to the one outlined in the discussion of the construct validity of the independent variable, because both the independent and the dependent variables often represent abstract constructs. For example, constructs such as learning, depression, helplessness, and aversion have been used as both independent and dependent variables. If one of these constructs were used as a dependent variable, your first task would be to identify the experimental operations that would represent the abstract construct. For example, Nezu (1986) investigated the efficacy of a problem-solving therapy approach for treatment of depression. In conducting this study, Nezu naturally had to operationalize the dependent variable, depression. In this study, *depression* was operationally defined as a score on the Beck Depression Inventory. The question that the researcher must ask is whether the operationalization of the dependent variable really represents the dependent variable construct—in this case, the abstract construct of depression. Cook and Campbell (1979) provide several suggestions for achieving construct validity of the dependent variable. The first suggestion is that the construct must be clearly defined, because the development of the dependent variable must arise from this definition. Only by having a clear definition of depression could Beck proceed to develop the items on the Beck Depression Inventory that would tap the construct of depression. Once the empirical operations used to measure the construct—in this case, the items measuring depression—are developed, convergent and divergent data can be collected to provide additional verification that the construct has been measured. Convergent data are data that indicate that the dependent variable is measuring the same thing as are other measures of the same construct. For example, if the Beck Depression Inventory was highly correlated with the Zung Self-Rating Scale for depression and the Hamilton Rating Scale for depression, then convergent validity data would have been obtained. Assessing the divergence of the dependent variable involves demonstrating that the dependent variable is not related to conceptually different constructs. For example, the Beck Depression Inventory should not be correlated with a measure of intelligence, or of anxiety, because depression is a construct that is conceptually different from intelligence or anxiety.

The problem that arises in producing convergent and divergent data is that many of the constructs that we want to measure are multidimensional, and the different dimensions may not be highly correlated. Strength of conditioning has, for example, been inferred from amplitude, latency, and resistance to extinction. Investigators have sometimes used these three measures interchangeably, suggesting that these measures should be highly correlated and should demonstrate convergent validity. However, Hall and Kobrick (1952) have shown that the correlations between some of these measures are in fact very low, suggesting that they measure conceptually different things. This leads to some ambiguity in assessing the validity of a given construct. Although situations like this exist, they do not invalidate the process of assessing convergent and divergent validity of a given dependent variable construct.

Reducing Subject Error

Once a decision has been made as to what the dependent variable will be, the experimenter using human subjects must make sure that the subject is taking the measure seriously and not trying to fake the responses. The difficulty of this task increases as the degree of commitment decreases. For example, in filling out a questionnaire, some subjects will undoubtedly race through it, reading questions in a haphazard manner and checking off answers without putting much thought into them. One way of decreasing such errors is to disguise the measure of the dependent variable. In addition to increasing the likelihood that the subject takes the measure seriously, the disguise also helps guard against the possibility that the subject will cooperate with the experimenter.

Aronson and Carlsmith (1968, p. 58) present a number of techniques useful for disguising the dependent variable. One technique is to assess the dependent variable outside the context of the experiment. Carlsmith, Collins, and Helmreich (1966) solicited the aid of a consumer research analyst to assess the dependent variable after the subject had supposedly completed the experiment and left the room. Another technique is to assess behavior of significance to the subject, such as selection of a roommate. A third technique is to construct the experiment in such a way that the subject does not realize that the dependent variable is being observed. Lefkowitz, Blake, and Mouton (1955) observed the frequency with which people jaywalked or disobeyed signs upon introduction of various levels of an independent variable. A fourth method that is often used in attitude-change experiments is to embed key items in a larger questionnaire in hopes that the key items will not be recognized and falsely reported on. A similar technique is to disguise the reason for interest in a particular dependent variable. Aronson (1961) was interested in finding out whether the attractiveness of several colors varied as the effort expended in getting them varied. For this study, he needed a measure of attractiveness. When he asked the subjects to rate the attractiveness of colors, he told them that he was investigating a relationship between color preference and a person's performance. Using unsuspicious subjects such as young children is also a very good way to reduce or eliminate cooperation. Young children are very straightforward and do not have devious motives. A sixth technique is to use what Aronson and Carlsmith call the family of "whoops" procedures. The typical procedure is to collect pretest data and then claim that something happened to them so that posttest data can be collected. Christensen (1968) used this method in collecting test–retest reliability data on ratings of a series of concepts. A seventh procedure is to have a confederate collect the data. Karhan (1973) had the learner, a member of the experiment who had supposedly received electric shock for errors made, make a request of the subject. The dependent variable was the subject's response to the confederate's request. Another technique is to use a physiological measure that is presumably not under the subject's conscious control. However, a number of individuals have presented data indicating that these measures may be consciously influenced by the subject.

The Number of Dependent Variables

Should more than one dependent variable be used in a psychological experiment? This is a very reasonable question, particularly when more than one dependent variable could be used to measure the effect of variation of a given independent variable. In a learning experiment using rats, the dependent variable could be the frequency, amplitude, or latency of response. Likewise, in an attitude experiment, the dependent variable could be measured by a questionnaire, by observing behavior, or by a physiological measure. When more than one dependent variable can be used, the scientist usually selects only one and proceeds with the experiment. If the scientist elects to use more than one dependent variable, certain problems arise. Assuming the scientist knows how to measure each of the dependent variables, he or she must be concerned with the relationship among them. If the various dependent variable measures are very highly correlated (for example, .95 or above), there is reasonable assurance that they are identical measures, and all but one can be dropped. If they are not so highly correlated, the experimenter must ask why not. The lack of correspondence could be due to unreliability of the measures or to the fact that they are not measuring the same aspect of the construct under study. Two different measures of learning may evaluate different aspects of the learning process. These difficulties must be resolved, but all too often the scientist does not have the necessary data. As a field advances, more and more of the aspects of a phenomenon are unraveled and problems like these are resolved. Such cases support the notion that multiple dependent variables should be used in some experiments because they contribute to the understanding of a phenomenon, which is the goal of science.

The fact that the constructs we are attempting to measure, such as learning or anxiety, can be measured by several different techniques indicates that they are multidimensional. In order to get a good grasp of the effect that our independent variable has on our multidimensional dependent variable, we must use several different measures. Multidimensional statistical procedures have been developed to handle the simultaneous use of several dependent variables in the same study. When several independent variables and several dependent variables are manipulated in one experiment, a more elaborate statistical technique called *multivariate analysis of variance* must be used to analyze the results. This approach allows us to take the correlation between the dependent variables into account (Kerlinger and Pedhazur, 1973). Analyzing each dependent variable separately would violate one of the underlying assumptions of the statistical test if a correlation did exist between the different dependent variable measures. Using several dependent variables in a study and appropriately analyzing them can, therefore, increase our knowledge of the complex relationship that exists between antecedent conditions and behavior.

SUMMARY

In seeking an answer to a research question, one must develop a design that will provide the necessary information. Two primary ingredients in a research design are the independent and dependent variables. Before the research design can be finalized, it is necessary to make a number of decisions about these two variables.

For the independent variable, the investigator must first specify not only the number of independent variables to be used but also the exact concrete operations that will represent the conceptual variable or variables. For some independent variables, this is easy because specific empirical referents exist; others, however, are not so easily translated. In either case, the conceptual independent variable must be operationally defined. In addition to translating the independent variable, the investigator must also specify how variation is to be established in the independent variable. Generally, variation is created by a presence-versus-absence technique or by varying the amount or type of the independent variable. The investigator must determine, within the framework of one of these techniques, the exact mode for creating the variation. Will the variation be created by manipulating instructions or events or by measuring the internal states of the organism? After these decisions have been made, it is advisable to reexamine the operational definition of the independent variable to determine if the operations really represent the specified construct.

There are a number of ways in which concrete information can be attained regarding this issue. If several ways of defining the independent variable all produce the same experimental results, there is added assurance that the conceptual variable was translated appropriately. Also, if several empirical representations of the independent variable produce the expected outcomes, then we are more confident that the correct translation was made. Additional techniques that can be used with human subjects are to conduct long, probing interviews and to obtain a behavioral indication that the appropriate translation was made.

STUDY QUESTIONS

1. Define an *independent variable* and a *dependent variable* and give an example of each.
2. Name the ways in which variation can be achieved in the independent variable. Give an example of how each of these modes might be used in a hypothetical experiment to achieve the necessary variation.
3. How does experimental manipulation of the independent variable differ from individual difference manipulation of the independent variable, and how can each be achieved?
4. What issues need to be considered when constructing the independent variable?
5. What is meant by the *construct validity* of the independent variable, and how can you be sure that construct validity has been attained?
6. What issues need to be considered when deciding on the number of independent and dependent variables to use in a research study?
7. What is the purpose of the dependent variable, and what issues must be considered when deciding on the dependent variable?
8. What is the difference between *reliability* and *validity*, and how do these topics relate to the dependent variable?
9. What role does construct validity play with respect to the dependent variable?

10. What procedures can be used to reduce subject error when assessing the dependent variable?

KEY TERMS AND CONCEPTS

Variable
Independent variable
Dependent variable
Discrete variable
Continuous variable
Qualitative variable
Quantitative variable
Experimental manipulation

Instructional manipulation
Individual difference manipulation
Construct validity
Convergent outcome
Divergent outcome
Interaction
Validity

CHAPTER 7

Control in Experimentation

LEARNING OBJECTIVES

1. To gain an understanding of the meaning of control in experimentation.
2. To understand the meaning of extraneous variables and why they need to be controlled.
3. To understand the major types of extraneous variables that can affect an experiment.
4. To understand how bias is produced by each of the extraneous variables discussed in this chapter.

Over the past two decades several theories have been advanced that relate diet to antisocial behavior. Until the 1980s, these theories appeared mostly in popular books and were supported primarily by food faddists. In recent years, however, these theories have penetrated the scientific profession. They are discussed at meetings of criminologists and in books and articles written for people who work in the corrections and criminal justice systems. Advocates of various dietary theories have stated that food allergies, or adverse reactions to food additives or specific foods such as milk or sugar, are the cause of the antisocial behavior exhibited by many prison inmates. In a number of states, correctional facilities have even changed inmates' diets, eliminating foods to which inmates may be allergic and providing vitamin supplements to correct supposed nutritional deficiencies.

As a result of the broad claims made about the impact of diet on antisocial behavior, some studies have been conducted in an attempt to document this presumed relationship. Stephen Schoenthaler at California State College, Stanislaus, has conducted a variety of studies using individuals in juvenile detention facilities as research subjects (Schoenthaler, 1983). In all but one of these studies, he changed the diet by replacing sugar with honey; soft drinks and Kool-Aid with fruit juice; and high-sugar cereals, desserts, and snacks with foods thought to be lower in simple sugars. In all of these studies Schoenthaler concluded that the dietary change decreased antisocial behavior and that the declines ranged from 21 to 54 percent. Such astounding results suggest that diet is a very important variable in antisocial behavior. Before these results can be accepted as true, however, the design of Schoenthaler's studies must be examined. In this case, unfortunately, there are a variety of other extraneous variables that could also account for the observed decline in antisocial behavior. For example, in one of the institutions investigated by Schoenthaler, the percentage of female inmates increased at the same time as the dietary change was being implemented. Assuming that females are less violent and antisocial than men, such a change in the composition of the inmate population should result in a decrease in antisocial behavior, in turn decreasing the certainty with which diet can be presumed to be the causal variable. In any research study, such extraneous variables must be controlled in order to be able to identify the effect of the independent variable. In this chapter I will discuss some of the major extraneous variables that may creep into an experiment.

INTRODUCTION

Internal validity
The extent to which the observed effect is caused only by the experimental treatment condition

In any experiment, the goal of the researcher is to attain internal validity (Campbell and Stanley, 1963). **Internal validity** refers to the extent to which we can accurately state that the independent variable produced the observed effect. When conducting an experiment, the scientist wants to identify the effect produced by the independent variable. If the observed effect, as measured by the dependent variable, is caused only by the variation in the independent variable, then internal validity has been achieved. The difficulty that arises is one of determining whether the observed effect is caused *only* by the independent variable, since the dependent variable could be influenced by variables other than the independent variable. For example, if we were investigating the influence of tutoring (independent variable) on grades (dependent variable), we would like to conclude that any improvement in the grades of the students who received the tutoring over that of those who did not was in fact a result of the tutoring. However, if the tutored students were brighter than those who were not tutored, the improvement in grades could be due to the fact that the tutored students were brighter. In such an instance, intelligence would represent an **extraneous variable,** which has the effect of confounding the results of the experiment. Once an extraneous variable creeps into an experiment, we can no longer draw any conclusion regarding the causal relationship that exists between the independent and the dependent variable. Therefore, it is necessary to control for the influence of such outside variables in order to attain internal validity. Since achieving control over the variation produced by extraneous variables is a prime component in the research process, it will be discussed in some detail.

Extraneous variable
Any variable other than the independent variable that influences the dependent variable

CONTROL OF EXTRANEOUS VARIABLES

The neophyte may think that controlling for the effect of extraneous variables means completely eliminating the influence of these variables from the experiment. It is possible to eliminate totally the influence of some variables. The effect of visual stimulation could, for example, be eliminated by conducting the experiment in a lightproof, blacked-out room, since such variables would be held constant at a magnitude of zero. However, most of the variables that could influence a psychological experiment—such as intelligence, past experience, and history of reinforcement—are not of a type that can be eliminated. Although it is not possible to eliminate the influence of such variables from the experiment, it *is* possible to eliminate any *differential* influence that these variables may have across the various levels of the independent variable. In other words, it is possible to keep the influence of these variables **constant** across the various levels of the independent variable; therefore, any differential influence noted on the dependent variable can be attributed to the levels of variation in the independent variable. Consider the study conducted by Wade and Blier (1974). They investigated the differential effect that two methods of learning have on the retention of lists of words (they actually used consonant-vowel-consonant trigrams). In this study, they had to control for the associations that subjects had with these words, because it has been shown that association value influences rate of learning. Therefore, they chose words that had previously been shown to have an average association value of 48.4 percent for subjects. In this manner, they held the association value of the words constant across the two groups of subjects and eliminated any differential influence that this variable might have had.

Constancy
Stability, or absence of change, in the influence exerted by an extraneous variable in all treatment conditions

Ideally, the scientist would like to keep the amount or type of each extraneous factor identical throughout the experiment. This means that the same amount and type of the factor must be present for all subjects. Variables that are noncontinuous (do not vary as a function of the independent variable or progress through the experiment) can meet the criterion of ideal constancy. Ideal constancy can be obtained for variables such as sex of the subject or rate of presentation of words, but not for factors such as interest or learning ability.

There are two reasons why variables such as learning ability cannot be held completely constant. First, constancy requires an exact measure, and our measuring devices are too crude to give more than an approximation of the amount of a characteristic such as interest, which is distributed along a continuum ranging from some very low point to some extremely high point. Second, some variables present in an experiment may change as the experiment progresses. Fatigue, motivation, interest, attention, and many other variables fall into this category. If these factors do vary, constancy dictates that the magnitude of these changes must be the same for all individuals. Degree of fatigue would have to increase and decrease for all subjects simultaneously. Even if it were possible to arrange this (which it probably is not), it is still possible that the waxing and waning of these factors would influence the behavior measured by the dependent variable. Increased fatigue frequently does affect performance. If the waxing and waning of a variable such as fatigue, even though it occurs simultaneously and in the same amount for all subjects, affects the dependent variable of performance, then constancy is not achieved, and it would be impossible to determine what caused the variation in the dependent variable of performance. The variation could be due entirely to the independent variable, entirely to the waxing and waning of fatigue, or to a combination of these two. As you can readily see, a situation like this would not allow you to identify unambiguously the cause of the variation noted in the dependent variable. Constancy requires an equal amount and type of a factor in all subjects throughout the course of the experiment.

Although the ideal level of constancy is not always achievable, in most cases it is possible to eliminate the differential effects of extraneous variables on the dependent variable, thereby allowing the scientist to relate unambiguously the variation in the independent variable to the dependent variable. How is constancy achieved? That is, how do we arrange factors in such a way that they have no differential influence on the result of the experiment? The only way is through control. Control means exerting a constant influence. Thus, if we wanted to hold constant the trait of dominance, we would attempt to make sure that this trait had an equal influence on all groups of subjects.

Control, or achieving constancy of potential extraneous variables, is often relatively easy to accomplish once the extraneous variables have been identified. The difficulty frequently lies in identifying these variables. Before determining techniques for controlling extraneous variables, we must identify the variables that need to be controlled. The variables identified in this chapter do not run the gamut of extraneous variables, but they do represent the more salient ones.

EXTRANEOUS VARIABLES TO BE CONTROLLED

In experimentation, the condition the scientist strives for is constancy of all variables except the one or more that are deliberately being manipulated. Campbell and Stanley (1963) state that experiments are internally valid when the obtained effect can be unambiguously attributed to the manipulation of the independent variable. In other words, if the effects obtained in the experiment are due only to the experimental conditions manipulated by the scientist and not to any other variables, the experiment has internal validity. To the extent that other variables may possibly have contributed to the observed effects, the experiment is not internally valid.

In any experiment there are always some variables other than the independent variable that could influence the observed effects. These potentially confounding variables must be identified and then dealt with or held constant. Cook and Campbell (1979) list a number of classes of variables that need to be controlled for internal validity to be attained.

History

History variable
An extraneous variable occurring between the pre- and the postmeasurement of the dependent variable

The **history variable** operates in an experiment that is designed in such a way as to have both a pre- and a postmeasurement of the dependent variable. *History* refers to the specific events, other than the independent variable, that occur between the first and second measurement of the dependent variable, as illustrated in Figure 7.1. These events, in addition to the independent variable, could have influenced the postmeasurement; therefore, these events become plausible rival hypotheses concerning the change that occurred between the pre- and the postmeasurement. Consider an attitude-change experiment. One typical procedure is to pretest subjects to identify their current attitudes. An experimental condition is then introduced in an attempt to change the subjects' attitudes, and a postattitude measurement is given to all subjects. The difference between the pre- and postattitude scores is typically taken as a measurement of the effect of the experimental variable on the attitude. The difficulty with making this assumption is that a certain amount of time elapses between the pre- and the postmeasure. It is possible that the subjects experienced events during this time that had effects on their attitudes—effects that were reflected in the postmeasurement. If such events actually took place, the history variable, in addition to the experimental variable, may have influenced the observed effects, creating *internal invalidity*, since the history events would be plausible rival hypotheses.

FIGURE 7.1   Illustration of extraneous history events.

A study by Schoenthaler (1983) can be used to illustrate history events. As mentioned earlier, he investigated the impact of a dietary change on violent and aggressive behaviors of institutionalized juveniles. A record of such behaviors was maintained for each inmate for three months prior to, as well as three months after, dietary change. The results of this study revealed that the mean number of violent and aggressive behaviors exhibited during the three months prior to the dietary change was significantly greater than the mean number of such behaviors exhibited after the dietary change. Schoenthaler concluded that the dietary change was responsible for the significant reduction in violent and aggressive behavior. Although this may be true, it is also important to realize that six months elapsed between the beginning of the pretesting and the completion of the posttesting, and many other events that took place during this time could have accounted for the alteration in behavior. Schoenthaler also realized this and considered a variety of alternative explanations, such as "system-wide changes" that coincided with the introduction of the dietary alteration. Although these rival hypotheses did not seem to be supported by facts and apparently could not explain the observed change in violent and aggressive behavior, there was one alternative explanation that was not controlled and that could explain the observed results. The juveniles who participated in the study were institutionalized for the entire six-month period and participated in both the pre- and postdietary phases. It is reasonable to assume that institutionalization itself should, over time, reduce the frequency of violent and aggressive behavior, and that a six-month period should be sufficient to induce such a reduction. Consequently, such a history event serves as a rival hypothesis for the effect that Schoenthaler attributed to the dietary change. This history event must be controlled before one can conclude with any degree of certainty that a dietary change can have a beneficial impact on violent and aggressive behavior.

Generally speaking, the longer the time lapse between the pre- and the posttest, the greater the possibility of history becoming a rival explanation. But even short time lapses can generate the history effect. If group data are collected and an irrelevant, unique event such as an obstreperous joke or comment occurs between the pre- and the posttest, this event can have an influence on the posttest, making it a rival hypothesis.

Maturation

Maturation
Changes in biological
and psychological
conditions that occur
with the passage of time

Maturation refers to changes in the internal conditions of the individual that occur as a function of the passage of time. The changes involve both biological and psychological processes, such as age, learning, fatigue, boredom, and hunger, that are not related to specific external events but reside within the individual. To the extent that such changes affect the individual's response to the experimental treatment, they create internal invalidity.

Consider a study that attempts to evaluate the benefits achieved from a Head Start program. Assume the investigator gave the subjects a preachievement measure at the beginning of the school year and a postachievement measure at the end of the school year. In comparing the pre- and postachievement measures, she found that significant increases in achievement existed and concluded that Head Start programs are very beneficial. Such a study is internally invalid because there was no control for the maturational influence. The increased achievement could have been due to the changes that occurred with the passage of time. A group of children who did not participate in Head Start may have progressed an equal amount. In order to determine the effect of a program such as Head Start, a control group that did not receive the treatment would also have to be included to control for the potential rival influence of maturation.

Liddle and Long (1958) conducted a study that did not seem to control for the maturational variable. They identified a group of "slow learners," culturally deprived children who had been unsuccessful in the first grade, and set up an experimental room in an attempt to motivate these children and to increase the amount that they learned. The investigators gave the children an initial intelligence test and assigned them a reading-grade placement score. Toward the end of the second year in the experimental room, the children were given Metropolitan Achievement Tests and showed "an improvement of about 1.75 years in less than two school years" (p. 145). On the basis of such evidence, the authors suggested that an experimental room like the one they had developed enhances the learning of slow learners. One of the problems with this study is that maturational influences were not ruled out. These slow learners were two years older at the end of the study and therefore were probably more mature. I suspect that some learning took place outside this experimental room. The investigators did not isolate the effect due *just* to the experimental room but allowed many other variables to enter in, one of which could have been maturation.

Instrumentation

Instrumentation Changes in the assessment of the dependent variable

Instrumentation refers to changes that occur over time in the measurement of the dependent variable. This class of extraneous variables does not refer to subject changes but to changes that occur during the process of measurement. Unfortunately, many of the techniques that we use to measure our dependent variable are subject to change during the course of the study. The measurement situation that is most subject to the instrumentation source of error is one that requires the use of human observers. Physical measurements show minor changes, but human observers are subject to influences such as fatigue, boredom, and learning processes. In administering intelligence tests, the tester typically gains facility and skill over time and collects more reliable and valid data as additional tests are given. Observers and interviewers are often used to assess the effects of various experimental treatments. As the observers and interviewers assess more and more individuals, they gain skill. The interviewers may, for example, gain additional skill with the interview schedule or with observing a particular type of behavior, producing shifts in the response measure that cannot be attributed to either the subject or the treatment conditions. This is why studies that use human observers to measure the behavioral characteristics of interest typically use more than one observer and have each of the observers go through a training program. In this way, some of the biases inherent in making observations can be minimized, and the various observers can serve as checks on one another to ensure that accurate data are being collected. Typically, the data collected by the various observers must coincide before they are considered valid.

Statistical Regression

Many psychological experiments (such as the attitude change experiment described in the discussion of the history factor) require pre- and posttesting on the same dependent variable measure or some other equivalent form for the purpose of measuring change. Additionally, these studies sometimes select only the two groups of subjects having the extreme scores, such as high and low attitude scores. The two extreme scoring groups are then given an experimental treatment condition, and a posttest score is obtained. A variable that could cause the pre- and posttest scores of the extreme groups to change is **statistical regression,** which "refers to the fact that the extreme scores in a particular distribution will tend to move— that is, regress—toward the mean of the distribution as a function of repeated testing" (Neale and Liebert, 1973, p. 38). The scores of the high groups may become lower, not because of any treatment condition introduced, but because of the statistical regression phenomenon. Low scorers could show increases upon retesting because of statistical regression and not because of any experimental treatment effect. This regression phenomenon exists because the first and second measurements are not perfectly correlated. In other words, there is some degree of unreliability in the measuring device.

Statistical regression
The lowering of
extremely high scores or
the raising of extremely
low scores during
posttesting

The regression toward the mean is illustrated in Table 7.1. A total of twelve subjects were pretested, and the scores of these subjects ranged from 46 to 123. A group of three extremely high scorers and four extremely low scorers was selected from the original twelve subjects. These high- and low-scoring subjects were then posttested. The scores of subjects with high pretest scores declined upon posttesting, whereas those of subjects with low pretest scores increased upon posttesting. In this example, an experimental treatment condition was not administered to the subjects to cause a change in their scores. Rather, the decline in the scores of the high-scoring group and the increase in the scores of the low-scoring group were caused entirely by statistical regression resulting from unreliability of the measuring device.

TABLE 7.1   Illustration of the Statistical Regression Effect

Subject	Pretest Score	Selected Subject	Pretest Score	Selected Subject	Posttest Score
S_1	110 ———→S_1		110 ———→S_1		103
S_2	46	→S_3	123 ———→S_3		116
S_3	123	S_8	105 ———→S_8		98
S_4	92				
S_5	59				
S_6	73				
S_7	99				
S_8	105				
S_9	67	S_2	46 ———→S_2		57
S_{10}	84	S_5	59 ———→S_5		63
S_{11}	61	→S_9	67 ———→S_9		70
S_{12}	96	→S_{11}	61 ———→S_{11}		65

Regression effects are different from those of maturation or history. They constitute a real source of possible internal invalidity, which must be controlled if any valid conclusion is to be drawn regarding the cause of the observed effects.

Selection

The **selection** bias exists when a differential selection procedure is used for placing subjects in the various comparison groups. Ideally, a sample of subjects is randomly chosen from a population, and then these subjects are randomly assigned to the various treatment groups. When this procedure cannot be followed and assignment to groups is based on some different procedure, possible rival hypotheses are introduced. Assume that you wanted to investigate the relative efficacy of a given type of therapy on various types of psychotic behaviors. For your subjects, you selected two groups of psychotic patients. After two months, progress in therapy was evaluated and you found that the subjects exhibiting one type of psychotic reaction improved significantly more than those with the other type. With these results, one is tempted to say that the therapy technique used is the agent that produced the difference in improvement between the two groups of psychotic patients. But there may be other differences between the two groups that would provide a better explanation of the observed difference. The psychotic patients who improved most may possess characteristics that predispose them toward more rapid improvement with almost any type of therapy. If this is the case, then it is these characteristics and not the type of therapy that caused the more rapid improvement. Such difficulties are encountered when subjects are differentially selected based on a criterion such as type of psychosis, because the independent variable manipulation represents an individual difference manipulation. Therefore, a study with a selection bias boils down to an ex post facto study, with all of its inherent difficulties.

Selection can also interact with maturation, history, or instrumentation to produce effects that appear to be results of treatment. Consider the selection by maturation interaction that occurs when the experimental groups selected are maturing at different rates. Kusche and Greenberg (1983) discovered that deaf children's understanding of the concepts of good and bad develops more slowly than that of hearing children, as depicted in Figure 7.2. If these maturational differences were

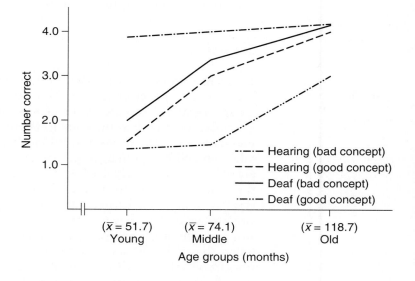

FIGURE 7.2 The evolution of good and bad concepts as a function of age and hearing status. (From "Evaluative Understanding and Role Taking Ability: A Comparison of Deaf and Hearing Children" by C. A. Kusche and M. T. Greenberg, 1983, *Child Development, 54,* pp. 141–147.)

not known, a study that attempted to teach the concepts of good and bad to deaf and to hearing children might conclude that the instructional program was more effective for the hearing children. However, the difference in response would be strictly a result of selection by maturation interaction and not a result of the instructional program. Similar treatment effects can occur if a selection by history or a selection by instrumentation interaction exists. For example, if a history effect influenced one of two treatment groups, a difference might exist between the two groups not because of the treatment effect but because of the impact of history on only one of the groups.

Mortality

Mortality
A differential loss of subjects from the various experimental groups

Mortality refers to the differential subject loss from the various comparison groups in an experiment. Most psychological experiments, both human and infrahuman, must contend with this potential source of bias at some time. Physiological experiments involving electrode implantation sometimes experience subject loss because of the complications that may arise from the surgical procedures. Human experiments must contend with subjects who do not show up for the experiment at the designated time and place or do not participate in all the conditions required by the study. The difficulty arises not just because subjects are lost but because the loss of subjects may produce differences in the groups that cannot be attributed to the experimental treatment. Consider the following example. Assume that you want to test the effect of a certain treatment condition on conformity. You know that past research has demonstrated that females conform to a greater degree than males do, so you control for this factor by assigning an equal number of males and females to two groups. When you actually run the experiment, however, half of the females assigned to the group that does not receive the treatment condition do not show up, and half of the males assigned to the other group (the one receiving the treatment condition) do not show up. Statistical analysis reveals that the group receiving the treatment condition conforms significantly more than does the group not receiving any treatment. Can you conclude that this significantly greater degree of conformity is caused by the independent variable administered? Such an inference would be incorrect, because more females were in the group receiving the experimental treatment and past research indicates that females exhibit a greater degree of conformity. This variable, and not the independent variable, may have produced the observed significant difference.

SUBJECT AND EXPERIMENTER EFFECTS TO BE CONTROLLED

Any human experiment involves an interaction between an experimenter and a subject. In this situation, two roles can be identified: that of the experimenter and that of the subject. Each role has specific behavioral requirements, and mutual expectations are held by the two role members. These expectations should define the behavior that is appropriate for each member. When agreeing to take part in an experiment, a person is making an implicit contract to play the role of the subject. Theoretically, this means that the subject will listen to the instructions and perform the tasks requested to the best of his or her ability and as truthfully as possible. In reality, such an ideal situation does not always exist because the subject has certain perceptions of the experiment that may alter behavior. The subject may want to comply and participate in the experiment but, because of certain perceptions and motives, may respond in several different ways. That is to say, there is an interaction between the way a person responds in an experiment and his or her motives and perception of the experiment.

Subject Effect

In an experiment, the researcher would like to have ideal subjects—subjects who would bring no preconceived notions to the laboratory. Once in the laboratory, they would accept instructions and be motivated to respond in as truthful a manner as possible. Although such a situation would be wonderful, it exists only in the experimenter's mind. When subjects enter the experiment, they are generally naive regarding its purpose or the task required of them. Once they appear for the experiment, however, they receive information from the way the experimenter greets them, from the instructions given regarding the experiment, from the task required of them, from the

Demand characteristics
Any of the cues
available in an
experiment, such as the
instructions, the
experimenter, rumors, or
the experimental setting

laboratory setting (including the available equipment), and from any rumors they have heard about the experiment. This information, called the **demand characteristics** of the experiment (Orne, 1962), defines the experiment from the subjects' point of view. It provides the subjects with the information from which they create their perceptions of the purpose of the experiment and the task required. Once the subjects identify this task, they are motivated to perform it. It is in the performance of the experimental task that the subjects' perceptions can influence the outcome of the experiment.

In the past, it was thought that subjects assumed a specific role (Orne, 1962; Rosenberg, 1969; Fillenbaum, 1966; Masling, 1966) and attempted to portray this role when performing the experimental task. Increasingly, this view is being rejected (for example, Carlston and Cohen, 1980; Carlopia, Adair, Lindsay, and Spinner, 1983). In its place is the notion that subjects respond to the experimental task as they perceive it. If the experiment involves a learning task, the subject will attempt to learn the material presented. However, subjects do not take an uninvolved, neutral approach, because often their performance implies something about them. For example, a learning task indirectly says something about the subjects' intelligence. If they learn the material rapidly, this suggests that they are intelligent. Most individuals have a desire to appear intelligent, so they will try to learn as rapidly as possible. Similarly, if the task suggests something about emotional stability, subjects will respond in such a way as to appear most emotionally stable (Rosenberg, 1969). Consequently, although subjects seem to approach an experiment with the motivation to perform the task requested, superimposed on this desire is the wish to make a **positive self-presentation** (Christensen, 1981). This means that subjects use their perceptions of the experiment to determine how to respond to the experimental task in such a way that they appear most positive.

Positive self-presentation
Subjects' motivation to
respond in such a way as
to present themselves in
the most positive manner

Consider the experiment conducted by Christensen (1977). In this experiment, an attempt was made to verbally condition the subjects—to increase the subjects' use of certain pronouns such as *we* and *they* by saying "good" whenever the subjects used one of them. Some subjects interpreted the experimenter's reaction of saying "good" as an attempt to manipulate their behavior. These subjects resisted any behavioral manifestation of conditioning. This resistance was caused by their viewing manipulation as negative. If they did not demonstrate any conditioning, then they would show that they could not be manipulated and, in this way, present themselves most positively. Similarly, Bradley (1978) has revealed that individuals take credit for desirable acts but deny blame for undesirable ones in order to enhance themselves.

Conditions Producing a Positive Self-Presentation Motive

In order to control the interactive effects that exist between the subjects' behavior and their role in an experiment, it would be advantageous to know the conditions that alter subjects' behavior in their attempt to attain favorable self-presentations. Only when such conditions are identified can one construct conditions that control for the confounding effect that may be produced.

Tedeschi, Schlenker, and Bonoma (1971) provide some insight into the general conditions that may determine whether or not the self-presentation motive will exist within an experiment. They state that this motive arises only when the behavior in which the subject engages is indicative of the subject's true intentions, beliefs, or feelings. If subjects believe that others view their behavior as being determined by some external source not under their control, then the positive self-presentation motive is not aroused. However, our experiments are seldom constructed so that the subjects believe that others think their behavior is externally determined. Thus it seems that the positive self-presentation motive would exist in most research studies.

Implication for Research

The implication of the positive self-presentation motive is that experimenters must take into consideration the influence of the subjects' perceptions of the experiment. No longer can we assume that the responses obtained from subjects are a function only of the physical or psychological stimuli representative of the independent variable. Consider the studies conducted by Carlston and Cohen (1980) and Carlopia et al. (1983). Carlston and Cohen contrasted the responses of subjects instructed to portray a good, faithful, negative, or apprehensive role with those of control subjects who were not given a role to portray. These investigators found that the responses of the role-play subjects differed from those of the control subjects. Based on these data, Carlston and Cohen concluded that subject roles probably do not produce experimental bias. Both Christensen (1982) and Carlopia et al. (1983) objected to this conclusion. Carlopia et al. hypothesized that Carlston and Cohen's conclusion was based on the differential perceptions that subjects had of the experimental hypothesis. These investigators then proceeded to replicate Carlston and Cohen's study. In doing so, they showed that subjects could be divided into those who perceived the experimental hypothesis to be true and those who perceived the experimental hypothesis to be false. Those who did not believe the experimental hypothesis supported Carlston and Cohen's conclusion, whereas those who believed the experimental hypothesis did *not* support that conclusion. Consequently, the responses produced by the subjects and the conclusions drawn by the experimenters were a function of how the subjects perceived the experiment and not a function of just the independent variable manipulation.

These studies indicate the necessity of trying to ensure that constant subject perceptions exist throughout all phases of the experiment. When such constancy is not maintained, artifactual confounding can be expected to occur from the interaction of the motive of positive self-presentation with the experimental treatment condition. There appear to be two types of interaction that can exist (Christensen, 1981): an intertreatment and an intratreatment interaction. An **intertreatment interaction** exists when subjects' perceptions of the different experimental treatment conditions suggest to them different ways of presenting themselves in a positive manner. For example, Sigall, Aronson, and Van Hoose (1970) found that in one treatment condition subjects increased their performance when told that they were responding to a dull, boring task under decreased illumination levels. However, in another treatment condition, subjects decreased their performance when told that rapid performance on the task indicated obsessive, compulsive performance. In the first experimental task, subjects apparently perceived that they could appear most positive by overcoming the obstacles of low illumination and a dull, boring task by performing rapidly. In the other treatment condition, subjects believed that they would appear most positive by performing slowly in order not to seem obsessive–compulsive.

An **intratreatment interaction** exists when different subjects in the same treatment condition perceive different ways

Intertreatment interaction
Perception by subjects in different treatment groups that they can fulfill the positive self-presentation motive by responding in different ways

Intratreatment interaction
Perception by subjects in the same treatment condition that they can fulfill the positive self-presentation motive by responding in different ways

of presenting themselves in the most positive light. For example, Turner and Solomon (1962) found that some subjects participating in an avoidance conditioning paradigm failed to learn to move a lever to avoid shock. Investigation of the perceptions of these subjects disclosed that they perceived the purpose of the study to be one of determining who could tolerate the shock. Consequently, these subjects apparently believed that they would appear most positive if they could demonstrate their capacity to endure the shock. When the instructions were changed to eliminate this perception, all subjects demonstrated avoidance conditioning. If these instructions had not been altered, the internal validity of the study would have been compromised, because the obtained results would have been due to the combined factors of subjects' perceptions and avoidance conditioning and not just the avoidance conditioning paradigm.

Experimenter Effect

We have just seen that the subjects who are used in psychological research are usually not apathetic or willing to passively accept and follow the experimenter's instructions. Rather, they have motives that can have an effect on the experimental results. In like manner, the experimenter is not just a passive, noninteractive observer but an active agent who can influence the outcome of the experiment. Friedman (1967, pp. 3–4) has appropriately stated that in the past psychology has

> implicitly subscribed to the democratic notion that all *experimenters* are created equal; that they have been endowed by their graduate training with certain interchangeable properties; that among these properties are the anonymity and impersonality which allow them to elicit from the same subject identical data which they then identically observe and record. Just as inches were once supposed to adhere in tables regardless of the identity of the measuring instrument, so needs, motives, traits, IQs, anxieties, and attitudes were supposed to adhere to patients and subjects and to emerge uncontaminated by the identity and attitude of the examiner or experimenter.

Such a conception of the experimenter is highly inappropriate, because research, as we shall see, has demonstrated biasing effects that are directly attributable to the experimenter. Take a look at the motives that the experimenter brings with him or her. First, the experimenter has a specific motive for conducting the experiment. The experimenter is a scientist attempting to uncover the laws of nature through experimentation. In performing this task, he or she develops certain perceptions of the experiment and the subject. Lyons (1964, p. 105) states that the experimenter wants subjects to be perfect servants—intelligent individuals who will cooperate and maintain their position without becoming hostile or negative. It is easy to see why such a desire exists. The scientist seeks to understand, control, and predict behavior. To attain this goal, the scientist must eliminate the subject effect, and so he or she dreams of the ideal subject, who does not have any bias. Also, the experimenter has expectations regarding the outcome of the experiment. He or she has made certain hypotheses and would like to see these confirmed. Although this aspect of science is legitimate and sanctioned, it can, as we shall see, lead to certain difficulties. Additionally, journals have a bias toward publishing primarily positive results, which essentially means that studies supporting hypotheses have a greater chance of being accepted for publication. Knowing this, the experimenter has an even greater desire to see the hypotheses confirmed. Can this desire or expectancy bias the results of the experiment so as to increase the probability of attaining the desired outcome? Consider the fascinating story of Clever Hans. Clever Hans was a remarkable horse that could apparently solve many types of arithmetic problems. Von Osten, the master of Clever Hans, would give Hans a problem, and then Hans would give the correct answer by tapping with his hoof. Pfungst (1911/1965) observed and studied this incredible behavior. Careful scrutiny revealed that von Osten would, as Hans approached the correct answer, look up at Hans. This response of looking up represented a cue for Hans to stop tapping his foot. The cue was unintentional and not noticed by observers, who attributed mathematical skills to the horse.

Observations such as those made by Pfungst of Clever Hans would seem to indicate that one's desires and expectancies can somehow be communicated to the subject and that the subject will respond to them. The research has suggested that subjects are motivated to present themselves in the most positive manner. If this is true, then the subtle cues presented by the experimenter in the experimental session may very well be picked up by the subjects and influence their performance in the direction desired by the experimenter. Consequently, the experimenter may represent a demand characteristic.

The experimenter, zealous to confirm his or her hypothesis, may also unintentionally influence the recording of data to support the prediction. Kennedy and Uphoff (1939) investigated the frequency of misrecording of responses as a function of subjects' orientation. Subjects, classified on the basis of their belief or disbelief in extrasensory perception (ESP), were requested to record the guesses made by the "receiver." The receiver was supposedly trying to receive messages sent by a transmitter. Kennedy and Uphoff found that 63 percent of the errors that were in the direction of increasing the telepathic scores were made by believers in ESP, whereas 67 percent of the errors that were in the direction of lowering the telepathic scores were made by disbelievers. Such data indicate that biased recording, unintentional as it may be, exists in some experiments.

Additionally, the experimenter is an active participant in social interaction with the subject. As stated earlier, the experiment may be considered a social situation in which two roles exist—that of the experimenter and that of the subject. The role behavior of the subject can vary slightly as a function of the experimenter's attributes. McGuigan (1963), for example, found that the results of a learning experiment varied as a function of the experimenter. Some of the nine researchers used to test the effectiveness of the same four methods of learning found significant differences, whereas the others did not. Such research shows that certain attributes, behavior, or characteristics of the experimenter may influence the subjects' responses in a particular manner.

The ways in which the experimenter can potentially bias the results of an experiment can be divided into two types: bias arising from the attributes of the experimenter and bias resulting from the expectancy of the experimenter.

Experimenter attributes
The physical and psychological characteristics of an experimenter that may create differential responses in subjects

Experimenter Attributes The term **experimenter attributes** refers to the physical and psychological characteristics of an experimenter that may interact with the independent variable to cause differential performance in subjects. There is a large body of data revealing the differential influence produced by various aspects of experimenters. Rosenthal (1966) has summarized a great deal of this research and has proposed that at least three categories of attributes exist. The first is *biosocial attributes,* which include factors such as the experimenter's age, sex, race, and religion. The second category is *psychosocial attributes.* These attributes include the experimenter's psychometrically determined characteristics of anxiety level, need for social approval, hostility, authoritarianism, intelligence, and dominance, and social behavior of relative status and warmth. The third category proposed by Rosenthal represents *situational factors,* including whether or not the experimenter and the subject have had prior contact, whether the experimenter is a naive or experienced researcher, and whether the subject is friendly or hostile. Additionally, the characteristics or physical appearance of the laboratory may influence the outcome of the research study.

A great many biasing factors have been identified. Does this mean that experimenter attributes will always affect the experiment and lead to artifactual results? McGuigan (1963) says that there are three possibilities.

1. *The attributes of the experimenters have absolutely no effect on the outcome of the experiment.* Ideally, this is the type of situation to strive for.
2. *The experimenter attributes affect the dependent variable, but the influence is identical for all subjects.* Such a case would be of no concern, because all subjects would be uniformly affected and therefore all experimenters would get the same experimental results.
3. *The experimenter attributes differentially affect subjects.* Here, the results of the experiment would be partially a function of the experimenter who conducted it. If, for example, a black experimenter found that subjects in an attitude-change experiment responded in a significantly less prejudiced manner, whereas a significant change was not found by a white experimenter, one would have to conclude that the biosocial attribute of race differentially affected subjects' responses.

Psychologists are working on the problem of the influence of the experimenters' characteristics. From this research, we should ultimately be able to identify what attributes will influence our experiments and when. Jung (1971, p. 49) has stated that "the extent to which the variable can affect results may vary with the type of experiment, being stronger in social and personality experiments and weaker with psychophysical, perceptual, and sensory experiments." Psychologists need to work toward the goal of being able to identify where, under what conditions, and in what type of experiments researcher attributes are confounding variables, because these variables (as well as others) may well account for some of the controversies that arise from failure to replicate previously published studies.

About the only experimenter attribute that has been investigated in sufficient depth to allow us to begin to answer these questions is the attribute of experimenter gender. Rumenik, Capasso, and Hendrick (1977) reviewed the literature relating to this attribute and found that "despite the sloppy methodological state of most research . . ." (p. 874), the data suggest that young children perform better for female experimenters on a variety of tasks. For adults, male experimenters seem to elicit better performance. Within a client–counselor relation, the studies suggest that male counselors elicit more information-seeking responses, whereas female counselors elicit more self-disclosure and emotional expression.

Before anything definitive can be stated about the influence of experimenter attributes on specific types of research, it is necessary that researchers conduct studies that are methodologically sound. Specifically, they must attempt to overcome three deficiencies identified by Johnson (1976). First, researchers must control for the confounding influence of attributes other than the ones being studied. If gender of the experimenter is being studied, researchers must control for other attributes such as age, race, warmth, and need for social approval. Second, it is necessary to take into account and study the interactive influence of subject and experimenter attributes. For example, it may be that male and female experimenters of one ethnic group obtain different responses from male and female subjects of another ethnic group. Third, it is necessary for researchers to sample a variety of types of tasks. A given experimenter attribute may have an influence on one task but not on another. Until such methodological refinements are made, about the only thing that can be said about experimenter attributes is that they "may at times affect how subjects perform in the experiment, but we can rarely predict beforehand what experimenter attributes will exert what kind of effects on subjects' performance on what kinds of tasks" (Barber, 1976).

Experimenter expectancies
The influence of the experimenter's expectations regarding the outcome of an experiment

Experimenter Expectancies The term **experimenter expectancies** refers to the biasing effects that can be attributed to the expectancies the experimenter has regarding the outcome of the experiment. As noted earlier, experimenters are motivated by several forces to see their hypotheses validated. Therefore, they have expectancies regarding the outcome of the experiment. These expectancies can lead the experimenter to behave unintentionally in ways that will bias the results of the experiment in the desired direction. These unintentional influences can operate on the experimenter to alter his or her behavior and on the subjects to alter their behavior.

Effect on the Experimenter. It has been well documented that the expectancies we have can color our perceptions of our physical and social worlds. Research in social perception has repeatedly demonstrated the biased nature of our perceptions of others. In light of this research, it would be naive to assume that the expectancies of the experimenter did not have a potential influence on his or her behavior. In fact, there are several documented ways in which these expectancies have actually influenced the outcome of experiments. The expectancies of the experimenter can lead him or her to record responses inaccurately in the direction that supports the expectancies, as was noted in the ESP experiment conducted by Kennedy and Uphoff (1939), discussed earlier. Rosenthal (1978) summarized the results of twenty-one studies relating to the expectancy issue. These studies showed that, on the average, 60 percent of the recording biases favored experimenter expectancies. In one study, 91 percent of the recording biases supported the experimenters' expectancies. Impressive as these percentages are, it is important to understand that these recording errors, both biased and unbiased, represent only a small portion of the overall number of observations made. Generally speaking, only about 1 percent of all observations are misrecorded, and of these about two-thirds support the experimenters' expectancies (Rosenthal, 1978). Such a rate of misrecordings, even if the majority of them support the expectancies of the experimenters, is so small as to seldom affect the conclusions reached in a given study. But just because recording errors occur infrequently does not mean that experimenters can stop worrying about them, because when we relax we run the risk of increasing such errors. Reviews like the one conducted by Rosenthal reveal that when we attempt to avoid recording errors we are relatively successful—at least successful enough to avoid reaching an unfounded conclusion that can be directly traced to recording errors.

A second type of bias falling under the category of "effect on the experimenter" involves the effects of expectancies on interpretation of the data collected. Once the data have been collected, the experimenter attempts to explain them. Practically any set of data can be interpreted in different ways, depending on the orientation of the person doing the interpreting. Robinson and Cohen (1954), for example, have reported finding differences in the psychological reports written for thirty patients by three examiners. Barber and Silver (1968) disagreed with the conclusion Rosenthal reached regarding experimenter bias. This disagreement does not revolve around the validity of the data collected but around the interpretation of these data. However, the interpreter effects, though real, are not considered to be a serious methodological problem because the interpretations of the recorded data are assumed to be a function of the experimenter and his or her specific orientation. This is supported by the fact that the debates occurring in the literature seldom involve another's observations but often involve the interpretations that are placed on the observed data.

Effect on the Subject. It is relatively easy to believe that the experimenters' expectancies may cause them to behave in ways that support their expectancies. It is harder to see how these same expectancies can influence the subject to behave in a way that would support them, yet there is a body of research that demonstrates just this phenomenon. Remember that subjects seem to be motivated toward positive self-presentation. How do they know what response will maximize the possibility of achieving such a positive self-presentation? Somehow they make use of the demand characteristics surrounding the experiment, one of which seems to be the experimenter. The researcher has certain expectancies that lead him or her unintentionally to behave in ways that convey these expectancies. Subjects pick up these subtle cues and respond accordingly. Von Osten, for example, conveyed to Clever Hans when he should stop tapping his foot. Rosenthal and his associates have devoted a great deal of attention to this source of bias. They have demonstrated in many studies that experimenters definitely can influence the results of the study in the direction of their hypotheses. To put it another way, the experimenter can influence the subjects' responses in such a way that they will support the experimenter's hypothesis. For example, Rosenthal and Fode (1963) found that researchers who expected to get high success ratings on photographs previously judged neutral actually got significantly higher ratings than did researchers who were led to expect that they would get low success ratings.

Do these biasing effects exist in different types of experiments in psychology? One might initially think that the biasing effects of the expectancies of the experimenter on the subjects' responses would be limited to human types of experimentation—more specifically, to human experiments in such areas as social and personality psychology. However, when Rosenthal and Rubin (1978) summarized the studies conducted on expectancy, they found, as shown in Table 7.2, that the expectancy effect had been demonstrated in eight different research areas. Perhaps the most revealing bit of information contained in Table 7.2 is that the expectancy effect is not confined to human experiments. Not only has it been demonstrated in animal experiments, but a greater proportion of animal studies display the expectancy effect.

Mediation of Expectancy

The evidence presented by Rosenthal and others is consistent in indicating that the problem of the biasing effects of the experimenter is serious and needs to be dealt with. To deal most effectively with such biases, we must know what is causing them. In other words, just how is the experimenter transmitting expectancies? In addressing this question, Rosen-

TABLE 7.2 Number of Experimenter Expectancy Studies Conducted in Eight Different Research Areas

Research Area	Number of Studies Conducted	Proportion Demonstrating Expectancy
Reaction time	9	0.22
Inkblot tests	9	0.44
Animal learning	15	0.73
Laboratory interviews	29	0.38
Psychophysical judgments	23	0.43
Learning and ability	34	0.29
Person perception	119	0.27
Everyday situations	112	0.40

Source: Based on Table 1 in "Interpersonal Expectancy Effects: The First 345 Studies" by R. Rosenthal and D. B. Rubin, 1978, *The Behavioral and Brain Sciences, 3,* pp. 377–415.

thal considers the possibility of recording errors, particularly those biased in the direction of the expectancy. This would seem to be an important issue, since several studies (for example, Johnson and Ryan, 1976) have indicated that recording biases can account for much of the so-called expectancy effects. Rosenthal (1976) directly confronted this issue and found that although recording errors can account for some of the expectancy effect in some studies, they cannot account for all of it. To further support the fact that the experimenter expectancy bias cannot be reduced to a recording bias, Rosenthal (1976) reviewed thirty-six studies that employed special techniques for the control of recording errors and deliberate cheating. He found that these studies were *more* rather than less likely to demonstrate the experimenter expectancy effect. Exactly why such studies should be more susceptible to an experimenter expectancy effect is not known. It may be that the investigators not only provided safeguards against cheating and recording errors but also reduced the influence of other errors, thereby creating a more powerful and precise test of the expectancy effect.

If recording errors and intentional biases are insufficient to account for expectancy effects, then how can they be explained? Lack of an explanation of expectancy effects is the primary shortcoming of this research area (Adair, 1978). In all probability, these effects do not work in a unitary fashion, since they have been demonstrated to exist in both animal and human research. In animal studies, differences in animal handling seem to be important (Rosenthal and Fode, 1963). In human studies, important factors seem to be nonverbal cues (Rosenthal, 1980), such as facial or postural signals (Barber, 1976); intonation, such as emphasizing different key sections of the instructions (Adair, 1973); or a nod, smile, or glance (Rosenthal, 1969). However, as Rosenberg (1980) stated and as I emphasized earlier, we must also consider the subject when discussing the mediating influence of expectancy effects, because it is the subjects' responses that ultimately demonstrate these effects. The subjects are motivated to present themselves in the most positive manner. In doing so, they use the demand characteristics of the experimental situation—including the behavior of the experimenter—to define the most appropriate way of responding to induce a positive self-presentation. Consequently, subjects apparently make use of many nonverbal cues transmitted by the experimenter to define the responses that will maximize the probability that they will present themselves in the most positive manner.

Although several theoretical models attempt to explain the communication of expectancies within a classroom environment (for example, Harris and Rosenthal, 1985), no conceptual integration or theoretical statement has been proposed to explain the communication of experimenter expectancies. Therefore, we do not know when expectancy effects might occur or what may mediate them; the evidence merely tells us that these effects are real possible sources of bias. However, it is precisely because we do not know when expectancy effects are likely to occur that we must always protect ourselves against them as a potential source of bias (Ellsworth, 1978).

Magnitude of the Expectancy Effects

The influence of experimenter expectancies has been demonstrated repeatedly in a wide variety of contexts. Table 7.2 illustrates that, of the studies conducted, about one-third showed a significant expectancy effect. This rate is about seven times greater than would have been expected if the effect did not exist (Rosenthal, 1976). However, if the effect of experimenter expectancies were extremely small (though real), it might not pose a significant threat to internal validity, which would mean that researchers need not concern themselves with this potential bias. Rosenthal (1978) reviewed five studies that directly addressed this issue. These studies compared the effect produced by the experimental treatment condition with the effect produced by expectancy. In three of these five studies, the expectancy effect was greater than the treatment condition. For example, Burnham (cited in Rosenthal and Rubin, 1978) compared lesioned rats with nonlesioned (sham surgery) rats on a discrimination learning task. Half of the lesioned and half of the nonlesioned rats were assigned to experimenters who were told that they had received lesioned rats. The remainder were assigned to researchers who were told that they had received nonlesioned rats. Figure 7.3 shows that all of the rats tested by experimenters who were told that the rats were nonlesioned performed better than did the rats tested by experimenters who were told that the rats were lesioned. This evidence indicates that expectancy effects can be quite large and that precautions should be taken against them.

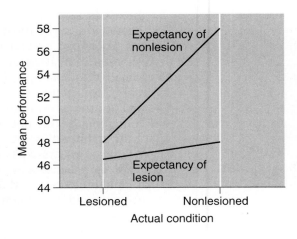

FIGURE 7.3 Discrimination learning of lesioned and nonlesioned rats as a function of experimenter expectancy. (Based on data from "Interpersonal Expectancy Effects: The First 345 Studies" by R. Rosenthal and D. B. Rubin, 1978, *The Behavioral and Brain Sciences, 3*, pp. 377–415.)

SEQUENCING EFFECT TO BE CONTROLLED

In an experiment, variation in the independent variable may be established by presence versus absence or by varying the amount or type of a variable. In addition to making a decision as to the manner in which the independent variable will be varied, the investigator must make a decision regarding how the subjects are to be used in the experiment. There are two choices here: the investigator can randomly assign subjects to the various treatment groups, or he or she can administer the different levels of variation of the independent variable to the same subjects. In a drug experiment having both a placebo and a drug condition, the experimenter could either randomly assign subjects to the placebo and the drug conditions or have all subjects respond under both the placebo and the drug conditions. Although there is a definite advantage (which will be discussed at a later time) to having the same subjects respond in both treatment conditions, there is also a definite disadvantage that involves a sequencing effect.

Sequencing effect
The influence on a subject's response of participation in a prior treatment condition

The **sequencing effect** occurs when participation in one condition affects the response that the subject will make in a subsequent treatment condition. If subjects participate first in the drug condition and then in the placebo condition, their responses in the placebo condition may be partially a function of carry-over effects of the drug. If so, then a sequencing effect has taken place. Sequencing effects occur any time a subject's response in one treatment condition is partially determined by participation in a prior treatment condition. Carry-over effects may be the result of such factors as practice received or fatigue incurred.

ADDITIONAL EXTRANEOUS VARIABLES TO BE HELD CONSTANT

Subject sophistication
The biasing influence that can arise from a subject's knowledge of or familiarity with psychology and psychological experiments

In addition to those already discussed, there are many other variables that could have an influence on a given experiment, including the subjects' motivation and ability and physical variables such as noise level and lighting. For example, a number of investigations (Page, 1968, 1969; Page and Kahle, 1976; and Page and Scheidt, 1971) have disclosed that **subject sophistication** can have a confounding influence on the results produced in psychological experiments. By subject sophistication I mean subjects' "familiarity with or sophistication in the subject matter, and methods of experimental psychology" (Page, 1968, p. 60). Investigators have revealed that a number of the effects previously identified could not be replicated unless sophisticated subjects were used. Use of novice subjects resulted in a failure to find the effect. For example, Schafer and Murphy (1943) found that subjects reported a pair of ambiguous human profiles that had previously been associated with winning money more frequently than they did a pair that had been associated with losing money. However, Page (1968) found this difference to exist only for the sophisticated group. Such findings suggest that in some of our experiments we must control for the influence of subject sophistication to eliminate the possibility that our results are due to this extraneous variable. When we gain better understanding of the reason that subject sophistication alters the results of our experiments, we will be in a better position to state which experiments should control for this extraneous variable and which ones can disregard it. At the present time, which experiments are subject to this effect is an empirical question.

The materials and apparatus used should also be constant for all subjects. If an apparatus breaks down, the experimenter should fix it rather than using a nonequivalent piece.

It would be impossible to list all of the variables that could possibly affect an experiment. A number of the more salient ones have been discussed. Beyond this, the experimenter must use his or her own knowledge and foresight to anticipate potential sources of error and build in controls for them.

SUMMARY

One of the most important tasks confronting the researcher is to ensure that the experiment is internally valid. To attain internal validity, the experimenter must control for the influence of extraneous variables that could serve as rival hypotheses for explaining the effects produced by the independent variable. Ideally, attaining the desired control involves completely eliminating the influence of all extraneous variables. In most cases, however, this is impossible. Therefore, control most frequently refers to holding the influence of the extraneous variables constant across the various levels of the independent variable. The task of maintaining constancy is difficult for some variables, because they may change as the subject progresses through the experiment.

Providing the desired control involves first identifying the variables that need to be controlled. Some of the more salient variables that could influence the experiment and serve as rival hypotheses are as follows:

History. Any of the many events other than the independent variable that occur between a pre- and a postmeasurement of the dependent variable.

Maturation. Any of the many conditions internal to the individual that change as a function of the passage of time.

Instrumentation. Any changes that occur as a function of measuring the dependent variable.

Statistical regression. Any change that can be attributed to the tendency of extremely high or low scores to regress toward the mean.

Selection. Any change due to the differential selection procedure used in placing subjects in various groups.

Mortality. Any change due to the loss of a subject from one or more of the various comparison groups.

Subject effect. Any change in performance that can be attributed to a subject's motives or attitude.

Experimenter effect. Any change in a subject's performance that can be attributed to the experimenter.

Sequencing. Any change in a subject's performance that can be attributed to the fact that the subject participated in more than one treatment condition.

Subject sophistication. Any change in a subject's performance as a function of sophistication or familiarity with the experimental procedures or subject matter of psychology.

STUDY QUESTIONS

1. What is meant by *internal validity*, and why is it an important ingredient in a psychological experiment?
2. How is control obtained over most extraneous variables?
3. Identify and define the major extraneous variables that need to be controlled within an experiment, and explain how each one operates to confound a psychological experiment.
4. Define *demand characteristics* and describe how they affect a psychological experiment.
5. Explain how subjects can bias the results of a psychological experiment and how this effect might manifest itself in the experiment.
6. What are the various ways in which the experimenter might affect a psychological experiment, and how are these effects exerted to produce bias?
7. How are expectancy effects mediated, and are they a significant source of bias in a psychological experiment?
8. What are sequencing effects?
9. Does the subject's familiarity with psychology and psychological experiments have any influence on the outcome of an experiment?

KEY TERMS AND CONCEPTS

Internal validity   Intertreatment interaction
Extraneous variable   Intratreatment interaction
Constancy   Experimenter effect
History variable   Experimenter attributes
Maturation   Biosocial attributes
Instrumentation   Psychosocial attributes
Statistical regression   Situational factors
Selection   Experimenter expectancies
Mortality   Sequencing effect
Demand characteristics   Subject sophistication
Positive self-presentation

CHAPTER 8

Techniques for Achieving Constancy

LEARNING OBJECTIVES

1. To learn the major control techniques that are used by researchers.
2. To understand how each of the control techniques operates to produce the necessary control.
3. To understand the type of variable that is controlled by each of the techniques discussed.

One of the areas that has attracted considerable attention over the past decade is the effect of diet on behavior. The increased attention and interest have come from both researchers and practitioners. For example, a colleague recently stopped me in the hall and inquired about the possible effect of breakfast on a student's classroom behavior. Apparently, this psychologist had been approached regarding the unmanageable behavior of a specific child. According to teacher reports, the child's behavior, when he arrived at school, was considered normal. The child was responsive to instructions, would do his schoolwork, and in general was like all the rest of the children in the classroom. But at about 10:00 A.M. his whole demeanor changed. Instead of being a pleasant, responsive child, he became extremely disruptive. He would get out of his seat and roam around the classroom, talk back to the teacher, and make strange sounds. Inquiry into the child's eating habits revealed that he typically ate a breakfast consisting of very high-carbohydrate foods such as pancakes and syrup. Could this high-carbohydrate diet have the effect of triggering disruptive behavior several hours after its consumption? Possibly, but many other variables, including other dietary substances, could have caused the disruptive behavior.

Let's look at another actual case. Several years ago the parents of a four-year-old child found that he would suddenly become extremely disobedient and even rather violent. He would start running around in circles and, when his parents tried to restrain him, would run away from them as fast as he could, run straight into a wall, bounce off it, pick himself up, and do the same thing again and again until his parents finally caught him and physically restrained him. Eventually it was established that this behavior was elicited when the child consumed Kool-Aid sweetened with Nutrasweet (the sugar substitute aspartame). Many parents and teachers believe that the diet their children receive and whether or not they have eaten a decent breakfast have an impact on their later behavior. In fact, a survey of pediatricians and family practitioners (Bennett and Sherman, 1983) has revealed that 45 percent of them periodically recommend a low-sugar diet when treating children with an attention-deficit disorder, despite the fact that there is little evidence to support the notion that such a diet is of any benefit. Does this mean that diet does not affect subsequent behavior or that breakfast makes lit-

tle or no difference in subsequent behavior? Little high-quality research has been done on this question, but the studies that have been conducted indicate that there is such a relationship. Any investigation into an area like this, however, is littered with variables that must be controlled. For example, a maturational variable such as the age of the subject may influence the outcome of a diet behavior study, because younger individuals who are still maturing may be more susceptible to the impact of dietary manipulations. Additionally, other variables—previous dietary habits, the food combinations ingested, the personality of the subjects, or their metabolic rate—may have an impact on the effect of a specific diet or even on whether skipping meals has any impact. How do we gain control over such variables so that a conclusion can be reached regarding the impact of a specific dietary variable? Many techniques have been developed over the years that enable the researcher to control the influence of such variables. In this chapter the control techniques that are most frequently used by researchers will be presented.

INTRODUCTION

In order to conduct an internally valid experiment, it is necessary to control for the influence of extraneous variables like those presented in Chapter 7. This means that some procedure must be incorporated into the study that will eliminate any differential influence that these variables may have on the dependent variable. There are three general methods of achieving the desired level of control. First, control can be attained through appropriate design of the experiment. In fact, one of the purposes of experimental design is to eliminate the differential influence of extraneous variables. (The control function of appropriate designs will become evident in Chapter 9.) A second means of attaining control involves making statistical adjustments by using techniques such as analysis of covariance. These techniques, however, are beyond the scope of the present text and therefore will not be discussed. A third means of acquiring the desired control is to incorporate one or more of the available control techniques into the design of the experiment. This method is intimately related to control through appropriate design of the experiment, because any control technique must be incorporated into the design of the experiment. However, control techniques are discussed separately in this chapter to allow more effective illustration of the variables that they control and the way they control these unwanted sources of variation.

All of the following techniques cannot and should not be incorporated into one study. Indeed, it would be impossible to do so. By the same token, it is often possible and advisable to use more than one of the techniques. The researcher must decide which of the possible extraneous variables could influence the experiment and, given this knowledge, select from the available techniques those that will allow the desired control. Failure to do so will create internal invalidity.

RANDOMIZATION

Randomization
A control technique that equates groups of subjects by ensuring every member an equal chance of being assigned to any group

Randomization, the most important and basic of all the control methods, is a statistical control technique designed to assure that extraneous variables, known or unknown, will not systematically bias the study results. It is the only technique for controlling unknown sources of variation. As Cochran and Cox (1957) have stated: "Randomization is somewhat analogous to insurance, in that it is a precaution against disturbances that may or may not occur and that may or may not be serious if they do occur. It is generally advisable to take the trouble to randomize even when it is not expected that there will be any serious bias from failure to randomize. The experimenter is thus protected against unusual events that upset his expectations" (p. 8).

How does randomization eliminate systematic bias in the experiment? Randomization refers to use of some clearly stated procedures such as tossing coins, drawing cards from a well-shuffled deck, or using a table of random numbers. To provide for maximum control of any systematic bias in the process of selecting the sample of subjects on which the study is conducted, one should select subjects randomly from a population. Population, if you recall, refers to all of something, such as all college students in the United States or all females. Randomly selecting subjects from a population provides maximum assurance that a systematic bias does not exist in the selection process and that you have selected for the study a sample that is representative of the total population. By **representative** I mean that the sample subjects have the same characteristic as the subjects in the population. If the average IQ in the population is 120, then the average IQ of the subjects in the sample should also be 120. Only in this way can the results of the study say something about the total population.

For example, assume that you wanted to conduct a study focusing on interpersonal attraction among college professors. In this instance the population of interest is all college professors and you would want the results of the study to say something about all college professors and not just those who participate in your study.

One way to ensure that the study results say something about all college professors is to include all of them in the study. However, including the total population is seldom, if ever, possible, so you must select a sample from the population on which to conduct the study. A sample, if you recall, is any number less than the population. Whenever you select a sample for a study, you run the risk of getting an unusual group of individuals who do not represent the population of interest. For example, if you conducted the study only on female college professors, you would have excluded the responses of male college professors. This would provide a biased representation of interpersonal attraction among college professors and the results would say little about the overall population—something you want to avoid. This issue of generalization of sample results to the population will be discussed in greater detail in Chapter 12.

How do you select a sample of subjects that is representative of the population? The only way in which this can be done is to select a random sample. A random sample is a sample of subjects that is selected in such a way that each subject has an equal chance of being selected and the selection of one does not affect the selection of another. Exhibit 8.1 illustrates the procedure involved in the selection of a random sample.

Once subjects have been randomly selected for a study, they should be randomly assigned to the same number of groups as there are experimental treatment conditions, as illustrated in Figure 8.1. The experimental treatment conditions should then be assigned to the experimental treatment groups. Although this is the ideal arrangement, one can seldom select

EXHIBIT 8.1 Procedure for Randomly Selecting a Sample of Subjects from a Population

The most frequently used method for randomly selecting a sample of subjects is to use a list of random numbers such as the following list. (Larger lists are contained in the appendixes of most statistics books.)

This list consists of rows and columns of numbers. The numbers in each position in the list are random in the sense that each of the numbers from 0 to 9 has an equal chance of occupying that position, and the selection of one number for a given position had no influence on the selection of another number for another

position. Consequently, every number was selected independently of every other number, and since each individual number is random, any combination of numbers is also random.

To use this list or any other table of random numbers to select a random sample from a population, you must first convert the people in the population to a series of numbers. This is easily accomplished by assigning each person a different number. For example, let's assume that you are concerned with the population of 5,000 female college professors in

	1	2	3	4	5	6	7	8	9	10	11	12	13	14
1	10480	15011	01536	02011	81647	91646	69179	14194	62590	36207	20969	99570	91291	90700
2	22368	46573	25595	85393	30995	89198	27982	53402	93965	34095	52666	19174	39615	99505
3	24130	48360	22527	97256	76393	64809	15179	24830	49340	32081	30680	19655	63348	58629
4	42167	93093	96243	61680	07856	16376	39440	53537	71341	57004	00849	74917	97758	16379
5	37570	39975	81837	16656	06121	91782	60468	81305	49684	60672	14110	06927	01263	54613
6	77921	06907	11008	42751	27756	53498	18602	70659	90655	15053	21916	81825	44349	42880
7	99562	72905	56420	69994	98872	31016	71194	18738	44013	48840	63213	21069	10634	12952
8	96301	91977	05463	07972	18876	20922	94595	56869	69014	60045	18425	84903	42508	32307
9	89579	14342	63661	10281	17453	18103	57740	84378	25331	12566	58678	44947	05585	56941
10	85475	36857	53342	53988	53060	59533	38867	62300	08158	17983	16439	11458	18593	64952
11	28918	68578	88231	33276	70997	79936	56865	05859	90106	31595	01547	85590	91610	78188
12	63553	40961	48235	03427	49626	69445	18663	72695	52180	20847	12234	90511	33703	90322
13	09429	93969	52636	92737	88974	33488	36320	17617	30015	08272	84115	27156	30613	74952
14	10365	61129	87529	85689	48237	52267	67689	93394	01511	26358	85104	20285	29975	89868
15	07119	97336	71048	08178	77233	13916	47564	81056	97735	85977	29372	74461	28551	90707
16	51085	12765	51821	51259	77452	16308	60756	92144	49442	53900	70960	63990	75601	40719
17	02368	21382	52404	60268	89368	19885	55322	44819	01188	65255	64835	44919	05944	55157
18	01011	54092	33362	94904	31273	04146	18594	29852	71585	85030	51132	01915	92747	64951
19	52162	53916	46369	58586	23216	14513	83149	98736	23495	64350	94738	17752	35156	35749
20	07056	97628	33787	09998	42698	06691	76988	13602	51851	46104	88916	19509	25625	58104
21	48663	91245	85828	14346	09172	30168	90229	04734	59193	22178	30421	61666	99904	32812
22	54164	58492	22421	74103	47070	25306	76468	26384	58151	06646	21524	15227	96909	44592
23	32639	32363	05597	24200	13363	38005	94342	28728	35806	06912	17012	64161	18296	22851
24	29334	27001	87637	87308	58731	00256	45834	15398	46557	41135	10367	07684	36188	18510
25	02488	33062	28834	07351	19731	92420	60952	61280	50001	67658	32586	86679	50720	94953
26	81525	72295	04839	96423	24878	82651	66566	14778	76797	14780	13300	87074	79666	95725
27	29676	20591	68086	26432	46901	20849	89768	81536	86645	12659	92259	57102	80428	25280
28	00742	57392	39064	66432	84673	40027	32832	61362	98947	96067	64760	64584	96096	98253
29	05366	04213	25669	26422	44407	44048	37937	63904	45766	66134	75470	66520	34693	90449
30	91921	26418	64117	94305	26766	25940	39972	22209	71500	64568	91402	42416	07844	69618

the United States. You want to draw a random sample of 10 individuals from this population of 5,000. Your first task is to assign each individual in the population a number from 0 to 4,999. Once you have assigned each professor a number, you are ready to use the list of random numbers to select your sample of 10. The first step is to block off the list of random numbers into columns of four, because you need four digits to represent the number for each professor. The first professor in the population would be represented by the number 0000, the second by the number 0001, and so forth.

To randomly select the 10 subjects, you would read down the first four columns until you encountered a number less than 5,000, since only 5,000 individuals exist in the population. The professor corresponding to the first number less than 5,000 would represent the first individual to be selected for the sample. You would then continue selecting numbers until you had selected 10 different numbers less than 5,000; the individuals corresponding to these numbers would represent your sample of 10 professors. In taking this random sample you must skip numbers greater than 4,999, and if a number is repeated you must also skip it. Using this procedure you would select the following 10 numbers, and the individuals corresponding to these numbers would represent the sample of 10 individuals.

1048	2891
2236	0942
2413	1036
4216	0711
3757	0236

subjects randomly from a population. For example, just think of the difficulty of randomly selecting a sample of college professors from all professors within the United States. This would be not simply difficult to do but next to impossible—unless you were independently wealthy or had a research grant to either go to all the various universities to test these subjects or have them come to your university. Consequently, random selection of subjects from a population is an ideal that is seldom achieved. Fortunately, random selection of subjects from a population is not the crucial element needed to achieve control over the influence of extraneous variables. Random assignment of the subjects to treatment conditions is essential. Random selection of subjects provides assurance that your sample is *representative* of the population from which it was drawn. It therefore has implications for generalization of the results of the experiment back to the population. Random assignment provides assurance that the extraneous variables are controlled.

The key word in this whole process of selecting and assigning subjects is *random.* The term *random* "may be used in

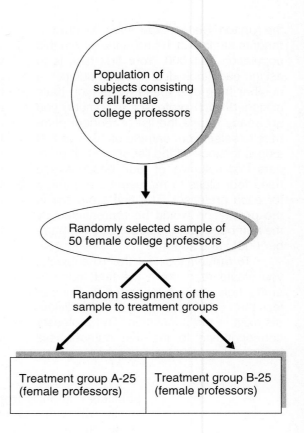

FIGURE 8.1   Illustration of the ideal procedure for obtaining subjects for an experiment.

a theoretical sense to refer to an assumption about the equiprobability of events. Thus, a random sample is one such that every member of the population has an equal probability of being included in it" (Ferguson, 1966, p. 133). In a random selection of a sample of one hundred subjects from a population of, say, college freshmen, every freshman has an equal chance of being included in the sample of one hundred. In like manner, random assignment of subjects to the experimental groups assures that each sample subject has an equal opportunity of being assigned to each group.

In order to provide equiprobability of events when randomly assigning subjects to treatment conditions, it is necessary to use a randomization procedure, like the one presented in Exhibit 8.2. When such a procedure is used, maximum assurance is provided that any systematic bias will be eliminated from the experimental results. This is because the ran-

---

EXHIBIT 8.2   Procedure for Randomly Assigning Subjects to Experimental Treatment

---

The most popular procedure for randomly assigning subjects to experimental treatment conditions is to use a list of random numbers, such as the following list of 200 numbers (Larger lists are contained in the appendixes of most statistics books.)

This list consists of a series of twenty rows and ten columns. The number in each position is random because each of the numbers from 0 to 9 had an equal chance of occupying that position and the selection of one number for a given position had no influence in the selection of another number for another position. Therefore, since each individual number is random, any combination of the numbers must be random.

Assume that you have fifteen subjects and you want to randomly assign them to three experimental treatment

groups. First, you would give each subject a number from 0 to 14. You would then block the list of random numbers into columns of two, to provide five pairs of columns, since two columns are necessary to represent the total sample of subjects.

Now you are ready to randomly assign the five subjects to each of the three treatment groups. The procedure that is usually followed is to randomly select the first five subjects from the group of fifteen and assign them to one treatment group. Then randomly select a second group of five subjects from the group of fifteen and assign them to another treatment group. Once these ten subjects have been randomly selected and assigned, only five subjects remain; these five subjects are assigned to the third treatment group.

	1	2	3	4	5	6	7	8	9	10
1	8	1	4	5	5	6	9	8	7	3
2	2	7	9	6	5	4	6	4	8	3
3	0	0	0	5	5	8	9	7	6	9
4	7	8	3	4	7	0	7	7	5	2
5	8	5	8	6	3	5	4	2	2	2
6	7	3	5	3	6	8	0	7	3	3
7	1	8	6	0	1	0	7	4	4	7
8	7	9	5	3	0	1	5	5	5	1
9	5	6	6	7	8	5	8	1	1	9
10	3	0	3	3	9	1	9	9	1	9
11	9	7	4	7	8	4	7	1	0	9
12	5	6	4	5	1	4	5	4	1	1
13	5	7	4	0	4	2	5	9	6	7
14	8	6	0	5	6	9	4	4	3	2
15	6	7	6	7	3	3	7	1	8	9
16	2	6	0	6	7	3	3	0	6	9
17	6	7	5	5	1	4	7	4	1	2
18	6	3	0	9	9	9	5	3	8	0
19	0	3	7	3	0	3	0	6	8	6
20	7	1	6	8	2	0	5	3	2	1

EXHIBIT 8.2   *(continued)*

To randomly select the first subject for the first group, read down the first two columns until you encounter a number less than 15. From the above list, we find that the first such number is 00. Consequently, the first randomly selected subject is the subject with the number 0. Proceed down the columns until you encounter the second number less than 15, which is 03. Subject number 3 represents the second randomly selected subject. Once you have reached the bottom of the first two columns, start at the top of the next two columns. With this procedure, the subject numbers 05, 06, and 09 are selected, which represent the remaining three of the first five randomly selected subjects. Note that if you encounter a number that has already been selected (as we did with the number 05), you must disregard it.

To randomly select the second group of five subjects, proceed down the columns and identify numbers less than 15 that have not already been chosen. Using this procedure, we find the numbers 10, 01, 14, 07, and 11. These numbers correspond to the second group of randomly selected subjects. The third group represents the remaining subjects.

We now have the following three randomly selected groups of subjects.

00	01	02
03	07	04
05	10	08
06	11	12
09	14	13

Once each of the three groups has been randomly selected, it must be randomly assigned to one of the three experimental treatment conditions. This is accomplished by using only one column of the table of random numbers, since there are only three groups of subjects. The three groups are numbered from 0 to 2. Proceed down the first column until you reach the first of these three numbers. In looking at column 1, you can see that the first number is 2. Consequently, group 2 (the third group of subjects) is assigned to the first treatment condition. The second number encountered is 0, so group 0 (the first group of subjects) is assigned to the second treatment condition. This means that group 1 (the second group of subjects) is assigned to the third treatment condition. Now we have randomly selected three groups of subjects and randomly assigned them to three treatment conditions.

Treatment Condition

A_1	A_2	A_3
Group 2	Group 0	Group 1

dom selection of subjects and the random assignment of subjects to treatment conditions are assumed to result in random distribution of all extraneous variables. Consequently, the distribution and influence of the extraneous variables should be about the same in all groups of subjects.

Consider the following example using only random assignment of subjects. Professor X was conducting a study on learning. Intelligence naturally is correlated with learning ability, so this factor must be controlled for, or held constant. Let us consider two possibilities—one that provides for the needed control through the use of random assignment and one that does not. Assume first that no random assignment of subjects existed (no control) but that the first ten subjects who showed up for the experiment were assigned to treatment Group A and the second ten subjects were assigned to treatment Group B. Assume further that the results of the experiment revealed that treatment Group B learned significantly faster than treatment Group A. Is this difference caused by the different experimental treatment conditions that were administered to the two groups or by the fact that the subjects in Group B *may* have been more intelligent than those in Group A? Suppose that the investigator also considers the intelligence factor to be a possible confounding variable and therefore gives all subjects an intelligence test. The left-hand side of Table 8.1 depicts the hypothetical distribution of IQ scores of these twenty subjects. From this table, you can see that the mean IQ score of the people in Group B is 10.6 points higher than that of those in Group A. Intelligence is, therefore, a potentially confounding variable and serves as a rival hypothesis for explaining the observed performance difference in the two groups. To state that the treatment conditions produced the observed effect,

TABLE 8.1 Hypothetical Distribution of Twenty Subjects' IQ Scores

Group Assignment Based on Arrival Sequence				*Random Assignment of Subjects to Groups*			
Group A		*Group B*		*Group A*		*Group B*	
Subjects	*IQ Scores*	*Subjects*	*IQ Scores*	*Subjects*	*IQ Scores*	*Subjects*	*IQ Scores*
1	97	11	100	1	97	3	100
2	97	12	108	2	97	4	103
3	100	13	110	11	100	6	108
4	103	14	113	5	105	12	108
5	105	15	117	13	110	7	109
6	108	16	119	9	113	8	111
7	109	17	120	15	117	14	113
8	111	18	122	10	118	16	119
9	113	19	128	19	128	17	120
10	118	20	130	20	130	18	122
Mean IQ score	106.1		116.7		111.5		111.3

Mean difference between the two groups: 10.6

Mean difference between the two groups: 0.2

researchers must control for potentially confounding variables such as the IQ difference.

One means of eliminating such a bias would have been to randomly assign the twenty subjects to the two treatment groups as they showed up for the experiment. The right-hand side of Table 8.1 depicts the random distribution of the twenty subjects and their corresponding hypothetical IQ scores. Now note that the mean IQ scores for the two groups are very similar. There is only a 0.2 point IQ difference as opposed to the prior 10.6 point difference. For the mean IQ scores to be so similar, both groups of subjects had to have a similar distribution of IQ scores, the effect of which is to control for the potential biasing effect of IQ. The IQ scores in Table 8.1 have been rank ordered to show this similar distribution.

Random assignment produces control by virtue of the fact that the variables to be controlled are distributed in approximately the same manner in all groups (ideally the distribution would be exactly the same). When the distribution is approximately equal, the influence of the extraneous variables is held constant, because they cannot exert any differential influence on the dependent variable. Does this mean that randomization will *always* result in equal distribution of the variables to be controlled? The control function of randomization stems from the fact that random selection and assignment of subjects also results in the random selection and assignment of most extraneous variables. Since every subject, and therefore the extraneous variables present, had an equal chance of being selected and then assigned to a particular group, the extraneous variables to be controlled are distributed randomly. But because chance determines the distribution of the extraneous variables, it is also possible that, by chance, these variables are not equally distributed among the various groups of subjects. In other words, bias can still exist when one uses the randomization procedure. The smaller the number of subjects, the greater the risk that this will happen. However, randomization still decreases the probability of creating a biased distribution, even if one has access only to a small group of subjects. Since the probability of the groups' being equal is so much greater with randomization, it is an extremely powerful method for controlling extraneous variables. And since it is really the *only* method for controlling unknown variables, it is necessary to randomize whenever and wherever possible, even when another control technique is being used.

There are, however, several extraneous variables that are not controlled for by randomization, including the subject effect and the experimenter effect. The potential influences of the subjects' motive of positive self-presentation and the experimenter's expectancies or attributes are not randomly distributed. Instead, these potential biasing effects are a function of how subjects perceive the experiment or the expectations researchers have regarding the outcome of the experiment. Consequently, these extraneous variables must be controlled by the use of techniques other than randomization.

MATCHING

Matching
Using any of a variety of techniques for equating subjects on one or more variables

Although randomization does provide the best guard against interpreting differences in the dependent variable as being the result of variables other than the independent variable, it is not the best technique for increasing the sensitivity of the experiment. In any study, it is desirable to demonstrate the influence of the independent variable, regardless of how small its effect may be. Suppose we want to isolate the potential effect of televised aggression on children's behavior. Assume that the effect is one of increasing aggressive behavior in children (this has been found in a number of studies) but the amount of increase is small. In order to isolate and detect this small effect, we need to construct an experiment that will be as sensitive as possible. The sensitivity of an experiment can be increased by **matching** the subjects in the various experimental treatment groups. An explanation of how matching accomplishes this requires a discussion of the way in which statistical techniques operate, which is beyond the scope of the text. For our purposes, you need only remember that one of the benefits of matching is that the sensitivity of the experiment is increased. A second benefit of matching is that the variables on which subjects are matched are controlled in the sense that constancy of influence is attained. If subjects in all treatment conditions are matched on intelligence, then the intelligence level of the subjects is held constant and therefore controlled for all groups.

Here we have two definite benefits that can accrue from matching. It is important to remember, however, that matching is no substitute for randomization. Randomization should still be incorporated whenever possible, because one cannot attain an exact match on most variables, and it is impossible to identify and match on all variables that could affect the results of the experiment.

The sections that follow present a number of ways in which matching can be accomplished.

Matching by Holding Variables Constant

One technique that can be used to increase the sensitivity of the experiment and control an extraneous variable is to hold the extraneous variable constant for all experimental groups. This means that all subjects in each experimental group will have the same degree or type of extraneous variable. If we are studying conformity, then sex of subjects needs to be controlled, because conformity has been shown to vary with the sex of the subject. As illustrated in Figure 8.2, the sex variable can be controlled simply by using only male subjects in the experiment. This has the effect of matching all subjects in terms of the sex variable, so that the sensitivity of the experiment is increased. Hauri and Ohmstead (1983) used only insomniacs in their investigation of estimates of the length of time required to fall asleep. This matching procedure creates a more homogeneous subject sample, because only subjects with a certain amount or type of the extraneous variable are included in the subject pool.

Although widely used, the technique of holding variables constant is not without its disadvantages. Two can readily be identified. The first disadvantage is that the technique restricts the size of the subject population. Consequently, in some cases, it may be difficult to find enough subjects to participate in the study. Consider a study that was conducted to investigate the influence of assistance given to single parents (assistance with child care or with household chores) on their attitudes and perceived interactions with their child or children. The study was limited to single parents, and researchers had to find volunteers. After two weeks of advertising, eighteen single parents had volunteered to participate in the study. If the study had not been limited to single parents, the subject pool from which researchers could have drawn would have been much larger, with the probable effect that more individuals would have volunteered their help.

The second disadvantage is more serious. The results of the study can be generalized only to the type of subject who

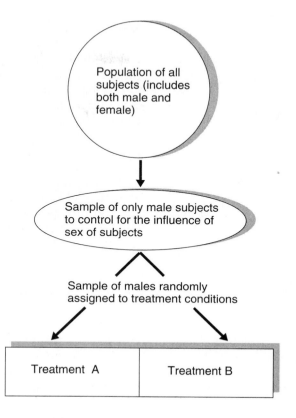

FIGURE 8.2 Illustration of matching by holding variables constant.

participated in the study. The results obtained from the single-parent study can be generalized only to other single parents. If someone wanted to know whether two-parent families would derive the same benefit from receiving this type of assistance, he or she would have to conduct a similar study using two-parent families. Conclusions from such a study might indeed be the same as those obtained from the single-parent study, but this is an empirical question. The only way we can find out if the results of one study can be generalized to individuals of another population is to conduct an identical study using representatives of the second population as subjects.

Matching by Building the Extraneous Variable into the Research Design

A second means of increasing the sensitivity of an experiment is to build the extraneous variable into the research design.

Assume that we were conducting a learning experiment and wanted to control for the effects of intelligence. Also assume that we had considered the previous technique of holding the variable constant by selecting only individuals with IQs of 110 to 120, but thought it unwise and inexpedient to do so. In this case, we could select several IQ levels (for example, 90 to 99, 100 to 109, and 110 to 120), as illustrated in Figure 8.3, and treat them as we would an independent variable. This would allow us to identify and extract the influence of the intelligence variable. Intelligence, therefore, would not represent a source of random fluctuation, and the sensitivity of the experiment would be increased.

To provide further insight into this control technique, I will cite a study conducted by Kendler, Kendler, and Learnard (1962) that investigated the influence of the subjects' age on

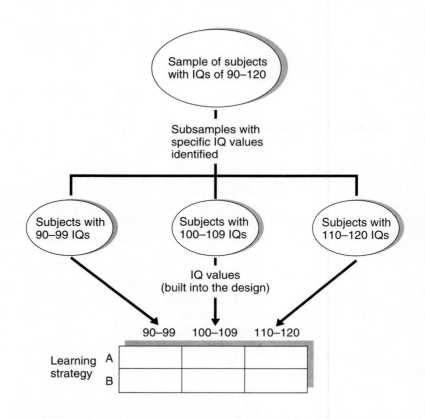

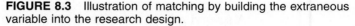

FIGURE 8.3 Illustration of matching by building the extraneous variable into the research design.

their use of internal mediating responses. Prior research conducted on this topic indicated that rats do not use mediational processes whereas college students do, and that three- and four-year-old children do not use the mediational process whereas nearly half of the children between ages five and seven do. Such data suggested that a developmental process was involved in the use of internal mediating responses, which must be controlled for in order to avoid attaining contradictory results from various studies. Realizing this, Kendler, Kendler, and Learnard controlled for age by building it into the design of their study. Children at five chronological age levels—three, four, six, eight, and ten years of age—were required to engage in a task that would elicit mediational responses. Thus the researchers controlled for the age factor by matching the five different age groups, and then used these age groupings as an independent variable. In this way, the age variable was controlled and its influence on mediational responses was displayed.

Building the extraneous variable into the research design seems like an excellent technique for achieving control and increasing sensitivity. But the technique is recommended only if one is interested in the differences produced by the various levels of the extraneous variable or in the interaction between the levels of the extraneous variable and other independent variables. In the hypothetical learning experiment, one might be interested in the differences produced by the three levels of intelligence and how these levels interact with the learning strategies. The primary reason Kendler, Kendler, and Learnard conducted their study was to investigate the differences produced by subjects of different ages. If they had not been interested in such conditions, then another control technique would probably have been more efficient. When such conditions are of interest, the technique is excellent because it isolates the variation caused by the extraneous variable. This control technique takes a factor that can operate as an extraneous variable, biasing the experiment, and makes it focal in the experiment as an independent variable.

Matching by Yoked Control

Yoked control
A matching technique that matches subjects on the basis of the temporal sequence of administering an event

The **yoked control** matching technique controls for the possible influence of the temporal relationship between an event and a response. Consider the widely quoted study conducted by Brady (1958) in which he investigated the relationship between emotional stress and development of ulcers. Brady trained monkeys to press a lever at least once during every 20-second interval to avoid receiving electric shock. The monkeys learned this task quite rapidly, and only occasionally would they miss a 20-second interval and receive a shock. In order to determine whether the monkeys developed ulcers from the psychological stress rather than the physical stress resulting from the cumulative effect of the shocks, Brady had to include a control monkey that would receive an equal number of shocks. This was easily accomplished, but there was still one additional variable that needed to be controlled—the temporal sequence of administering the shocks. It may be that one temporal sequence produces ulcers whereas another does not. If the experimental and the control monkeys received a different temporal sequence of shocks, this difference and not the stress variable could be the cause of the ulcers. Consequently, both monkeys had to receive the same temporal sequence to control this variable. Brady placed the experimental and the control monkeys in yoked chain, whereby both monkeys would receive shock when the experimental monkey failed to press the lever during the 20-second interval. However, the control animal could not influence the situation and essentially had to sit back and accept the fact that sometimes the shock was going to occur. The only apparent difference between these animals was the ability to influence the occurrence of the shock. If only the experimental monkey got ulcers, as was the case in this experiment, the ulcers could be attributed to the psychological stress.

The yoked control technique appears to be an excellent way to control the biasing effects of the temporal distribution of events, but there is some controversy over its effectiveness. Church (1964) believes that this control technique may introduce a source of bias in the results of a study; however, Kimmel and Terrant (1968) believe that Church bases his arguments on unwarranted assumptions. In spite of Kimmel and Terrant's arguments, Church does seem to be correct in some instances. This means that, in some instances, the yoked control technique is appropriate for controlling the influence of the temporal relationship between an event and a response. In other instances it may introduce rather than eliminate error. Therefore, before using this control technique it would be advisable to refer to both Church (1964) and Kimmel and Terrant (1968) to determine whether the yoked control technique is appropriate for your purpose.

Matching by Equating Subjects

A third technique for controlling extraneous variables and also increasing the sensitivity of the experiment is to equate subjects on the variable or variables to be controlled. If intelligence needs to be controlled, then the investigator must make sure that the subjects in each of the treatment groups are of the same intelligence level.

Matching by equating subjects is very similar to matching by building the extraneous variable into the study design: both techniques attempt to eliminate the influence of the extraneous variable by creating equivalent groups of subjects. The difference lies in the procedure for creating the equivalent groups. The previously discussed method creates equivalent groups by establishing categories of the extraneous variable into which subjects are placed, thereby creating another independent variable. The present method does not build the extraneous variable into the design of the study but matches subjects on the variable to be controlled, where the number of subjects is always some multiple of the number of levels of the independent variable. There are two techniques that are commonly used to accomplish this matching, which Selltiz et al. (1959) labeled the precision control technique and the frequency control technique.

Precision control
A matching technique in which each subject is matched with another subject on selected variables

Precision Control The technique of **precision control** requires the investigator to match subjects in the various treatment groups on a case-by-case basis for each of the selected extraneous variables. Scholtz (1973) investigated the defense styles used by individuals who attempted suicide versus those used by individuals who did not attempt suicide. All subjects were neuropsychiatric patients. The suicide subjects were identified as those individuals who, among other things, had attempted suicide during the past year. The other subjects had evidenced "no history of a suicide attempt nor marked suicidal ideation" (p. 71). For a non–suicide attempter to be included in the study, the subject had to be of the same age, sex, race, marital status, diagnosis, and education as a suicide attempter. Matching on these variables on a case-by-case basis resulted in thirty-five pairs of subjects.

The Scholtz study illustrates the various advantages and disadvantages of the precision control matching technique. Before discussing them, I should point out that the Scholtz study was an ex post facto study, since the subjects assigned themselves to the various groups; they could not be randomly assigned after being paired. In a truly experimental study, subjects would be matched and then randomly assigned to the different groups, as illustrated in Figure 8.4. As stated before, matching is never a substitute for random assignment.

The principal advantage of the precision control technique is that it increases the sensitivity of the study by ensuring that the subjects in the various groups are equal on at least the paired variables. If sensitivity is to be increased, the variables on which subjects are matched must be correlated with the dependent variable. How much of a correlation should exist? Kerlinger (1973) states that matching is a waste of time unless the variables on which subjects are matched correlate greater than 0.5 or 0.6 with the dependent variable (this criterion holds only for linearly related variables). This corresponds to the data Billewicz (1965) obtained from his simulation experiments.

The precision control technique has three major disadvantages. First, it is difficult to know which are the most

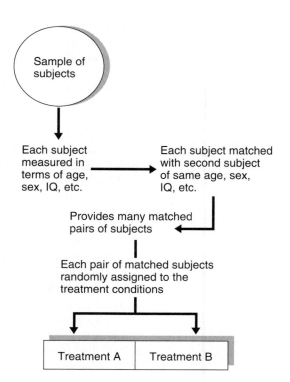

FIGURE 8.4   Illustration of matching by the precision control technique.

important variables to match. In most instances, there are many potentially relevant variables. In his study, Scholtz selected age, sex, race, marital status, diagnosis, and education, but many other variables could have been selected. Variables selected should be those that show the lowest intercorrelation but the highest correlation with the dependent variable.

A second problem encountered in precision control matching is that the difficulty in finding matched subjects increases disproportionately as the number of variables increases. Scholtz matched on six variables, which must have been very difficult. His task would have been much easier if matching had been attempted on only two variables, such as sex and age. In order to match individuals on many variables, one must have a large pool of subjects available in order to obtain a few who are matched on the relevant variables. Fortunately, the relevant variables are generally intercorrelated, so the number that can be used successfully to increase precision is limited. Matching also limits the generality of the results of the study. Assume that you are matching on age and education and that the subjects in your final sample of matched subjects are between the ages of twenty and thirty and have only high school educations. Since this is the type of subject included in the study, you can generalize the results only to other individuals having the same characteristics.

A third disadvantage is that some variables are very difficult to match. If having received psychotherapy was considered a relevant variable, an individual who had received psychotherapy would have to be matched with another person who had also received psychotherapy. A related difficulty is the inability to obtain adequate measures of the variables to be matched. If we wanted to equate individuals on the basis of the effect of psychotherapy, we would have to measure such an effect. Matching can only be as accurate as available measurement.

Frequency Distribution Control The precision control technique of matching is excellent for increasing sensitivity, but many subjects must be eliminated because they cannot be matched. **Frequency distribution control** attempts to overcome this disadvantage while retaining some of the advantages of matching. This technique, as the name implies, matches groups of subjects in terms of overall distribution of the selected variable or variables rather than on a case-by-case basis. If IQ were to be matched in this fashion, the two or

Frequency distribution control
A matching technique that matches groups of subjects by equating the overall distribution of the chosen variable

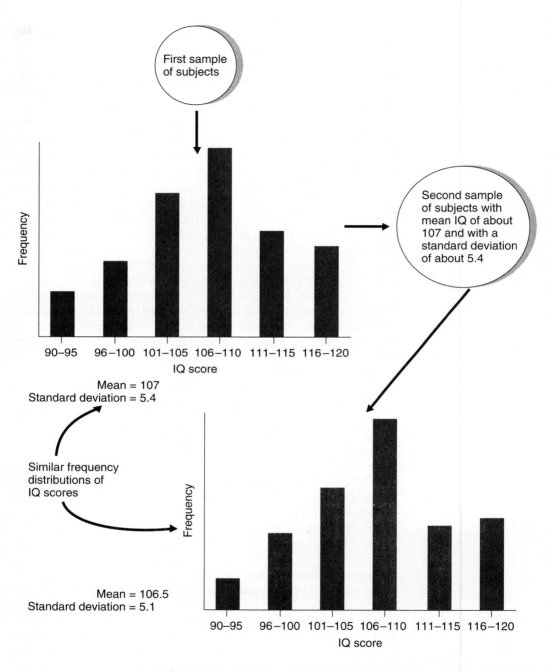

FIGURE 8.5   Illustration of the frequency distribution control technique.

more groups of subjects would have to have the same average IQ, as well as the same standard deviation and skewness of IQ scores, as illustrated in Figure 8.5. This means that, generally speaking, the investigator would select the first group of subjects and determine the mean, standard deviation, and so forth of their IQ scores. Then another group having the same statistical measures would be selected. If more than one variable was considered to be a relevant variable on which to match subjects, the groups of subjects would have to have the same statistical measures on both of these variables. The number of subjects lost using this technique would not be as great as the number lost using the precision control method, because each additional subject would merely have to contribute to producing the appropriate statistical indexes rather than be identical to another subject on the relevant variables. Consequently, this technique allows more flexibility in terms of being able to use a particular subject.

The major disadvantage of matching by the frequency distribution control method is that the combinations of variables may be mismatched in the various groups. If age and IQ were to be matched, one group might include old subjects with high IQs and young subjects with low IQs, whereas the other group might be composed of the opposite combination. In this case, the mean and distribution of the two variables would be equivalent but the subjects in each group would be completely different. This disadvantage obviously exists only if matching is conducted on more than one variable.

COUNTERBALANCING

Counterbalancing
A technique used to control sequencing effects

Order effect
A sequencing effect arising from the order in which the treatment conditions are administered to subjects

Counterbalancing is the technique used to control for sequencing effects. Sequencing effects can occur when the investigator elects to construct an experiment in which all subjects serve in each of several experimental conditions. (See Figure 8.6.) Under these conditions, there are two types of effects that can occur. The first is an **order effect,** which arises from the order in which the treatment conditions are administered to the subjects. Suppose that you are conducting a verbal learning experiment in which the independent variable is rate of presentation of nonsense syllables. Nonsense syllables with a 50

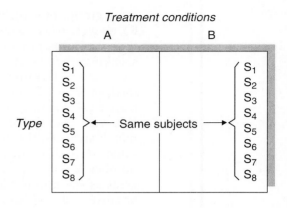

FIGURE 8.6   Illustration of the type of design that may include sequencing effects.

percent level of meaningfulness are randomly assigned to three lists. The subject has to learn sequentially list S (the slow list, in which the syllables are presented at 6-second intervals), then list M (the moderate list, in which the syllables are presented at 4-second intervals), and finally list F (the fast list, in which the syllables are presented at 2-second intervals). In such an experiment, there is the possibility that practice with the equipment, learning the nonsense syllables, or just general familiarity with the surroundings of the experimental environment may enhance performance. Let us assume that one or more of these variables does enhance performance and that the increment due to the order effect is four units of performance for subjects progressing from list S to M and two units of performance for subjects progressing from list M to list F. The left half of Table 8.2 depicts these order effects. As you can see, order effects could affect the conclusions reached, because performance increments occurred in the learning of these two lists that were entirely the result of order effects. You should also be aware that when the increment in performance is caused by order effects, the particular sequence of the list is irrelevant. If the order of the lists were reversed, the increments in performance would still occur in the same ordinal position, as is shown in the right half of Table 8.2. Increments due to order effects are strictly the result of subjects' increased familiarity and practice with the whole experimental environment. Other experimental factors, such as the time of testing (morning, noon, or night), may produce the order effect. Such effects must be controlled to avoid reaching false conclusions.

TABLE 8.2 Hypothetical Order Effects

	List Learned			Reversed Order of List		
	S	M	F	F	M	S
Increment in performance	0	4	2	0	4	2

Carry-over effect
A sequencing effect that occurs when performance in one treatment condition affects performance in another treatment condition

The second type of sequencing effect that can occur is a carry-over effect. A **carry-over effect** occurs when performance in one treatment condition is partially dependent on the conditions that precede it. D'Amato (1970) provides an excellent example of carry-over effects in his simulated experiment designed to investigate the influence of monetary reward (5, 10, or 15 cents) on performance. In this type of study, it is possible that when subjects serve in all three conditions, performance in a particular treatment condition may be partially a function of the conditions that precede it (a dime may be more rewarding when it is preceded by 5 cents than when it is preceded by 15 cents). "Let us simplify the analysis by assuming that the carry-over effects from one condition to another will be directly proportional to the difference in the monetary rewards of the two conditions. We will assume that going from A (5 cents) to B (10 cents) or from B to C (15 cents) results in a positive carry-over (increment in performance) of two units, whereas traveling in the reverse direction results in the same amount of carry-over effect but negative in sign, i.e., leads to a decrement in performance of two units. Transitions from A to C and from C to A both result in four units of carry-over effects, positive and negative, respectively" (p. 53). Table 8.3 illustrates that the carry-over effects for any one treatment condition are a function of the preceding treatment conditions. Such effects need to be controlled to identify unambiguously the effects due to the independent variable.

The order effects and the carry-over effects are potential sources of bias in any studies in which the subject partakes of several treatment conditions. In such cases, the sequencing effects need to be controlled, and researchers often resort to counterbalancing. Counterbalancing techniques have disadvantages associated with statistical analysis; students having a good statistical background are referred to Gaito (1958, 1961) for a discussion of these dangers.

TABLE 8.3 Calculation of Assumed Carry-over Effects in Six Sequences

Sequence	Value of the Independent Variable		
	A (5 cents)	B (10 cents)	C (15 cents)
ABC	0	2	2
ACB	0	−2	4
BAC	−2	0	4
BCA	−4	0	2
CAB	−4	2	0
CBA	−2	−2	0
Total	−12	0	12

Source: From *Experimental Psychology Methodology: Psychophysics and Learning* by M. R. D'Amato. Copyright 1970, McGraw-Hill. Used with permission of McGraw-Hill Book Co.

Intrasubject Counterbalancing: The ABBA Technique

Intrasubject counterbalancing Administering the treatment conditions to each individual subject in more than one order

Intrasubject counterbalancing controls for sequencing effects by having each subject take the treatment conditions first in one order and then in the reverse order. Suppose that you were conducting a Pepsi challenge experiment to find out if people preferred Pepsi over Coke. In this experiment, the treatment conditions would consist of the two colas. Subjects would make an assessment of liking after tasting first the Pepsi (cola A) and then the Coke (cola B), in AB order. Subjects would then taste the Coke (cola B) a second time, followed by the Pepsi (cola A), and make a liking assessment after each, making the sequence ABBA. In other words, each subject would taste each cola drink twice and make an assessment of liking after each tasting. The results of the liking assessment obtained from the two Pepsi tastings would be combined for each subject, as would the results of the two Coke tastings, once again making the study a two-treatment-condition experiment. The Pepsi liking assessment would then be compared to the Coke liking assessment to determine whether one cola was liked more than the other. Any observed difference would not be attributable to carry-over or order effects, because they would have been equalized, or held constant, across groups.

To illustrate how intrasubject counterbalancing controls for carry-over and order effects, let us assume that each subject increments his or her liking assessment by one unit for each treatment condition in which he or she participates solely because of sequencing effects. If the ABBA technique is employed, these sequencing effects will be constant across treatment conditions and will therefore be controlled. This constant influence is illustrated in the top half of Table 8.4. For both the A and the B treatment condition, liking was increased by a constant amount of three units. Therefore, sequencing was controlled. Note, however, that the sequencing effect was linear in the sense that a constant increment was added to liking in each successive position in the sequence.

Would the ABBA technique control for carry-over and order effects if they were not linear? The answer is no. The ABBA method is based on the assumption that the sequencing effects are linear, or constant for each successive position in the sequence. (This assumption is also made for the incomplete counterbalancing method discussed later.) If a constant effect is not attained, the sequencing effect will differentially affect the results, as shown in the bottom half of Table 8.4. In this case, the sequencing effects were not controlled, because liking increased by ten units for condition B but only eight units for condition A. This is because the sequence effect was twice as powerful for progression from the first A condition to the first B condition as it was for progression through the remainder of the conditions. Can such differential sequence effects be controlled? The answer depends on whether you are considering carry-over effects or order effects.

Differential order effects can be held constant by having each treatment condition appear in every possible position in the sequence. This means that in addition to an ABBA se-

TABLE 8.4   Sequencing Effects for the ABBA Technique

	Treatment Condition			
	A	*B*	*B*	*A*
Linear sequencing effect				
Sequence effect	0	1	2	3
A sequence effect 0 + 3 = 3				
B sequence effect 1 + 2 = 3				
Nonlinear sequencing effect				
Sequence effect	0	4	6	8
A sequence effect 0 + 8 = 8				
B sequence effect 4 + 6 = 10				

TABLE 8.5   Control for Order Effects Using Intrasubject
Counterbalancing

	Sequence I				Sequence II			
	A	B	B	A	B	A	A	B
Order effect	0	2	3	4	0	2	3	4
Total A order effect $0 + 4 + 2 + 3 = 9$								
Total B order effect $2 + 3 + 0 + 4 = 9$								

quence a BAAB sequence must be included to control nonlinear order effects. Half of the subjects can then be assigned
to each sequence. Let us assume that each subject increments
two units of liking after tasting the first cola and one unit of
liking after tasting each subsequent cola just because of order
effects. If both the ABBA and the BAAB sequences are employed, the results in Table 8.5 will occur. The total order
effects for both treatment conditions are equal, which means
that the effect is held constant. This actually represents a
combining of intrasubject with intragroup counterbalancing.

It is not as easy to control for differential carry-over effects. These effects frequently defy control because the carry-
over may vary as a function of the preceding treatment
conditions. Such a condition was illustrated earlier in the example of a simulated experiment designed to test the influence
of monetary reward on performance. Table 8.3 shows the assumed carry-over effects. Note that the carry-over effect for
any one treatment condition varies as a function of the particular treatment conditions that precede it. Also note that the
total carry-over effects for the treatment conditions are not
identical. Here, then, is a case in which carry-over effects are
not controlled. When carry-over effects are linear, they can be
controlled by the ABBA sequence, but in nonlinear cases such
as this, they cannot. The investigator who suspects such a
situation should consider using some other technique, such as
precision control; otherwise, the carry-over effects serve as a
rival hypothesis.

Intragroup Counterbalancing

A primary disadvantage of using the intrasubject counterbalancing technique is that each treatment condition must be

presented to each subject more than once. As the number of treatment conditions increases, the length of the sequence of conditions each subject must take also increases. For example, with the three treatment conditions A, B, and C, each subject must take a sequence of six treatment conditions—ABCCBA. Intragroup counterbalancing allows the experimenter to avoid this time-consuming process. **Intragroup counterbalancing** differs from intrasubject counterbalancing in that groups of subjects rather than individuals are counterbalanced. Because the intragroup technique attempts to control sequencing effects over groups, it represents a more efficient technique, particularly when more than two treatment conditions exist.

Intragroup counterbalancing Administering the treatment conditions to various members of each group of subjects in more than one order

Incomplete Counterbalancing The intragroup counterbalancing technique used most frequently is **incomplete counterbalancing,** which derives its name from the fact that all possible sequences of treatment conditions are not enumerated. The first criterion that incomplete counterbalancing must meet is that, for the sequences enumerated, each treatment condition must appear an equal number of times in each ordinal position. Also, each treatment condition must precede and be followed by every other condition an equal number of times.

Incomplete counterbalancing Enumerating fewer than all possible sequences, and requiring different groups of subjects to take each of the enumerated sequences

Assume that you are conducting an experiment to determine whether caffeine affects reaction time. You want to administer 100, 200, 300, and 400 mg of caffeine (conditions A, B, C, and D, respectively) to subjects to see whether reaction time increases as the amount of caffeine consumed increases. You know that if each subject takes all four doses of caffeine, sequencing effects could alter the results of your experiment, so you want to counterbalance the order in which the dosages are administered to the subjects. Whenever the number of treatment conditions is even, as is the case with the four caffeine dosages, then the number of counterbalanced sequences equals the number of treatment conditions. The sequences are established in the following way. The first sequence takes the form 1, 2, n, 3, $(n - 1)$, 4, $(n - 2)$, 5, and so forth, until we have accounted for the total number of treatment conditions. In the case of the caffeine study, with four treatment conditions, the first sequence would be ABDC, or 1, 2, 4, 3. If an experiment consisted of six treatment conditions, the first sequence would be ABFCED, or 1, 2, 6, 3, 5, 4. The remaining sequences of the incomplete counterbalancing tech-

nique are then established by incrementing each value in the
preceding sequence by 1. For example, for the caffeine study,
where the first sequence is ABDC, the second sequence would
be BCAD. Naturally, to increment the last treatment condition, D, by 1, you do not proceed to E but go back to A. This
procedure results in the following set of sequences for the
caffeine study.

Subject	Sequence			
1	A	B	D	C
2	B	C	A	D
3	C	D	B	A
4	D	A	C	B

If the number of treatment conditions is odd, as with five
treatment conditions, the criterion that each value must precede and follow every other value an equal number of times is
not fulfilled if the above procedure is followed. For example,
the foregoing procedure would give the following set of sequences:

Sequence				
A	B	E	C	D
B	C	A	D	E
C	D	B	E	A
D	E	C	A	B
E	A	D	B	C

In this case, each treatment condition appears in every possible
position; but, for example, D is immediately preceded by A
twice but never by B. To remedy this situation, we must enu-
merate five additional sequences that are exactly the reverse of
the first five sequences. In the five-treatment-condition example, the additional five sequences would appear as follows:

Sequence				
D	C	E	B	A
E	D	A	C	B
A	E	B	D	C
B	A	C	E	D
C	B	D	A	E

When these ten sequences are combined, the criteria of
incomplete counterbalancing are met. Consequently, the in-

complete counterbalancing technique provides for control of order effects.

How well does the incomplete counterbalancing technique control for sequencing effects? The influence of order effects is controlled, as every treatment condition occurs at each possible position in the sequence. In other words, every condition (A, B, C, and D) precedes and follows every other condition an equal number of times. However, carry-over effects are controlled only if they are linear for all sequences. If they are not, then incomplete counterbalancing is inadequate.

CONTROL OF SUBJECT EFFECTS

We have seen that subjects' behavior in an experiment can be influenced by the perceptions and motives they bring with them. It seems as though subjects are motivated to present themselves in the best possible light. If the demand characteristics suggest that a particular type of response will allow subjects to fulfill this motive, the subjects' responses will be a function of this motive in addition to the experimental treatment conditions. Such a situation will produce internal validity if the demand characteristics that operate in the experiment suggest to the subjects that the self-presentation motive can be fulfilled in different ways. For internal validity to be created, there must be constancy in the subjects' perceptions of the way in which the positive self-presentation motive can be fulfilled. Only then can we state with certainty that the independent variable has caused the variation in the subjects' responses to the dependent variable.

The experimenter can use a number of control techniques to try to ensure identical perceptions in all subjects. The following techniques cannot be used in all types of experiments; they are presented so that the experimenter can choose the most appropriate one for the particular study being conducted.

Double blind placebo model
A model in which neither the experimenter nor the subject is aware of the treatment condition administered to the subject

Double Blind Placebo Model

One of the best techniques for controlling demand characteristics is the **double blind placebo model.** This model requires that the experimenter "devise manipulations that appear essentially identical to subjects in all conditions"[1] and that the experimenter not know which group received the placebo condition or the experimental manipulation.

If you were conducting an experiment designed to test the effect of aspartame on disruptive behavior in young children, you would have to administer this sweetener to one group of children and a placebo to another group. Since both groups would think that they had received the aspartame, expectancies would be held constant. The experimenter also must not know whether a given subject received the aspartame or the placebo in order to avoid communicating the expectancy of generating disruptive behavior. Therefore, the experimenter as well as the subject must be "blind" to the treatment condition a given subject received. For some time, drug research has recognized the influence of patients' expectations on their experiences subsequent to taking a drug. Thus drug research consistently uses this model to eliminate subject bias.

Using this technique, Beecher (1966) found no difference in pain alleviation between a placebo group that was administered a weak saline solution and a drug group that was administered a large dose of morphine. Such results ran counter to a large body of previous research. However, Beecher communicated with another experienced drug researcher, who revealed that demand characteristics probably existed in the prior studies. This researcher said that he "found that as long as he knew what the subject had received, he could reproduce fine dose-effect curves; but when he was kept in ignorance, he was no more able than we were to distinguish between a large dose of morphine and an inert substance such as saline" (p. 841). In the former cases, the subject knew the correct response and acted accordingly.

Use of the double blind placebo model is a way to eliminate the development of differential subject perceptions, because all subjects are told that they are given (and appear to be given) the same experimental treatment. And since the researcher does not know which subjects have received the experimental treatments, he or she cannot communicate this information to the subjects. Therefore, the demand charac-

---

[1]Reprinted by special permission from "Experimentation in Social Psychology" by E. Aronson and J. M. Carlsmith, in *The Handbook of Social Psychology*, 2nd edition, volume 2, edited by G. Lindzey and E. Aronson, 1968, p. 62. Reading, Mass.: Addison-Wesley.

teristics surrounding the administration of the treatment conditions are controlled by the double blind placebo model.

Unfortunately, many types of experiments cannot use such a technique because all conditions cannot be made to appear identical in all respects. In such cases, other techniques must be employed.

Deception

One of the more common methods used to solve the problem of subject perceptions is the use of deception in the experiment. **Deception** involves providing all subjects with a hypothesis that is unrelated to or orthogonal to the real hypothesis. Almost all experiments contain some form of deception, ranging from minor deceit (an omission or a slight alteration of the truth) to elaborate schemes. Christensen, Krietsch, White, and Stagner (1985), in their investigation of the impact of diet on mood disturbance, told subjects that the double blind challenges in which they had participated had isolated certain food substances as the causal factors in their mood disturbance. But the particular foods mentioned to the subjects were ones that actually had *not* been investigated in the challenges to which they had been subjected. Subjects were given this bogus information to induce the perception that the offending foods had been isolated and that the remaining foods could be eaten without inducing any detrimental effect on mood states. At the other end of the continuum, there are experiments in which subjects are given unrelated or bogus hypotheses to ensure that they do not discover the real hypothesis.

Deception
Giving the subject a bogus rationale for the experiment

Aronson and Mills (1959) repeatedly used deception in their study; at just about every stage of the experiment, some type of cover for the real purpose was given. For example, rather than telling the subjects that the experiment was investigating the effect of severity of initiation, the researchers said that the study was investigating the "dynamics of the group discussion process." Is it better to use such deception or simply to refrain from giving any rationale for the tasks to be completed in the experiment? It seems as though providing subjects with a false, but plausible, hypothesis is the preferred procedure, because the subjects' curiosity may be satisfied so that they do not try to devise their own hypotheses. If different subjects perceive the study to be investigating different hypotheses, their responses may create a source of bias.

The rationale underlying the deception approach is "to provide a cognitive analogy to the placebo."[2] In a placebo experiment, all subjects think they have received the same independent variable. In the deception experiment, all subjects receive the same false information about what is being done, which should produce relatively constant subject perceptions of the purpose of the experiment. Therefore, deception seems to be an excellent technique for controlling the potential biasing influence that can arise from subjects' differential perceptions regarding the hypothesis of the experiment. The one problem with deception is that it frequently prompts objections on ethical grounds (see Chapter 5).

Disguised Experiment

Disguised experiment
A study that is conducted without communicating to the subject that he or she is part of the experiment

The **disguised experiment** is conducted in a context that does not communicate to the subjects that they are participating in an experiment. This means that a procedure must be established so that the independent variable as well as the dependent variable can be administered without telling the subjects that they are in an experiment. Abelson and Miller (1967) conducted such a study in their investigation of the influence of a personal insult on persuasion. The experimenter, disguised as a roving reporter, approached a subject seated on a park bench. The experimenter explained to the subject that he was conducting a survey on a particular issue. The individual was asked to give an opinion about the issue, and then the person seated next to him or her—an experimental confederate—was asked for his or her views on the same topic. In one treatment condition, the confederate derogated the subject before expressing an opposite point of view. The experimenter then obtained a second measure of the subject's opinion to assess the influence of the confederate's insult. This whole experiment was disguised in the sense that the subjects had no way of knowing that they were participants, so demand characteristics were minimal.

---

[2]Reprinted by special permission from "Experimentation in Social Psychology" by E. Aronson and J. M. Carlsmith, in *The Handbook of Social Psychology*, 2nd edition, volume 2, edited by G. Lindzey and E. Aronson, 1968, p. 63. Reading, Mass.: Addison-Wesley.

Use of the disguised experiment has much to recommend it and is an excellent way of controlling experimental demand characteristics. But this method is not without its limitations. First, many studies cannot be disguised in this way. Second, disguised experiments generally have to be field studies, and associated with field studies is the difficulty of controlling extraneous variables. The third problem is an ethical one: subjects are not informed that they are participating in an experiment and therefore are not given the option of declining.

Independent Measurement of the Dependent Variable

Independent measurement of the dependent variable Assessment of the dependent variable in a situation that is removed from the experiment

In **independent measurement of the dependent variable,** the experimenter measures the dependent variable in a context that is completely removed from the manipulation of the independent variable. One way to accomplish this is by manipulating the independent variable within the context of one experiment and measuring the dependent variable at some later time within the context of an unrelated experiment. Carlsmith, Collins, and Helmreich (1966) conducted a study that illustrates this procedure. They were investigating the influence of one's attitude toward a task when one was paid various amounts of money to state that the boring task performed was actually interesting. All subjects performed the tasks required of them and then, when they thought that they had completed the experiment, were asked to participate in another study conducted by a different group of individuals. This second study was actually a bogus one set up especially to measure the subjects' attitudes toward the boring task they had completed in the first experiment. Such a situation minimizes subject bias because the subjects think they are participating in another study (assuming there is nothing about the procedure that arouses the subjects' suspicions) and form hypotheses relative to this new study. Consequently, any biasing effects should not systematically influence one group over another.

This technique is good when it can be used—in cases in which the dependent variable can be independently measured. In many studies, however, this cannot be accomplished because the variables are interdependent. There is also an ethical issue involved here, because subjects are not told the true purpose of the experiment.

Procedural Control, or Control of Subject Interpretation

The four techniques just discussed are excellent for controlling some of the demand characteristics of the experiment. "However, these control techniques seem to be limited to ensuring that subjects have a unified perception of the treatment condition they are in, whether or not they receive a given treatment, and the purpose of the experiment" (Christensen, 1981, p. 567). There is little recognition of the fact that the subjects' perceptions are also affected by the many demand characteristics surrounding the whole procedure. For example, it has been demonstrated that subjects respond differently to a verbal conditioning task depending on how they interpret the verbal reinforcer (Christensen, 1977). To provide adequate control of subject perceptions and the positive self-presentation motive, we need to know the types of situations and instructions that will alter subjects' perceptions of how to create the most positive image. The literature on this issue, however, is in its infancy. At the present time, therefore, it is necessary to consider each experiment separately and try to determine if subjects' perceptions of the experiment might lead them to respond differentially to the levels of variation in the independent variable.

Retrospective verbal report
An oral report in which the subject retrospectively recalls aspects of the experiment

Postexperimental inquiry
An interview of the subject after the experiment is over

A variety of techniques that can be used to gain insight into subjects' perceptions of the experiment are summarized in Christensen (1981) and Adair and Spinner (1981). These methods can be grouped into two categories: retrospective verbal reports and concurrent verbal reports. A **retrospective verbal report** consists of a technique such as the **postexperimental inquiry,** which is exactly what it says it is: questioning the subject regarding the essential aspects of the experiment after completion of the study. What did the subject think the experiment was about? What did he or she think the experimenter expected to find? What type of response did the subject attempt to give, and why? How does the subject think others will respond in this situation? Such information will help to expose the factors underlying the subject's perception of his or her response.

Concurrent verbal report
A subject's oral report of the experiment that is obtained as the experiment is being performed

Sacrifice groups
Groups of subjects that are stopped and interviewed at different stages of the experiment

Concurrent probing
Obtaining a subject's perceptions of the experiment after completion of each trial

Think-aloud technique
A method that requires subjects to verbalize their thoughts as they are performing the experiment

Concurrent verbal reports include such techniques as Solomon's "sacrifice" group (Orne, 1973), concurrent probing, and the "think-aloud" technique (Ericsson and Simon, 1980). In Solomon's **sacrifice groups,** each group of subjects is "sacrificed" by being stopped at a different point in the experiment and probed regarding the subjects' perceptions of the experiment. **Concurrent probing** requires subjects to report, at the end of each trial, the perceptions they have regarding the experiment. The **think-aloud technique** requires subjects to verbalize any thoughts or perceptions they have regarding the experiment as they are performing the experimental task. Ericsson and Simon (1980) consider this the most effective technique, because it does not require the subject to recall information and hence eliminates distortions in reporting due to failure to remember or due to the biasing influence that may result from the experimenter's probing.

None of these techniques is foolproof or without disadvantages. However, use of these methods will provide some evidence regarding subjects' perceptions of the experiment and will enable you to design an experiment in such a way as to minimize the differential influence of the subjects' motive of positive self-presentation.

CONTROL OF EXPERIMENTER EFFECTS

Experimenter effects
The biasing influence that can be exerted by the experimenter

Experimenter effects have been defined as the unintentional biasing effects that the experimenter can have on the results of the experiment. The experimenter is not a passive, noninfluential agent in an experiment, but an active potential source of bias. This potential bias seems to exist in most types of experiments, although it may not be quite as powerful as Rosenthal maintains.

Page and Yates (1973) have shown that 90 percent of the respondents they surveyed felt that the implications of experimenter bias for psychology were serious. Additionally, 81 percent of the respondents felt that the presence of experimenter-related controls should be a major criterion for publishability of studies. Such data suggest that psychologists in general consider the experimenter bias effect to be of importance in psychological research and see the need to incorporate techniques to control for such potential effects. According to Wyer, Dion, and Ellsworth (1978), problems such as experimenter bias are widely understood in social psychology, and it is assumed "that most persons who submit papers to JESP avoid these problems as a matter of course" (p. 143). However, Silverman (1974) concluded from his survey that "despite all of the rhetoric and data on experimenter effects, it appears that psychologists show little more concern for their experimenters as sources of variance than they might for the light fixtures in their laboratories" (p. 276). His findings indicate that it is important to present and emphasize the use of controls for experimenter bias.

Control of Recording Errors

Errors resulting from the misrecording of data can be minimized if the person recording the data remains aware of the necessity of making careful observations to ensure the accuracy of data transcription. An even better approach would be to use multiple observers or data recorders. If, for example, three individuals independently recorded the data, discrepancies could be noted and resolved to generate more accurate data. Naturally, all data recorders could err in the same direction, which would mask the error, but the probability of this occurring is remote. This procedure could be improved even further if the data recorders were kept blind regarding the experimental conditions in which the subject was responding (Rosenthal, 1978).

The best means for controlling recording errors, though not possible in all studies, is to eliminate the human data recorder and have responses recorded by some mechanical or electronic device. In some research laboratories, the subjects' responses are automatically fed into a computer.

Control of Experimenter Attribute Errors

At first glance, there seems to be a simple and logical solution to the problem created by experimenter attributes. Throughout much of this text, I have referred to control in terms of constancy. Because most extraneous variables cannot be eliminated, they are held constant so that a differential influence is not exerted on the subjects' responses in the various treatment groups. In like manner, the influence of experimenter attributes could be held constant across all treatment conditions. Some experimenters, because of their attributes, may obtain more of an effect than other experimenters. But this increased effect should be constant across all treatment groups. Therefore, the influence of experimenter attributes should not significantly affect the *mean differences* among treatment groups. Assume that a cold and a warm experimenter independently conduct the same learning study and that the warm experimenter obtains an average of 3 more units of learning from subjects in each of the two treatment groups than does the cold experimenter, as shown in the top half of Table 8.6. Note that the mean difference between Groups A and B is identical for both experimenters, indicating that they would have reached the same conclusions even though each obtained different absolute amounts of learning. In such a situation, the effects of the experimenter attributes would not have had any influence on the final conclusion reached.

Control through the technique of constancy does imply that the variable being held constant—experimenter attributes, in this case—produces an equal effect on all treatment groups. If this assumption is not accurate or if the experimenter's attributes interact with the various treatment effects, control has not been achieved. If, in the foregoing example, a warm experimenter obtained an average of 8 units of performance from subjects in Group A and 21 units of performance from subjects in Group B, whereas the cold experimenter obtained identical performance from subjects in both treatment groups (as shown in the bottom half of Table 8.6), we would not have controlled for the influence of experimenter attributes. In this

TABLE 8.6 Hypothetical Data Illustrating the Mean Difference in Learning Obtained from a Warm and a Cold Experimenter

| | Experimental Group | | Mean |
Experimenters	*A*	*B*	*Difference*
Experimenter attributes controlled			
Warm	10	20	10
Cold	7	17	10
Experimenter attributes not controlled			
Warm	8	21	13
Cold	17	17	0

case, the two experimenters would have produced conflicting results. Unfortunately, we do not know which attributes interact with numerous independent variables that exist in psychology. Since we do not know how much difference is exerted by various experimenters, a number of individuals (for example, McGuigan, 1963; Rosenthal, 1966) have suggested that several experimenters be employed in a given study. (The ideal but impractical recommendation is that a random sample of experimenters be selected to conduct the experiment.)

If more than one experimenter were employed, evidence could be acquired as to whether there was an interaction between the treatment conditions and an experimenter's attributes. If identical results were produced by all experimenters, we would have increased assurance that the independent variable and the experimenter attributes did not interact. If the experimenters produced different results, however, we would know that an interaction existed and could perhaps identify the probable cause of the interaction.

Since such interaction effects do occur in some studies, several individuals (for example, McGuigan, 1963) have recommended that the experimenter be studied as an independent variable. Lyons (1964), however, feels this merely complicates the issue, since an experimenter with given attributes still has to study the influence of other experimenters' attributes, and certain investigators may find an influence of certain attributes whereas others may not. How far back can we push the problem? The solution that Lyons proposes is to automate the experiment and thus get rid of the experimenter. But even if this solution is employed, some human contact is still necessary in the form of recruiting subjects and greeting them before turning them over to the automated section. Also, automation is not always feasible and is often expensive.

Aronson and Carlsmith (1968) believe that the presence of an experimenter, as well as potentially producing bias, is frequently necessary in an experiment to eliminate bias. They argue that the experimenter can help standardize the extent to which all subjects understand the instructions. Jung (1971) also states that the experimenter may be necessary to detect the occurrence of unanticipated phenomena that could affect the outcome of the experiment and to identify ways of improving the experiment. In the final analysis, the possible gains of having a live experimenter must be weighed against the possible bias that he or she may produce.

As you can see, we do not yet have a good means for controlling the potential artifactual influence of the experimenter in cases where automation cannot be used. To learn more about the influence of experimenter attributes, we must conduct experiments that systematically vary experimenter attributes and types of psychological tasks as well as subject attributes. It may be that subject and experimenter attributes interact in some fashion to produce artifactual results (Johnson, 1976). But stating that additional research is needed to identify the situations that require control of experimenter attributes provides little direction or assistance to the investigator who must use live experimenters. Based on his review of the literature, Johnson (1976) has found that the experimenter attributes effect can be minimized if one controls for "those experimenter attributes which correspond with the psychological task" (p. 75). In other words, if the experimenter attribute is correlated with the dependent variable, then it should be controlled. On hostility-related tasks, it is necessary to hold the experimenters' hostility level constant. In a weight reduction experiment, the weight of the therapist may be correlated with the success of the program. Therefore, to identify the relative effectiveness of different weight reduction techniques, it would be necessary, at the very least, to make sure the therapists were of approximately the same weight. Such an attribute consideration may not, however, have an artifactual influence in a verbal learning study. At the present time, it is necessary for the investigator to use his or her judgment as well as any available research to ascertain whether the given attributes of the experimenters may have a confounding influence on the study.

Control of Experimenter Expectancy Error

Rosenthal and his associates have presented a strong argument for the existence of experimenter expectancy effects in most types of psychological research. Despite the fact that certain individuals, notably Barber and Silver (1968), have presented counterarguments against Rosenthal, it seems important to devise techniques for eliminating bias of this type. There are a number of techniques that can be used for eliminating or at least minimizing expectancy effects. Generally, they involve automating the experiment or keeping the experimenter ignorant of the condition the subject is in so that appropriate cues cannot be transmitted. Rosenthal (1966) discusses such techniques, several of which will now be presented.

The Blind Technique The **blind technique** actually corresponds to the experimenter's half of the double blind placebo model. In the blind technique, the experimenter knows the hypothesis but is blind as to which treatment condition the subject is in. Consequently, the experimenter cannot unintentionally treat groups differently.

Rosenthal (1966) has suggested that we need a professional experimenter—a trained data collector analogous to the laboratory technician. This person's interest and emotional investment would be in collecting the most accurate data possible and not in attaining support of the hypothesis. The scientist would not attempt to keep the hypothesis from this individual, since it would be very difficult to do so (Rosenthal, Persinger, Vikan-Kline, and Mulry, 1963) and in any case the experimenter would probably just develop his or her own. However, because this person's primary interest would be in collecting accurate data, he or she would have less incentive to bias the results and therefore would probably not be as much of a biasing agent. As Rosenthal pointed out, this idea has already been implemented with survey research and may have merit for experimental psychology. However, Page and Yates (1973) have indicated that most psychologists are not favorably disposed toward this alternative.

At present, the blind technique is probably the best procedure for controlling experimenter expectancies. But there are many studies in which it is impossible to remain ignorant of the condition the subject is in, and in those cases the next best technique should be employed—the partial blind technique.

The Partial Blind Technique In cases where the blind technique cannot be employed, it is sometimes possible to use the **partial blind technique,** whereby the experimenter is kept ignorant of the condition the subject is in for a portion of the study. The experimenter could remain blind while initial contact was made with the subject and during all conditions prior to the actual presentation of the independent variable. When the treatment condition was to be administered to the subject, the experimenter could use some technique (such as pulling a number out of a pocket) that would designate which condition the subject was in. Therefore, all instructions and conditions

Blind technique
A method whereby knowledge of each subject's treatment condition is kept from the experimenter

Partial blind technique
A method whereby knowledge of each subject's treatment condition is kept from the experimenter through as many stages of the experiment as possible.

prior to the manipulations would be standardized and expectancy minimized. Aronson and Cope (1968) used this procedure in investigating the attraction between two people who share a common enemy. The experimenter explained the purpose of the study and instructed each subject in the performance of a task. After the task had been completed, the subject was randomly assigned to one of two experimental conditions. This was accomplished by having the experimenter unfold a slip of paper—given to him or her just prior to running the subject—that stated the subject's experimental condition. Only at this point did the experimenter learn the subject's experimental condition.

Although this procedure is only a partial solution, it is better than the experimenter's having knowledge of the subject's condition throughout the experiment. If the experimenter could leave the room immediately following administration of the independent variable and allow another person (who was ignorant of the experimental manipulations administered to the subject) to measure the dependent variable, the solution would come closer to approaching completeness. Again, in many experiments this is not possible because the independent and dependent variables cannot be temporally separated.

Automation The technique of totally automating the experimental procedures so that no experimenter–subject interaction is required

Automation A third possibility for eliminating expectancy bias in animal and human research is total **automation** of the experiment. Indeed, numerous animal researchers currently use automated data collection procedures. Many human studies could also be completely automated by having instructions written, tape recorded, filmed, televised, or presented by means of a microcomputer, and by recording responses via timers, counters, pen recorders, computers, or similar devices. These procedures are easily justified to the subject on the basis of control and standardization, and they minimize the subject–experimenter interaction. Johnson and Adair (1972) have provided some evidence that automation can reduce expectancy effects for male experimenters. Videbeck and Bates (1966) have demonstrated that the computer can be used to replace the experimenter altogether.

Psychological experiments are becoming increasingly automated. With each passing year, we find more electronic devices manufactured for use in our experiments. At present, however, few of them totally remove the researcher from the experimental environment. Complete automation, through such approaches as the use of computers, is restricted by such practical considerations as cost of equipment and programming. In most animal research, the experimenter must transport the animals to and from the home cages as well as feed and care for them; seldom is this operation totally automated. With human research, Aronson and Carlsmith (1968) make the point that the experimenter sometimes eliminates bias rather than acting as a biasing agent. Rosenthal (1966) states that when the experimenter's participation is considered vital, his or her behavior should be as constant as possible and experimenter–subject contact and interaction should be minimal.

LIKELIHOOD OF ACHIEVING CONTROL

So far we have looked at several categories of extraneous variables that need to be controlled and a number of techniques for controlling them. Do these methods allow us to achieve the desired control? Are they effective? The answer to these questions seems to be both yes and no. The control techniques are effective, but not 100 percent effective. Actually, we do not know exactly how effective they are. If we are controlling by equating subjects on some characteristic, then the effectiveness of the control is dependent on such factors as the ability to measure (for example, the ability of an intelligence test to measure intelligence). Likewise, the effectiveness of control through randomization depends on the extent to which the random procedure equated the groups. Since subjects were randomly assigned to groups, it is also possible that the factors affecting the experiments were unequally distributed among the groups, which would result in internal invalidity.

The point is that we can never be certain that complete control has been effected in the experiment. All we can do is increase the probability that we have attained the desired control of the extraneous variables that would be sources of rival hypotheses.

SUMMARY

In conducting an experiment that attempts to identify a causal relationship, the experimenter must accomplish one important task: controlling for the influence of extraneous variables. This is usually accomplished by using an available control technique. The technique of randomization is extremely valuable because it provides control for unknown as well as known sources of variation by distributing them equally across all experimental conditions so that the extraneous variables exert a constant influence.

Matching is a control technique that is less powerful than randomization in its ability to equate groups of subjects on all extraneous variables. The prime advantage of the matching technique is that it increases the sensitivity of the experiment while providing control on those extraneous variables that are matched. There are four basic matching techniques. One technique, matching by holding variables constant, produces control by including in the study only subjects with a given amount or type of an extraneous variable. Certain extraneous variables are therefore excluded from the study, which means they cannot influence the results. A second matching technique involves building the extraneous variable into the design of the experiment. In this case, the extraneous variable actually represents another independent variable, so its effect on the results is noted and isolated from the effects of other independent variables. The yoked control matching technique is very restrictive in that it controls only for the temporal relationship between an event and a response. It accomplishes this by having a yoked control subject receive the stimulus conditions at exactly the same time as does the experimental subject. The last matching technique involves equating subjects in each of the experimental groups either on a case-by-case basis (precision control) or by matching the distribution of extraneous variables in each experimental group. Regardless of which approach is used, the matching technique represents an attempt to generate groups of subjects that are equated on the extraneous variables considered to be of greatest importance.

The counterbalancing technique represents an attempt to control for both order and carry-over sequencing effects. Order effects exist where a change in performance arises from the order in which the treatment conditions are administered, whereas carry-over effects refer to the influence that one treatment condition has on performance under another treatment condition. Two counterbalancing techniques that can provide some control over sequencing effects are intrasubject counterbalancing, which involves counterbalancing subjects, and intragroup counterbalancing, which involves counterbalancing groups of subjects. These techniques are effective in controlling for all sequencing effects except nonlinear carry-over effects.

Subjects and experimenters have also been shown to be potential sources of bias in psychological experiments. The biasing influence of subjects is the result of their differential perceptions regarding the most effective mode for presenting themselves in the most positive manner. Use of the double blind placebo model, deception, disguising the experiment, and obtaining an independent measurement of the dependent variable are all effective ways of creating constant perceptions of the experimental hypothesis, the purpose of the experiment, and knowledge of being in the experiment. However, differential subject perceptions can be caused by other procedural aspects of the experiment. To determine whether these other procedures create differential perceptions, we must use a technique such as the retrospective or concurrent verbal report. Experimenter effects can be minimized by using some technique that either conceals from the experimenter the treatment condition that the subject is in or else eliminates experimenter–subject interaction. Such techniques include automation, the blind technique, and the partial blind technique.

Even after all of these control techniques have been considered for a given study and the appropriate ones have been used, we still cannot be completely sure that all extraneous variables have been controlled. The only sure thing that can be said is that more control is gained with the use of these techniques than would be without their use.

STUDY QUESTIONS

1. What are the three general ways in which control is achieved in experimentation?
2. What is randomization, how does it achieve the desired control, and why is it considered the most important control technique?
3. Identify the various matching control techniques and explain how each achieves control, the limitations of each, and when each would be used.
4. What is counterbalancing, when should it be used, and what are the various counterbalancing techniques?

5. What is the difference between an order effect and a carry-over effect?

6. What is the difference between a linear and a non-linear sequence effect, and how do you control for each of these effects?

7. What are the various techniques that can be used to control for subject effects? Explain how each one produces the necessary control.

8. What are the various techniques that can be used to control for experimenter effects? Explain how each one produces the necessary control.

KEY TERMS AND CONCEPTS

Randomization

Population

Representative sample

Sample

Random

Random assignment

Matching

Yoked control

Precision control

Frequency distribution control

Counterbalancing

Order effect

Carry-over effect

Intrasubject counterbalancing

Intragroup counterbalancing

Incomplete counterbalancing

Double blind placebo model

Deception

Disguised experiment

Independent measurement of the dependent variable

Retrospective verbal report

Postexperimental inquiry

Concurrent verbal report

Sacrifice groups

Concurrent probing

Think-aloud technique

Experimenter effects

Blind technique

Partial blind technique

Automation

CHAPTER 9

Experimental Research Design

LEARNING OBJECTIVES

1. To gain an understanding of the requirements of a good research design.
2. To be able to distinguish the ways in which good research designs differ from faulty research designs.
3. To learn the various types of good research designs.
4. To understand the difference between a between-subjects design and a within-subjects design.
5. To gain an understanding of the concept of *interaction*.

Consumer Reports *(1974) has revealed that new-car dealers, in particular, use a specific procedure to induce a customer to buy a car. This procedure involves making the customer an extremely attractive offer—perhaps offering the car at a price several hundred dollars below any competitor's price. The object is to make the customer an offer that can't be refused, so that he or she will decide to buy the car. Once the customer has decided to purchase the car, the salesperson begins filling out the appropriate forms. The next step in the procedure is for the salesperson to withdraw the sales advantage in one of several ways. The salesperson might tell the customer that the quoted price did not include an expensive option such as the stereo system that the customer assumed was part of the offer. More commonly, however, after most of the paperwork has been completed, the salesperson tells the customer that the "boss" must make the final approval of the transaction. When the salesperson returns from a discussion with the boss, the customer learns that the boss refused to accept the offer because the dealership would be losing money. The salesperson then attempts to persuade the customer to agree to pay more money for the car. The assumption on the part of the car dealership is that customers will stick to their decision to buy the car even though the final price is higher than the original price agreed on—and that more customers will decide to make a purchase than if this procedure were not used.*

Does this approach really work? Does it induce more people to make the ultimate decision to purchase an automobile? If Consumer Reports *is correct, this tactic is widely used by car dealers. Evidently the dealers are convinced of its effectiveness. To determine if the car dealers are correct, it is necessary to determine experimentally whether this procedure actually induces increased compliance. This is precisely what Cialdini, Cacioppo, Bassett, and Miller (1978) did. In experimentally testing the efficacy of this compliance tactic, however, these researchers had to formulate a research design that could be used to answer their research question. On the basis of the material presented in the previous chapters, you know that these researchers first had to identify the independent and dependent variables, the variables that must be controlled in order to ensure internal validity, and the techniques that must be used to control for the influence of these extraneous variables.*

Only after these decisions were made could Cialdini et al. construct a design that would incorporate not only the independent and dependent variables but also the control techniques that would provide a strategy for them to use in collecting data that would give them an answer to their research question. This chapter will present the basic research designs used in most experimental studies.

INTRODUCTION

*Research design
The outline, plan, or
strategy used to
investigate the research
problem*

Research design refers to the outline, plan, or strategy specifying the procedure to be used in seeking an answer to the research question. It specifies such things as how to collect and analyze the data. One purpose of the design is to control unwanted variation, which is accomplished by incorporating one or more of the control techniques discussed in Chapter 8 or by incorporating a control group. The significance of the control group will be discussed in detail later in the chapter, and the manner in which it assists in achieving control will be discussed in conjunction with the various research designs.

To illustrate the purposes of research design, let us evaluate the study conducted by Ossip-Klein et al. (1983) in which they attempted to determine if switching to low tar/nicotine/carbon monoxide cigarettes actually decreases a smoker's level of carbon monoxide. Forty adult smokers were recruited through advertising in newspapers, television, radio, and posters. The subjects were randomly assigned to two groups: control and experimental. Members of the control group were told to continue smoking their usual brand of cigarettes, whereas those in the experimental group were instructed to smoke a low tar/nicotine/carbon monoxide brand. Study results revealed that the level of carbon monoxide in the body was not altered by switching to a low tar/nicotine/carbon monoxide cigarette.

The procedure specified in the design selected by Ossip-Klein et al., depicted in Figure 9.1, is quite simple. First, the forty subjects were to be randomly assigned to the two groups, and then each group was to be assigned a different brand of cigarettes to smoke. All subjects were to be tested for carbon monoxide levels before and after smoking their designated brand. The design also suggests which statistical test to use in analyzing the data. Since there were to be two groups and these two groups were assessed twice, a factorial design based on a mixed model was called for. (This design will be discussed later in this chapter.) Note the intimate connection between research design and statistics.

Researchers sometimes design an experiment and collect data according to the specifications of the design without attempting to determine whether the design will permit statistical analysis. To their dismay, these individuals frequently find either that their data cannot be analyzed (so the research problem cannot be tested) or that analysis would not be worthwhile. This difficulty can be traced to the fact that the studies were not appropriately designed. As a rule of thumb, never conduct an experiment until you have determined if your research design permits analysis that will answer your research questions.

The design of the experiment also suggests the conclusions that can be drawn. With the design illustrated in Figure 9.1, a statistical test could be computed to determine if differences existed between the two groups of subjects, if the pre- and postmeasurements of the carbon monoxide level varied, and if the differences between the pre- and postmeasurements depended on the group being considered (an interaction effect that will be discussed in more detail later).

The design also shows how the controls for extraneous variables are incorporated. In the Ossip-Klein et al. experiment, the randomization control technique was incorporated by randomly assigning subjects to the two groups. Before being assigned, subjects were matched on several variables, such as number of cigarettes smoked per day.

Because the design suggests the observations that will be made and how these observations will be analyzed, it deter-

	Preresponse measure	Treatment condition	Postresponse measure
Experimental group	Carbon monoxide measure	Smokes low tar/nicotine cigarettes	Carbon monoxide measure
Control group	Carbon monoxide measure	Smokes usual brand of cigarettes	Carbon monoxide measure

FIGURE 9.1   Design of the Ossip-Klein et al. study.

mines whether or not valid, objective, and accurate answers to research questions will be obtained. Whether designs are good or bad depends on whether they enable one to attain the answers sought. It is usually much easier to design an experiment inappropriately, because careful thought and planning are not required. To the extent that the design is faulty, however, the results of the experiment will be faulty. How does one go about conceiving a good research design that will provide answers to the questions asked? It is no simple task, and there is no set way of instructing others in how to do it. Designing a piece of research requires thought—thought about the components to include and pitfalls to avoid. We will look first at some faulty research designs and then at some appropriate research designs.

FAULTY RESEARCH DESIGN

In seeking solutions to questions, the scientist conducts experiments, devising a certain strategy to be followed. Unfortunately, research has been conducted using designs that are inappropriate. The purpose in presenting some examples of defective designs is to demonstrate the types of extraneous variables that produce internal invalidity.

One-Group After-Only Design

One-group after-only design
A faulty research design in which the influence of a treatment condition on only one group of individuals is investigated

In the **one-group after-only design,** a single group of subjects is measured on a dependent variable after having undergone an experimental treatment (see Figure 9.2). Consider a hypothetical situation in which an institution starts a training program X (the treatment condition). The institution wants to evaluate the effectiveness of the program, so on completion of the program it assesses behaviors, the Y measure (for example, the opinions, attitudes, and perhaps performance of the individuals who went through the program). If the Y measures are positive and if the individuals' performances are good, then the validity of the program is thought to have been established.

For yielding scientific data, the design in Figure 9.2 is, as Campbell and Stanley (1963) state, of almost no value, because without some sort of comparison, it is impossible to determine whether a change in performance occurred as a result of the

FIGURE 9.2   One-group after-only design. (Adapted from *Experimental and Quasi-Experimental Designs for Research* by D. T. Campbell and J. C. Stanley, 1963. Chicago: Rand McNally and Company. Copyright 1963, American Educational Research Association, Washington, D.C.)

Treatment
X

Response measure
Y

experimental treatment. At a minimum, subjects should be pretested to determine whether they changed their responses after receiving the experimental treatment. Also, an equated comparison group that did not receive the treatment condition must be included. Without an equated comparison group or multiple pretests, it is impossible to determine whether change resulted from the experimental treatment or from some extraneous variable, such as a history, maturation, or statistical regression effect.

One-Group Before–After (Pretest–Posttest) Design

Most researchers recognize the deficiencies in the one-group after-only design and attempt to improve on it by including a pretest. For an evaluation of a curriculum or training program, some measure of improvement is necessary. Some individuals, however, assume they need only include a pretest that can be compared with a test taken after administration of some treatment condition. Figure 9.3 depicts such a plan, which corresponds to the **one-group before–after design.**

One-group before–after design
A faulty research design in which a treatment condition is interjected between a pre- and posttest of the dependent variable

A group of subjects is measured on the dependent variable, *Y*, prior to administration of the treatment condition. The

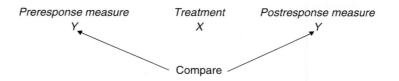

FIGURE 9.3   One-group before–after design. (Adapted from *Experimental and Quasi-Experimental Designs for Research* by D. T. Campbell and J. C. Stanley, 1963. Chicago: Rand McNally and Company. Copyright 1963, American Educational Research Association, Washington, D.C.)

independent variable, X, is then administered, and Y is again measured. The difference between the pre- and posttest scores is taken as an indication of the effectiveness of the treatment condition. In evaluation of a new curriculum, an attitude scale and an achievement test might be given at the beginning of the school year (pretest Y). The new curriculum—X—is then introduced to the students. At the end of the school year, the attitude scale and the achievement test (posttest Y) are again administered. The pre- and posttest scores on the attitude scale and the achievement test are examined for change. A significant change between these two scores is attributed to the new curriculum.

The Liddle and Long (1958) study represents an example of the use of this design. Liddle and Long selected eighteen slow learners, who were administered an intelligence test and assigned a reading grade placement score (pre-Y) prior to being placed in the experimental classroom. After students had spent approximately two years in the experimental classroom, the Metropolitan Achievement Tests were administered (post-Y) and these scores were compared with the previously assigned placement scores. This comparison indicated "an improvement of about 1.75 years in less than 2 school years" (p. 145). Such a study has intuitive appeal and at first seems to represent a good way to accomplish the research purpose—a change in performance can be seen and documented. In actuality, this design represents only a small improvement over the one-group after-only study because of the many uncontrolled rival hypotheses that could also explain the obtained results.

In the Liddle and Long study, almost two years elapsed between the pre- and posttests. Consequently, the uncontrolled rival hypotheses of history and maturation could account for some, if not all, of the observed change in performance. In order to determine conclusively that the observed change was caused by the treatment effect (the experimental classroom) and not by any of these rival hypotheses, researchers should have included an equated group of slow learners who were not placed in the experimental room. This equated group's performance could have been compared with the performance of the children who received the experimental treatment. If a significant difference had been found between the scores of these two groups, it could have been attributed to the influence of the experimental classroom, because both groups would have experienced any history and maturation effects that had occurred and, therefore, these variables would have been controlled. The design of the study was inadequate, not so much because the sources of rival hypotheses *can* affect the results, but because we do not know *if* they did.

Although the one-group before–after design does not allow us to control or to test for the potential influence of these effects, it is not totally worthless. In situations in which it is impossible to obtain an equated comparison group, the design can be used to provide some information. However, one should remain constantly aware of the possible confounding extraneous variables that can jeopardize internal validity.

Nonequivalent Posttest-Only Design

Nonequivalent posttest-only design A faulty research design in which the performance of an experimental group is compared with that of a nonequivalent control group

The primary disadvantage of the previous two designs is the impossibility of drawing any unambiguous conclusions as to the influence of the treatment condition. The **nonequivalent posttest-only design** makes an inadequate attempt to remedy this deficiency by including a comparison group. In this design, one group of subjects receives the treatment condition (X) and is then compared, on the dependent variable (Y), with a group that did not receive this treatment condition. Figure 9.4 depicts the design.

Brown, Wehe, Zunker, and Haslam (1971) conducted a study that illustrates the use of this scheme. They wanted to evaluate the influence of a student-to-student counseling program on potential college freshman dropouts. One group of potential dropouts received the student-to-student counseling, and another matched group—the comparison group—did not. Following the series of counseling sessions, all students were administered several tests designed to evaluate the effects of the program. First-semester grade point averages were also obtained. Results revealed that, on all dependent variable measures, the group receiving the counseling performed in a superior manner.

The design of this study appears to be adequate. A comparison group was included to evaluate the influence of the treatment condition, and subjects in both groups were matched. Why, then, is this design included as an example of one that is faulty? The reason is that the two groups are *assumed* to be equated on variables other than the independent variable. Granted, Brown et al. did match on a number of variables such as age, sex, and ACT composite scores. Match-

FIGURE 9.4 Nonequivalent posttest-only design. (Adapted from *Experimental and Quasi-Experimental Designs for Research* by D. T. Campbell and J. C. Stanley, 1963. Chicago: Rand McNally and Company. Copyright 1963, American Educational Research Association, Washington, D.C.)

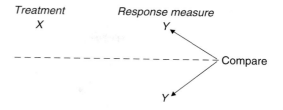

ing, however, is no assurance of having attained equated groups. As Campbell and Stanley (1963, p. 12) have stated, "Matching on background characteristics other than *O* [the dependent variable] is usually ineffective and misleading. . . ." The only way one can have any assurance that the groups are equated is to assign subjects randomly to the two groups. As indicated by the dashed lines in Figure 9.4, random assignment is not included in the nonequivalent posttest-only design. In studies where it is not possible to assign subjects randomly, the next best technique is to match on relevant variables. However, matching is no substitute for random assignment because it does not control for other variables such as motivation.

REQUIREMENTS OF TRUE RESEARCH DESIGNS

The designs just presented are considered faulty because, in general, they do not provide a way of isolating the effect of the treatment condition; rival hypotheses are not excluded. What, then, is a true research design? Kerlinger (1973) discusses three criteria that need to be met in a research design. The first criterion is whether the design answers the research questions, or adequately tests the hypothesis. Periodically, one encounters a situation in which an investigator designs a study and collects and analyzes the data, only to realize when he or she attempts to interpret the data that there is no answer to the research question. Such instances could have been avoided if, after the study was designed, the researcher had asked, "What conclusion or conclusions can I draw from this experiment?" Remember that the design of the study suggests the statistical tests that can be performed on the data, which in turn determine the conclusions that can be drawn. If the design allows us to conduct statistical tests that will provide an answer to the research question, the first criterion has been met.

The second measure of a true research design is whether extraneous variables have been controlled. This criterion relates to the concept of internal validity. If the observed effects can be attributed to the independent variable, the experiment is internally valid. In order to achieve internal validity, we must eliminate potential rival hypotheses. This can be accomplished by two means: control techniques or a control group.

Of the control techniques discussed in Chapter 8, the most important is randomization. The importance of this technique cannot be overemphasized; it is the only means by which unknown variables can be controlled. Also, statistical reasoning is dependent on the randomization procedure, so I emphasize again, *randomize whenever and wherever possible.*

Control group
The group of subjects that serves as a standard of comparison for determining if the treatment condition produced any effect

Experimental group
The group of subjects that receives the treatment condition

The second means for effecting control is inclusion of a control group. A **control group** is a group of subjects that does not receive the independent variable, receives zero amount of it, or receives a value that is in some sense a *standard* value, such as a typical treatment condition. An **experimental group** is a group of subjects that receives some amount of the independent variable. In the study conducted by Aronson and Mills on severity of initiation, the group that did not have to take the embarrassment test represented the control group, whereas the other two groups, which had to read either embarrassing or not very embarrassing material, represented the two experimental groups. In a drug study, the subjects receiving a placebo would represent the control group and the subjects receiving the drug would represent the experimental group.

A control group serves two functions. First, it serves as a source of comparison. The one-group after-only and the one-group before–after designs were considered faulty primarily because there was no way to tell if the treatment condition, *X*, caused the observed behavior, *Y*. To arrive at such a conclusion, we must have a comparison group or a control group that did not receive the treatment effect. Only by including a control group—assuming all other variables are controlled—can we get any concrete indication of whether or not the treatment condition produced results different from those that would have been attained in the absence of the treatment. Consider a hypothetical case of a father whose daughter always cries for candy when they go into a store. The parent does not like the behavior so, in order to get rid of it, he decides to spank the child whenever she cries for candy in the store and also to refuse to let her have any candy. After two weeks, the child has stopped the crying behavior, and the parent concludes that the spanking was effective. Is he correct? Note that the child also did not receive any candy during the two weeks, so a rival hypothesis is that crying was extinguished. To determine whether it was the spanking or extinction that stopped the behavior, we would also have to include a control child who did not receive the spanking. If both stopped crying in two weeks, then we would know that the spanking was not the variable causing the elimination of the crying behavior.

This hypothetical example also demonstrates the second function of a control group—that is, to serve as a control for rival hypotheses. All variables operating on the control and experimental groups must be identical, except for the one being manipulated by the experimenter. In this way, the influence of extraneous variables is held constant. The extinction variable was held constant across the child who did receive the spanking and the child who did not, and therefore did not confound the results. In the one-group before–after design, extraneous variables such as history and maturation can serve as rival hypotheses unless a control group is included. If a control group is included, these variables will affect the performance of both the control subjects and the experimental subjects, effectively holding their influence constant. It is in this way that a control group also serves a control function.

Before we leave the topic of the control group, one additional point needs to be made. A necessary requirement of the control group is that the subjects in the group be similar to those in the experimental group. If this condition does not exist, the control group cannot act as a baseline for evaluating the influence of the independent variable. The responses of the control group must stand for the responses that members of the experimental group would have given if they had not received the treatment condition. The subjects in the two groups must be as similar as possible so that theoretically they would yield identical scores in the absence of the introduction of the independent variable.

The third criterion of a true research design is generalizability, or external validity, as presented by Campbell and Stanley (1963). Generalizability asks the question "Can the results of this experiment be applied to individuals other than those who participated in the study?" If the answer is yes, then we need to follow with the question "To whom do the results apply?" Can we say that results should apply to everyone, only to females, or just to females who are attending college? In all cases, we would like to be able to generalize beyond the confines of the actual study. Whether or not we can generalize and how far we can generalize our results, though, is never completely known.

The foregoing three criteria represent the ideal. Naturally, the first criterion must be met by all studies, but the degrees to which the second and third are met will vary from one study to another. Basic research focuses primarily on the criterion of internal validity because its foremost concern is the examination of the relations among variables. Applied research, on the other hand, places equal emphasis on external and internal validity, since the central interest of such research is to apply the results to people and to situations.

PRETESTING SUBJECTS

One means of obtaining information about the pretreatment condition of the organism is to pretest subjects, as was done in the one-group before–after design. The experimenter can then directly observe change in the subjects' behavior as a result of the treatment effect. But one may legitimately question the need to pretest. Is it not sufficient and appropriate to assign subjects randomly to experimental and control groups and forget about pretesting? One can then assume comparability of the subjects in the two groups, and those in the control group provide the comparison data. Hence a pretest is unnecessary. However, there are several reasons (Selltiz et al., 1959; Lana, 1969) for including a pretest in the experimental design. These are as follows:

1. *Increased sensitivity.* One can increase the sensitivity of the experiment by matching subjects on relevant variables. Such matching requires pretesting (see Chapter 8).

2. *Ceiling effect.* Another reason for pretesting is to determine if there is room for the treatment condition to have an effect. Suppose you were investigating the efficiency of a particular persuasive communication for improving attitudes toward environmental protection. If, by chance, all subjects in

the experiment already had extremely positive attitudes toward environmental protection, there would be no room for the treatment condition to have an effect. Such a case could exist if, on a 10-point rating scale (with 10 being the positive end), all subjects were evaluated as being 8, 9, or 10. In such a situation, the effect of the persuasive communication cannot be assessed. Pretesting enables the investigator to identify the existence of a possible ceiling effect and take it into consideration when evaluating the effects of the independent variable.

3. *Initial position.* Many psychological studies are conducted in which it is necessary to know a person's initial position on the dependent variable because it may interact with the experimental condition. A treatment condition that tries to induce hostility toward a minority group may find that the effectiveness of this treatment condition is a function of the subjects' initial level of hostility. The treatment may be very successful with individuals having little hostility but unsuccessful with extremely hostile individuals. With such conditions, it is very helpful to pretest subjects. LeUnes, Christensen, and Wilkerson (1975), for example, pretested subjects on their attitudes toward various components of mental retardation. Subjects were then separated into a positive and a negative group in order to determine if subjects' initial attitudes affected whether an institutional tour provoked a change in attitude toward the various components of mental retardation. These investigators found that the subjects' initial attitudes were a significant factor.

4. *Initial comparability.* Another reason for pretesting is to assure that subjects are initially comparable on relevant variables. Ideally, subjects are randomly assigned to conditions. Although random assignment provides the greatest assurance possible of comparability of subjects, it is not infallible. Should randomization fail to provide comparability, comparison of the subgroups' pretest mean scores would tell us so.

In field research, we cannot always assign subjects randomly; rather, they must be taken as intact groups. Educational experiments, for example, are sometimes restricted to using one intact class for one group of subjects and another class for another group of subjects. In such instances, it is advisable to make sure that subjects do not differ initially on the independent variable. This kind of compromise occasionally has to be made. We must also recognize that the results of the experiment could be caused by group differences on characteristics other than the pretested variables. The pretest does, however, give some indication that the observed differences result from the treatment condition.

5. *Evidence of change.* Perhaps the most common reason for pretesting is to gain an empirical demonstration of whether the treatment condition succeeded in producing a change in the subjects. The most direct way of gaining such evidence is to measure the difference obtained before and after a treatment is introduced.

As you can see, there are several legitimate reasons for including a pretest in the study design. Unfortunately, there are also some difficulties that accompany pretesting (Oliver and Berger, 1980). First, pretesting may increase the amount of time or money required to complete the investigation. A more serious problem is that it may sensitize subjects to the experimental treatment condition. For example, pretesting subjects' opinions may alert them to the fact that they are participating in an attitude experiment, and this knowledge could heighten their sensitivity to the independent variable. Pretested subjects may therefore produce results that are not representative of those that would be obtained from a population that had not been pretested. This potential error is considered by Campbell and Stanley (1963) to be a factor jeopardizing external validity.

Lana (1969), however, has summarized the research that seeks to document this potential source of bias and reached some interesting conclusions. When the pretest involves a learning process such as requiring subjects to recall previously learned materials, the posttest score may very well be affected. "Ordinarily, if the task of the recall demanded by the pretest procedure is properly understood by the subject, the effect on the posttest should be facilitative" (p. 132). However, the conclusion regarding attitude research is somewhat different: "In attitude research pretest measures, if they have any impact at all, depress the effect being measured; any differences which can be attributed to the experimental treatment probably represent strong treatment effect" (p. 139).

Thus, although pretesting may influence the subjects' responses to the experimental treatment, the nature of this influence appears to depend on the type of study being conducted.

TRUE RESEARCH DESIGNS

*True experimental design
A good experimental design in which the influence of extraneous variables is controlled for while the influence of the independent variable is tested*

In this section we will consider some "true" experimental research designs. To be a **true experimental design,** a research design must enable the researcher to maintain control over the situation in terms of assignment of subjects to groups, in terms of who gets the treatment condition, and in terms of the amount of the treatment condition that subjects receive. In other words, the researcher must have a controlled experiment in order to have confidence in the relations discovered between the independent variables and the dependent variable. There are two basic types of true research designs: the after-only design and the before–after design.

After-Only Design

*After-only design
A true experimental design in which the experimental and the control groups' posttest scores are compared to assess the influence of the treatment condition*

The **after-only design** contains the basic components of most research plans used in the field of psychology. Its name is derived from the fact that the dependent variable is measured only once and this measurement occurs after the experimental treatment condition has been administered to the experimental group, as depicted in Figure 9.5. From this figure, you can see that the responses obtained from an experimental condition are compared with the responses obtained from a control condition after the treatment has been administered. However, the format illustrated here represents only the basic structure of the after-only design. The exact structure of the final design depends on several factors, such as the number of independent variables included in the investigation, the number of levels of variation of each independent variable, and whether the same

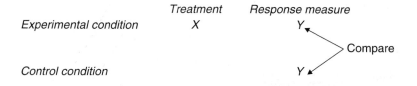

FIGURE 9.5 After-only design. (Adapted from *Experimental and Quasi-Experimental Designs for Research* by D. T. Campbell and J. C. Stanley, 1963. Chicago: Rand McNally and Company. Copyright 1963, American Educational Research Association, Washington, D.C.)

or different subjects are to be used in each treatment condition. After-only designs are usually dichotomized in terms of this last factor. If different subjects are used in each experimental treatment condition, then the after-only design is typically labeled a *between-subjects* design. If the same subjects are used in each experimental condition, then the after-only design is labeled a *within-subjects* design.

Between-subjects after-only design
A type of after-only research design in which subjects are randomly assigned to the experimental and control groups

Between-Subjects After-Only Design   In the **between-subjects after-only design,** the subjects are randomly assigned to as many groups as there are experimental treatment conditions. For example, if a study was investigating only one independent variable and the presence-versus-absence form of variation was being used with this independent variable, subjects would be randomly assigned to two treatment groups, as illustrated in Figure 9.6. This design is similar in appearance to the nonequivalent posttest-only design, but with one basic and important difference. Remember that the nonequivalent posttest-only design was criticized primarily from the standpoint that it does not provide any assurance of equality among the various groups. This between-subjects after-only design provides the necessary equivalence by randomly assigning subjects to the two groups. If enough subjects are included to allow randomization to work, then, theoretically, all possible extraneous variables are controlled (excluding those such as experimenter expectancies).

The study conducted by Cialdini, Cacioppo, Bassett, and Miller (1978) illustrates the use of the between-subjects after-only research design. The experiment that these investigators conducted represented an attempt to determine whether the tactic used by new-car dealers to entice a customer to pay more for an automobile really worked. In order to investigate

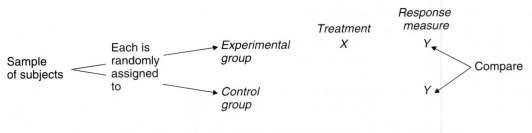

FIGURE 9.6   Two-group between-subjects after-only design.

this tactic, the researchers had to simulate, in the laboratory, the same type of psychological ploy that the car dealers used. This simulation involved contacting students by phone, explaining a psychological experiment to them, and asking them if they would be willing to participate in it. If they agreed to participate, they were told that the experiment was being conducted at 7:00 A.M. on Wednesday and Friday mornings. The experimenter then asked if it was possible to schedule the subject for one of these mornings. In this way the experimenter would simulate the tactic used by the new-car dealers: he or she would have obtained a commitment from the subject to participate in the experiment prior to increasing the cost of participation. Once the commitment was received, the cost was increased. The key issue was whether the subject would still agree to participate, as with the assumption of the new-car dealers that the customer, once committed to buying the car, would still agree to buy it even when the price was raised. Consequently, the dependent variable was the percentage of individuals agreeing to participate in the experiment. To determine if this tactic worked, however, it was necessary to include a control group that was not subjected to this tactic. This control group consisted of a group of subjects who were contacted by phone and asked to participate in a psychological experiment that began at 7:00 A.M. Note that these subjects were told that the experiment began at 7:00 A.M. *before* being asked to participate, like a customer who is told the real price of a car and then allowed to make a decision to purchase. Consequently, this experiment involved two groups, a control and an experimental group, and the subjects for each group were randomly selected from class rolls in order to control for the influence of extraneous variables. Therefore, the experiment was a between-subjects after-only research design. The results of this experiment revealed that the tactic worked: 56 percent of the subjects in the experimental group agreed to participate, whereas only 31 percent of the subjects in the control group did so.

Two difficulties can be identified in the design just presented. First, randomization is used to produce equivalence between the two groups. Although this is the best control technique available for achieving equivalence, it does not provide complete assurance that the necessary equivalence has been attained. (This is particularly true when the group of subjects being randomized is small.) In the Cialdini et al. experiment there were over thirty subjects randomly selected for each group. This is a large enough group to provide reasonable assurance that randomization produced the necessary group equivalence. If there is any doubt, it is advisable to combine matching with the randomization technique. Second, it is not the most sensitive design for detecting an effect caused by the independent variable. As discussed earlier, matching is the most effective technique for increasing the sensitivity of the experiment, and so these difficulties with the after-only design can be eliminated by matching subjects prior to randomly assigning them to the experimental treatment groups. However, the benefits of matching should always be weighed against the accompanying disadvantages, such as limitation of the available subject pool.

When, in the opinion of the investigator, the advantages of matching outweigh the disadvantages, a matched between-subjects after-only design should be used. As illustrated in Figure 9.7, this design requires that each member of the sample of subjects be matched with another subject on the variable or variables that are correlated with the dependent variable. The matched subjects are then randomly assigned to the experimental groups. Note that matching takes place *in addition to* randomization; it does not replace randomization. Using the two techniques increases both the sensitivity of the experiment and the probability that the groups are equivalent on the extraneous variables that must be controlled.

The two-group between-subjects after-only design just discussed illustrates the basic conceptual structure of the between-subjects design. However, experiments are seldom confined to two levels of variation of one independent variable. Instead, most studies use several levels of variation of one or more independent variables, and their schemes are extensions of the between-subjects after-only design. The two primary

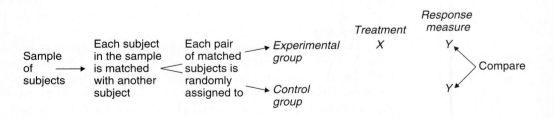

FIGURE 9.7   Matched between-subjects after-only design.

extensions are represented by the simple randomized subjects design and the factorial design.

Simple randomized subjects design
A between-subjects design in which the influence of several levels of variation on the independent variable is investigated

Simple Randomized Subjects Design. The **simple randomized subjects design** is a between-subjects after-only type of design that has been extended to include more than one level of the independent variable. There are many situations in which it is desirable to give varying amounts or degrees of an independent variable to different groups of subjects. In drug research, the investigator may want to administer different amounts of a drug to see if they produce differential reactions to the dependent variable. In such a case, subjects would be randomly assigned to the various treatment groups. If there were three experimental groups and one control group, subjects would be randomly assigned to the four groups, as shown in Figure 9.8. A statistical test would then be used to determine if a significant difference existed in the average responses of the four groups of subjects to the dependent variable.

Sigall, Aronson, and Van Hoose (1970) used the simple randomized subjects design in attempting to determine whether subjects are motivated to look good or whether they are motivated to cooperate with the experimenter to produce the results that he or she wants. To investigate these motives, they had a control group and three experimental groups. In one experimental group, the subjects were led to believe that they should increase performance; in another group, the subjects were led to believe that they should decrease performance; and in a third group, evaluative apprehension was generated by telling the subjects that increased performance was indicative of obsessive–compulsive behavior. The design of this experiment is depicted in Figure 9.8.

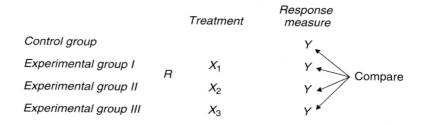

FIGURE 9.8 Simple randomized subjects design, with four levels of variation of the independent variable. *R* indicates that the four groups of subjects were randomly selected.

Analysis by Sigall et al. of their data revealed significant differences among the various groups. Additional statistical tests showed that the three experimental groups differed significantly from the control group and that the evaluative apprehension group differed significantly from the other two experimental groups. From this they concluded that a subject's primary motive is to look good rather than to cooperate with the experimenter. They arrived at this conclusion because the obsessive–compulsive group performed more slowly than any other group, whereas the other two experimental groups performed better than either the control or the obsessive–compulsive experimental group.

The simple randomized subjects design considers only one independent variable. In psychological research, as in other types of research, we are frequently interested in the effect of several independent variables acting in concert. In research on instructional effectiveness, researchers are interested in methods of instruction (for example, tutorial, discussion, lecture) as well as in other factors such as instructor attitude or experience. The simple randomized subjects design does not enable us to investigate several independent variables simultaneously, but a factorial design does.

Factorial design
A between-subjects design that enables us to investigate the independent and interactive influences of more than one independent variable

Factorial Design. In a **factorial design,** two or more independent variables are simultaneously studied to determine their independent and interactive effects on the dependent variable. Let us look at a hypothetical example that considers the effect of two independent variables, A and B. Assume that variable A has three levels of variation (A_1, A_2, and A_3) and that variable B has two levels of variation (B_1 and B_2). Figure 9.9 depicts this design, in which there are six possible combinations of the two independent variables—A_1B_1, A_1B_2, A_2B_1, A_2B_2, A_3B_1, and A_3B_2. Each one of these treatment combinations is referred to as a **cell.** There are six cells within this design to which the subjects would be randomly assigned. The subjects randomly assigned to A_1B_1 would receive the A_1 level of the first independent variable and the B_1 level of the second independent variable. In like manner, the subjects randomly assigned to the other cells would receive the designated combination of the two independent variables.

Cell
A specific treatment combination in a factorial design

Main effect
The influence of one independent variable in a factorial design

In an experiment that uses the design shown in Figure 9.9, two types of effects need to be analyzed: main effects and interaction effects. A **main effect** refers to the influence of one independent variable. The term *main effect* did not arise in the

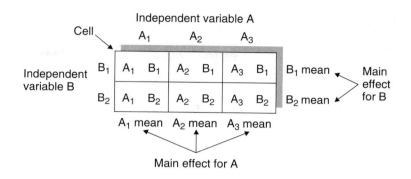

FIGURE 9.9 Factorial design with two independent variables.

simple randomized experiment or in the two-group between-subjects after-only experiment, because only one main effect or one independent variable existed. However, more than one independent variable exists in a factorial design, and the separate effects of each independent variable must be identified. To distinguish the influence of the different independent variables, we refer to each one as a separate main effect. In Figure 9.9, the two independent variables A and B each have a main effect. The main effect for A simply tells us if A produced a significant influence on behavior or if there was a significant difference among the three A mean scores. Similarly, the main effect for B tells us if B had a significant impact on behavior or if there was a significant difference between the two B mean scores.

Interaction effect
The influence of one independent variable on a second independent variable

An **interaction effect** refers to the influence that one independent variable has on another. The concept of interaction is rather difficult for most students to grasp, so I will digress in order to clarify this idea. First, I will present a number of possible outcomes that could accrue from an experiment having the design shown in Figure 9.9. Some of the outcomes will represent interactions and others will not, so that you can see the difference in the two situations. I will set up a progression from a situation in which one main effect is significant to a situation in which both main effects and the interaction are significant. The letter A will always represent one independent variable, and the letter B will always represent a second independent variable. Table 9.1 and Figure 9.10 depict these various cases. For the sake of clarity, the hypothetical scores in the cells will represent the mean score for the subjects in each cell.

TABLE 9.1 Tabular Presentation of Hypothetical Data Illustrating Different Kinds of Main and Interaction Effects

	A_1	A2	A_3	Mean		A_1	A_2	A_3	Mean
B_1	10	20	30	20	B1	20	20	20	20
B_2	10	20	30	20	B_2	30	30	30	30
Mean	10	20	30		Mean	25	25	25	

(a) A is significant; B and the interaction are not significant

(b) B is significant; A and the interaction are not significant

	A_1	A2	A_3	Mean		A_1	A_2	A_3	Mean
B_1	30	40	50	40	B1	10	20	30	20
B_2	50	40	30	40	B2	40	50	60	50
Mean	40	40	40		Mean	25	35	45	

(c) Interaction is significant; A and B are not significant

(d) A and B are significant; interaction is not significant

	A_1	A2	A_3	Mean		A_1	A_2	A_3	Mean
B_1	20	30	40	30	B1	10	20	30	20
B_2	30	30	30	30	B2	50	40	30	40
Mean	25	30	35		Mean	30	30	30	

(e) A and the interaction are significant; B is not significant

(f) B and the interaction are significant; A is not significant

	A_1	A2	A_3	Mean
B_1	30	50	70	50
B_2	20	30	40	30
Mean	25	40	55	

(g) A, B, and the interaction are significant

Parts (a), (b), and (d) of Figure 9.10 represent situations in which one or both of the main effects were significant. In each case, the mean scores for the level of variation of at least one of the main effects differ. This can readily be seen from both the numerical examples presented in Table 9.1 and the graphs in Figure 9.10. Note also from Figure 9.10 that the lines for levels B_1 and B_2 are parallel in each of these three cases. In such a situation an interaction cannot exist, because an interaction means that the effect of one variable, such as B_1, depends on the level of the other variable being considered, such as A_1, A_2, or A_3. In each of these cases, the B effect is the same at all levels of A.

Part (c) depicts the classical example of an interaction. Neither main effect is significant, as indicated by the fact that the three-column means are identical and the two-row means are identical and reveal no variation. However, if the A treatment effect is considered only for level B_1, we note that the scores systematically increase from level A_1 to level A_3. In like

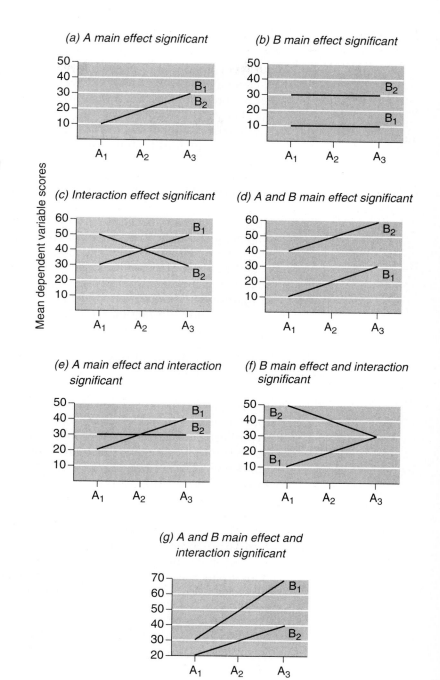

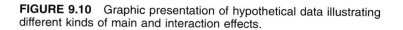

FIGURE 9.10 Graphic presentation of hypothetical data illustrating different kinds of main and interaction effects.

manner, if only level B_2 is considered, then there is a systematic decrease from level A_1 to A_3. In other words, A is effective but in opposite directions for levels B_1 and B_2, or the effect of A depends on which level of B we are considering. This is the definition of *interaction*. I have found graphs to be more helpful than tables in depicting interaction, but you should use whichever mode better conveys the information.

Parts (e) and (f) show examples of situations in which a main effect and an interaction are significant; part (g) represents a case in which both main effects and the interaction are significant. These illustrations exhaust the possibilities that exist in a factorial design having two independent variables. The exact nature of the main effects or the interaction may change, but one of these types of conditions will exist. Before we leave this section, one additional point needs to be made regarding the interpretation of significant main and interaction effects. Whenever either a main or an interaction effect *alone* is significant, you naturally have to interpret this effect. When *both* main and interaction effects are significant, however, and the main effect is contained in the interaction effect, then only the interaction effect is interpreted, because the significant interaction effect qualifies the meaning that would arise from the main effect alone.

A design similar to that just discussed was used by Swann, Wenzlaff, Krull, and Pelham (1992) to investigate the type of feedback preferred by depressed individuals. It was hypothesized that depressed people preferentially seek to interact with others who provide unfavorable feedback, whereas the nondepressed seek out others who provide positive feedback. In investigating this hypothesis, Swann et al. used two independent variables: level of depression and type of feedback. Three levels of depression were established for the first independent variable: nondepressed, dysphoric, and depressed. Similarly, three levels of feedback were established for the second independent variable: positive, neutral, and negative. This gives a 3×3 factorial design (three levels of one independent variable times three levels of the other) with nine cells, as shown in Figure 9.11.

Figure 9.12 displays a graph of the data collected from one portion of this study. This figure shows that an interaction exists between the type of evaluator and the level of depression. Depressed individuals have a greater desire to interact with evaluators or others who provide unfavorable feedback,

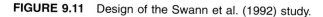

		Level of depression		
		Nondepressed (A₁)	Dysphoric (A₂)	Depressed (A₃)
Type of feedback	Positive (B₁)	A_1B_1	A_2B_1	A_3B_1
	Neutral (B₂)	A_1B_2	A_2B_2	A_3B_2
	Negative (B₃)	A_1B_3	A_2B_3	A_3B_3

FIGURE 9.11   Design of the Swann et al. (1992) study.

whereas nondepressed and dysphoric individuals have a greater desire to interact with others who provide favorable feedback.

So far, the discussion of factorial designs has been limited to those with two independent variables. There are times when it would be advantageous to include three or more independent variables in a study. Factorial designs enable us to include as many independent variables as we consider impor-

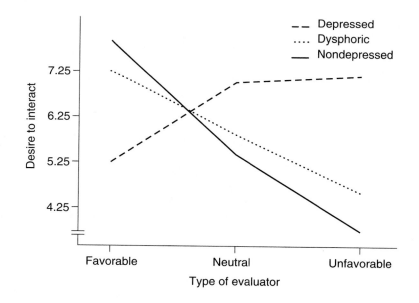

FIGURE 9.12   Self-verification and depression. (From "Allure of Negative Feedback: Self-verification Strivings among Depressed Persons" by W. B. Swann, Jr., R. M. Wenzlaff, D. S. Krull, and B. W. Pelham, 1992. *Journal of Abnormal Psychology, 101,* pp. 293–306.

tant. Mathematically or statistically, there is almost no limit to the number of independent variables that can be included in a study. Practically speaking, however, there are several difficulties associated with increasing the number of variables. First, there is an associated increase in the number of subjects required. In an experiment with two independent variables, each of which has two levels of variation, a 2×2 arrangement is generated, yielding four cells. If ten subjects are required for each cell, the experiment requires a total of forty subjects. In a three-variable design, with two levels of variation per independent variable, a $2 \times 2 \times 2$ arrangement exists, yielding eight cells, and eighty subjects are required in order to have ten subjects per cell. Four variables mean that sixteen cells and 160 subjects are required. As you can see, the required number of subjects increases rapidly with an increase in the number of independent variables. This difficulty, however, does not seem to be insurmountable; many studies are conducted with large numbers of subjects.

A second problem with factorial designs incorporating more than two variables is the increased difficulty of simultaneously manipulating the combinations of independent variables. In an attitude study, it is harder to simultaneously manipulate credibility of the communicator, type of message, sex of the communicator, prior attitudes of the audience, and intelligence of the audience (a five-variable problem) than it is just to manipulate credibility of the communicator and prior attitudes of the audience.

A third complication arises when higher-order interactions are significant. In a design with three independent variables, it is possible to have a significant interaction among the three variables A, B, and C. Consider a study that includes the variables of age, sex, and intelligence. A three-variable interaction means that the effect on the dependent variable is a joint function of the subjects' ages, sex, and intelligence levels. The investigator must look at this triple interaction and interpret its meaning, deciphering what combinations produce which effect, and why. Triple interactions can be quite difficult to interpret, and interactions of an even higher order tend to become unwieldy. Therefore, it is advisable to restrict the design to no more than three variables.

In spite of these problems, factorial designs are very popular because of their overriding advantages when appropriately used. The following four advantages of factorial designs were adapted from Kerlinger (1973, p. 257).

The first advantage is that more than one independent variable can be manipulated in an experiment, and therefore more than one hypothesis can be tested. In a one-variable experiment, only one hypothesis can be tested: Did the treatment condition produce the desired effect? In an experiment with three independent variables, however, seven hypotheses can be tested: one regarding each of the three main effects— A, B, and C—and one regarding each of the four interactions— $A \times B$, $A \times C$, $B \times C$, and $A \times B \times C$.

A second positive feature is that the researcher can control a potentially confounding variable by building it into the design. This, as noted in Chapter 7, is a mechanism for eliminating the influence of an extraneous variable. Naturally, the decision as to whether to include the extraneous variable (such as sex) in the design will partially be a function of how many independent variables are already included. If three or four are already included, it may be wise to effect control in another manner (perhaps by including only females). If only one or two independent variables exist, then the decision in most cases should be to include the extraneous variable in the design. Including the extraneous variable not only controls it but also may provide valuable information about its effect on the dependent variable.

The third advantage of the factorial design is that it produces greater precision than does an experiment with only one variable, for reasons discussed earlier.

The final benefit of the factorial design is that it enables the researcher to study the interactive effects of the independent variables on the dependent variable. This advantage is probably the most important because it enables us to hypothesize and test interactive effects. Testing main effects does not require a factorial design, but testing the interactions does. It is this testing of interactions that lets us investigate the complexity of behavior and see that behavior is caused by the interaction of many independent variables. Lana (1959), for example, specifically set out to test an interactive hypothesis put forth by Solomon (1949) and Campbell (1957). They stated that pretests have a potentially sensitizing effect on attitudes. Using the four-group design to test this interactive hypothesis, Lana found no significant interaction, so pretesting apparently did not have the hypothesized sensitizing effect on attitudes.

Within-subjects after-only design A type of after-only design in which the same subjects are repeatedly assessed on the dependent variable after participating in all experimental treatment conditions

Within-Subjects After-Only Design In the **within-subjects after-only design,** the same subjects participate in all experimental treatment conditions (see Figure 9.13). Actually, this is a repeated measures design, since all subjects are repeatedly measured under each treatment condition. Haslerud and Meyers (1958) used this scheme in their investigation of the transfer value of individually derived principles. All subjects were first trained on problems in which rules were given and on problems in which the subjects had to derive their own rules. After this training, *all* subjects solved problems using both the rules they had been given and the rules they had derived. In other words, subjects served under both conditions.

Among the benefits of using the within-subjects after-only design is the fact that the investigator need not worry about creating equivalence in the participating subjects, because the same subjects are involved in each treatment condition. In other words, subjects serve as their own control, and variables such as age, sex, and prior experience remain constant over the entire experiment. Since the subjects serve as their own control, the subjects in the various treatment conditions are perfectly matched, which increases the sensitivity of the experiment. Therefore, the within-subjects design is maximally sensitive to the effects of the independent variable.

Also, the within-subjects design does not require as many subjects as does the between-subjects design. In the former, with all subjects participating in all treatment conditions, the number of subjects needed for an entire experiment is equal to the number of subjects needed in one experimental treatment condition. In the between-subjects design, the number of subjects needed equals the number of subjects required for one

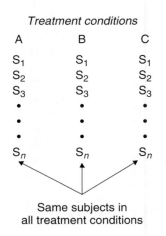

FIGURE 9.13   Within-subjects after-only design.

treatment condition times the number of treatment conditions. If ten subjects are needed in each treatment condition and there are three treatment conditions, then ten subjects would be needed in a within-subjects design, whereas thirty subjects would be needed in a between-subjects design.

With all these advantages, one might think that the within-subjects design would be used more than the between-subjects design. Actually, the reverse is true because of the disadvantages that also accompany the within-subjects design. The most serious handicap of this design is the confounding influence of a sequencing effect. Remember that a sequencing effect can occur when subjects participate in more than one treatment condition. Since the primary characteristic of a within-subjects design is that all subjects participate in all experimental treatment conditions, a sequencing rival hypothesis is a real possibility. In order to overcome the sequencing effect, investigators frequently use one of the counterbalancing techniques discussed earlier. However, counterbalancing controls only linear sequencing effects; if the sequencing effects are nonlinear, then a confounding sequencing influence exists even if counterbalancing is used. Even if the sequencing effect can be overcome through counterbalancing, in a within-subjects design there is no way to identify any impact of the extraneous variables of history, maturation, and statistical regression.

As you can see, there are some serious problems associated with the within-subjects design, and they are generally more difficult to control than those in the between-subjects design. As a result, the within-subjects design is not the most commonly used.

Combining Between- and Within-Subjects Designs Many times in psychological research there are several variables of interest, of which one or more would fit into a between-subjects design and the others would fit into a within-subjects design. Does this mean that two separate studies must be conducted, or can they be combined? As you probably suspected, they can be incorporated into one design, called a **factorial design based on a mixed model.** The simplest form of such a design involves a situation in which two independent variables have to be varied in two different ways. One independent variable requires a different group of subjects for each level of variation. The other independent variable is constructed in such a way that all subjects have to take each level

Factorial design based on a mixed model A factorial design that represents a combination of the within-subjects and the between-subjects designs of variation. Consequently, the first independent variable requires a between-subjects design, and the second independent variable requires a within-subjects design. When these two independent variables are included in the same scheme, it becomes a factorial design based on a mixed model, as illustrated in Figure 9.14.

In this design, subjects are randomly assigned to the different levels of variation of the between-subjects independent variable. All subjects then take each level of variation of the within-subjects independent variable. Therefore, we have the advantage of being able to test for the effects produced by each of the two independent variables, as well as for the interaction between the two independent variables. Additionally, we have the advantage of needing fewer subjects, because all subjects take all levels of variation of one of the independent variables. Therefore, the number of subjects required is only some multiple of the number of levels of the between-subjects independent variable.

The discussion of the factorial design based on a mixed model has been limited to consideration of only two independent variables. In no way is this meant to imply that the design cannot be extended to include more than two independent variables. As with the factorial designs, we can include as many independent variables as are considered necessary. We could include any combination of the between-subjects type of independent variable with the within-subjects

		Within-subjects independent variable		
		A_1	A_2	A_3
	B_1	S_1	S_1	S_1
		S_2	S_2	S_2
		S_3	S_3	S_3
		S_4	S_4	S_4
Between-subjects independent variable		S_5	S_5	S_5
	B_2	S_6	S_6	S_6
		S_7	S_7	S_7
		S_8	S_8	S_8
		S_9	S_9	S_9
		S_{10}	S_{10}	S_{10}

FIGURE 9.14   Factorial design based on a mixed model, with two independent variables.

type of independent variable. If we were conducting a study with three independent variables, two of which required all subjects to take each level of variation of both the independent variables, our design would include two independent variables of the within-subjects variety and one of the between-subjects variety.

Before–After Design

Before–after design
A true experimental design in which the treatment effect is assessed by comparing the difference between the experimental and control groups' pre- and posttest scores

The before–after design differs from the one-group before–after design in two important respects: the former incorporates a control group and randomization. In a **before–after design,** subjects are randomly assigned to groups and then pretested on the dependent variable, *Y.* The independent variable, *X,* is administered to the experimental group, and the experimental and control groups are posttested on the dependent variable, *Y.* The differences between the pre- and posttest scores for the experimental and control groups are then tested statistically to assess the effect of the independent variable. Figure 9.15 depicts this design.

The before–after design is also a good experimental design and does an excellent job of controlling for rival hypotheses such as history and maturation. Whereas the similar but faulty one-group before–after design was said to have been contaminated by extraneous variables, including history and maturation, the before–after plan neatly controls for many of these rival hypotheses. The history and maturation variables are clearly controlled, since any history events that may have produced a difference in the experimental group would also have produced a difference in the control group. Note, how-

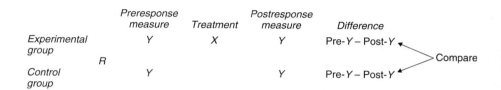

FIGURE 9.15 Before–after design. (Adapted from *Experimental and Quasi-Experimental Designs for Research* by D. T. Campbell and J. C. Stanley, 1963. Chicago: Rand McNally, and Company. Copyright 1963, American Educational Research Association, Washington, D.C.)

ever, that an intragroup history effect could exist in this or any design that includes more than one group of subjects. If all the subjects in the experimental group were treated in one session and all the subjects in the control group were treated in another session, it is possible that events took place in one group that did not take place in the other. If a differential event did take place (for example, laughter, a joke, or a comment about the experimental procedure), there would be no way of eliminating its influence, and it may have produced an effect that would be picked up by the dependent variable. Such an event would have to be considered a possible cause for any significant difference noted between the groups.

The intragroup history effect can be controlled by individually testing subjects who are randomly assigned to the treatment groups and by randomly determining when a control and an experimental treatment will be administered. If group administration of each level of variation of the independent variable is essential, then each separate group potentially has a different intrasession history. In this case, it would be necessary to test the various groups statistically to determine whether differences exist as a function of intrasession history. In other words, groups would have to be included as another independent variable.

Maturation and instrumentation are also controlled in this design because they should be equally manifested in both the experimental group and the control group. Equal manifestation of the testing effect in experiments that use observers or interviewers to collect the dependent variable data does, however, assume that the observers are randomly assigned to individual observation sessions. This ensures that the instrumentation effect is randomly distributed across groups. When this assumption cannot be met, a double blind model should be used, with each available observer used in both experimental and control sessions.

Regression and selection variables are controlled by virtue of the fact that subjects are randomly assigned to both the experimental and the control groups. Randomization ensures initial equality of groups as well as equality in the extent to which each group regresses toward the mean. Since subjects are randomly assigned, each group should have the same percentage of extreme scores and, therefore, should demonstrate the same degree of regression toward the mean. Selection is naturally ruled out, because random assignment has assured equality of the experimental and control groups at the time of randomization. As stated earlier, randomization does not provide 100 percent assurance, and one will occasionally be wrong. It is, however, our *best* protection against the selection rival hypothesis.

In the past (Cronbach and Furby, 1970; Kerlinger, 1973), it has been stated that a statistical analysis computed on the differences between pretest and posttest scores was inappropriate because such gain scores may have low reliability. However, Nicewander and Price (1978) ascertained that the reliability of our dependent variable measures do not have any general relationship to the power of the statistical test. This means that we cannot conclude that the experiment is faulty if the difference score has low reliability. Therefore, it seems appropriate to statistically analyze differences in the gain scores achieved by the experimental and control groups to determine if the experimental groups' gains were significantly greater than those achieved by the control group.

CHOICE OF A RESEARCH DESIGN

It is your task to choose which type of research design is most appropriate for a particular research study. There are some straightforward factors to consider in making the design selection. The choice requires a thorough knowledge of the research problem, of the extraneous variables to be controlled, and of the advantages and disadvantages inherent in the alternative designs available.

Research Question

First and foremost, you must select a design that will give you an answer to your problem. There are times when investigators try to force a problem into a specific research design. This is an example of the tail wagging the dog and seldom allows you to arrive at an appropriate answer. Thus the primary criterion in design selection is whether the design will enable you to arrive at an answer to the research question.

Control

The second factor to consider in selecting a research design is whether you can incorporate control techniques that will allow you to arrive unambiguously at a conclusion. If you have the choice of several designs that would enable you to answer your research question, then you should select the design that will provide maximum control over variables that could also explain the results. Control, therefore, is the second most important criterion.

Between- versus Within-Subjects Design

The third factor to consider is the nature of the research design. In some cases, the problem dictates the type of design. For example, if you are engaged in a learning study, you must give subjects a number of trials to enable them to learn the material. Where trials must be incorporated into the design, a within-subjects design is necessary. Where this is not the case, the investigator can manipulate the independent variable or variables with either a between-subjects or a within-subjects design.

When the options are a between- or a within-subjects design, the decision is usually made in favor of the latter. This is because, as mentioned earlier, the within-subjects design provides a more sensitive test of the independent variable.

SUMMARY

The design of a research study is the basic outline of the experiment, specifying how the data will be collected and analyzed and how unwanted variation will be controlled. The design determines to a great extent whether the research question will be answered. Studies based on inappropriate designs like the one-group after-only design, the one-group before–after design, and the nonequivalent posttest-only design do not provide the desired answers because they do not control for the influence of the many extraneous variables that can affect the results of an experiment.

A true research design satisfies three criteria. First, the design must test the hypotheses advanced. Second, extraneous variables must be controlled so that the experimenter can attribute the observed effects to the independent variable. Third, it must be possible to generalize the results. These three criteria represent the ideal; seldom will a study satisfy them all.

In designing a study that attempts to meet the conditions just stated, many investigators use a pretest. There are a number of good reasons for administering a pretest. It can be used to match subjects and thereby increase the sensitivity of the experiment. It can also be used to determine if a ceiling effect exists or to test a subject's initial position on a variable to see if the variable interacts with the independent variable. Other reasons include testing for initial comparability of subjects and establishing that subjects actually changed as a result of the independent variable.

The after-only and the before–after designs are true, or effective, general research designs because they have the ability to eliminate the influence of extraneous variables that serve as sources of rival hypotheses for explaining the observed results. These designs can control for unwanted variation, because they include a comparison control group and because subjects are randomly assigned to the experimental and control groups. The after-only design represents the prototype of most research designs. Although the basic after-only design is not commonly used, its variants—the between-subjects after-only design and the within-subjects after-only design—are very popular. The between-subjects type is used when subjects must be randomly assigned to the various experimental treatment groups. The within-subjects type is used when subjects must participate in all treatment groups. When different subjects must participate in some experimental treatment conditions and all subjects must participate in other experimental treatment conditions, a combination of the between- and the within-subjects designs is called for.

STUDY QUESTIONS

1. Define research design and identify its purpose.
2. Identify the faulty research designs and explain why each is faulty.
3. What are the requirements of a true research design?
4. What is a control group, and what functions does it serve?

5. Why would a researcher want to pretest subjects, and what difficulties can be caused by pretesting?
6. What is the difference between a between-subjects and a within-subjects design?
7. Identify the different types of between-subjects designs and diagram each one.
8. In what ways does a factorial design differ from a simple randomized design?
9. Define *interaction* and give an example.
10. In what situations would a researcher use a within-subjects design and a factorial design based on a mixed model?

KEY TERMS AND CONCEPTS

Research design

One-group after-only design

One-group before–after design

Nonequivalent posttest-only design

Control group

Experimental group

Ceiling effect

True experimental design

After-only design

Between-subjects after-only design

Simple randomized subjects design

Factorial design

Cell

Main effect

Interaction effect

Within-subjects after-only design

Factorial design based on a mixed model

Before–after design

CHAPTER 10

Quasi-Experimental Designs

LEARNING OBJECTIVES

1. To learn how quasi-experimental designs differ from true experimental designs.
2. To learn the basic characteristics of each of the quasi-experimental designs.
3. To understand how rival hypotheses are ruled out in each of the quasi-experimental designs.

During the 1970s drug use among teenagers and young adults rose dramatically. In 1972, for example, only 14 percent of young people between the ages of twelve and seventeen had tried marijuana. By 1979 this figure had more than doubled to 31 percent. Among eighteen- to twenty-five-year-olds there had been a similar trend, with 48 percent reporting that they had tried marijuana in 1972 and 68 percent in 1979. Although these percentages leveled off in the 1980s, the figures indicate that a large percentage of the population has experimented with drugs. Although these percentages focused on marijuana use, a similar trend has occurred with other, more addictive drugs such as heroin and cocaine. With the increased use of these drugs, more attention and effort have been directed toward controlling access to them. The most widely publicized attempts have been aimed at the manufacture, transportation, and distribution of drugs. However, there have also been increasing efforts toward the rehabilitation of individuals who have become addicted to drugs. Of the numerous drug abuse clinics created during the 1970s, many have continued into the 1990s. One question asked about these clinics is whether they have been effective—a legitimate question, because they are expensive to operate. If all drug addicts required treatment from a drug abuse clinic in order to eliminate their drug addiction, answering this research question would require only that the drug abuse clinics compute the percentage of individuals who have participated in the program and been cured of their addiction. But we all know that some addicts break their drug habit by themselves or with the help of family or friends. Consequently, in order to determine if the drug abuse programs are effective, it is necessary to include a comparison group of addicts who did not participate in a drug abuse program. On the basis of the material that you have learned in previous chapters, you should realize that this research question could be answered using a good experimental design. A sample of addicts seeking treatment could be randomly assigned to a control group that did not receive treatment or to an experimental group that did receive treatment. Then you could compare the percentage of individuals in the two groups who effectively eliminated their drug habit. The problem with using such an approach is that most drug abuse clinics will not allow a researcher to come in and determine randomly whether a person can or cannot receive treatment.

Rather, these clinics state that their mission is to treat drug addicts, and they will accept anyone who requests treatment: it would be unethical to do otherwise. This is one of the primary difficulties encountered in moving out of the laboratory and into the real world. Outside the laboratory setting, it is more difficult to use control techniques, and therefore harder to control for the influence of extraneous variables. But in such cases investigators need not throw up their hands and abandon the research. Rather, they must turn to the use of quasi-experimental designs—designs that enable researchers to investigate problems that preclude the use of procedures required by a true experimental design.

This chapter will present a variety of quasi-experimental designs and will discuss the way in which the influence of rival hypotheses must be considered when these designs are used.

INTRODUCTION

Quasi-experimental design
A research design in which an experimental procedure is applied but all extraneous variables are not controlled

A **quasi-experimental design** is an experimental design that does not meet all the requirements necessary for controlling the influence of extraneous variables. In most instances the requirement that is not met is that of random assignment of subjects to groups. For example, several years ago the head of the probation department in a large metropolitan area approached a group of investigators, of which I was a part, and asked us to design a study to investigate the validity of the hypothesis that the food eaten by juvenile delinquents is causally related to their delinquent behavior. This administrator said that we could use the juveniles who had been committed to one of the detention facilities as our subject pool. Since these youngsters were required to spend all of their time at this detention facility and the food available to them was prepared there, this was an ideal setting in which to test the nutrition–behavior hypothesis. Once we started designing the experiment, however, we encountered some of the constraints that investigators may find when moving out of the laboratory and into the real world. We were told that we could not randomly assign the juveniles into experimental and control groups; they all had to be treated in the same manner. Consequently, we realized at the outset that it would be impossible to conduct a true experiment and that we had to settle for a design that would not provide maximum assurance that the experimental and control groups were equated. In other words, we had to settle for a quasi-experimental design.

You may ask whether it is possible to draw causal inferences from studies based on a quasi-experimental design, since such a design does not rule out the influence of all rival hypotheses. Many causal inferences are made without using the experimental framework; they are made by rendering other rival interpretations implausible. If a friend of yours unknowingly stepped in front of an oncoming car and was pronounced dead after being hit by the car, you would probably attribute her death to the moving vehicle. Your friend might have died as a result of numerous other causes (a heart attack, for example), but such alternative explanations are not accepted because they are not plausible. In like manner, the causal interpretations arrived at from quasi-experimental analysis are those that are consistent with the data in situations where rival interpretations have been shown to be implausible. Of course, the identification of what is and is not plausible is not always as apparent as this illustration suggests. If it were, we would not need to conduct the experiment. I am simply demonstrating the type of procedure that must be used within the framework of quasi-experimental designs.

NONEQUIVALENT CONTROL GROUP DESIGN

Nonequivalent control group design
A quasi-experimental design in which the results obtained from nonequivalent experimental and control groups are compared

A number of designs have been identified by Cook and Campbell (1979) as being **nonequivalent control group designs.** This kind of design includes both an experimental and a control group, but subjects are not randomly assigned. The fact that subjects in the control and experimental groups are not equivalent on all variables may affect the dependent variable. These uncontrolled variables operate as rival hypotheses to explain the outcome of the experiment, making these designs quasi-experimental designs. But where a better design cannot be used, some form of a nonequivalent control group design is frequently recommended. The basic scheme, depicted in Figure 10.1, consists of giving an experimental group and a con-

	Preresponse measure	Treatment	Postresponse measure	Difference
Experimental group	Y_1	X	Y_2	$Y_1 - Y_2$
Control group	Y_1		Y_2	$Y_1 - Y_2$

Compare

FIGURE 10.1   Nonequivalent control group design. (Adapted from *Experimental and Quasi-Experimental Designs for Research* by D. T. Campbell and J. C. Stanley, 1963. Chicago: Rand McNally and Company. Copyright 1963, American Educational Research Association, Washington, D.C.)

trol group first a pretest and then a posttest (after the treatment condition is administered to the experimental group). The pre- to posttest difference scores of the two groups are then compared to determine if significant differences exist. The design appears identical to the before–after experimental design. However, there is one basic difference that makes one a *true* experimental design and the other a *quasi*-experimental design. In the before–after design, subjects are randomly assigned to the experimental and control groups, whereas in the nonequivalent control group design they are not. The absence of random assignment is what makes a design quasi-experimental.

Consider the study conducted by Becker, Rabinowitz, and Seligman (1980), which was concerned with the impact of the billing procedure on energy consumption. Because of the large energy bills resulting from increased energy costs, a number of utility companies have given their customers the option of using an "equal monthly payment plan." This scheme requires the utility company to bill the resident for one-twelfth of the yearly utility cost each month, as opposed to billing for the actual amount consumed. Although such a plan apparently produces a great deal of customer satisfaction, it runs the risk of increasing energy use, because there is no direct connection between energy used and the size of the monthly bill. The study conducted by Becker et al. was designed to determine if the equal monthly payment plan actually led to an increased use of energy.

In conducting such a study, we would ideally assign subjects randomly to either the equal monthly payment plan or the conventional payment plan (in which energy is paid for as it is consumed). For a variety of reasons, however, the utility companies contacted would not allow such random assign-ment, so the investigators had to formulate two groups with-out randomization. This meant that a quasi-experimental design had to be used, and Becker et al. selected the nonequiva-lent control group type. Figure 10.2 shows that the design of the Becker et al. study consisted of pretesting both groups on consumption of electricity prior to the implementation of the equal payment plan. Following this pretesting (which occurred during the summer months), the treatment plan was imple-mented for the experimental group, and consumption of elec-tricity was measured for both groups during the following summer.

In formulating the experimental and control groups, Becker et al. did not have the opportunity to assign subjects randomly, although they were aware of the need to have equated groups. Consequently, they devised a system that seemed to match the subjects on the variables that would influence electrical consumption. One of the companies whose customers were used in the study maintained records in such a way that it was possible to identify next-door neighbors. The investigators reasoned that next-door neighbors would be more likely to have similar-sized homes and to be more similar on other variables that may affect electrical consumption than would a random sample of individuals not on the equal monthly payment plan. Therefore, the control group consisted of next-door neighbors of those individuals who were on the equal monthly payment plan.

The results of this study for one company are depicted in Figure 10.3, which shows that a difference in electrical con-sumption existed between the two groups at pretesting time. However, the change in consumption between pretesting and

	Pretest response	Treatment conditions	Posttest response
Experimental group	Magnitude of electricity consumed	Equal monthly payment plan	Magnitude of electricity consumed
Control group	Magnitude of electricity consumed	Conventional payment plan	Magnitude of electricity consumed

FIGURE 10.2 The design of the Becker, Rabinowitz, and Seligman study.

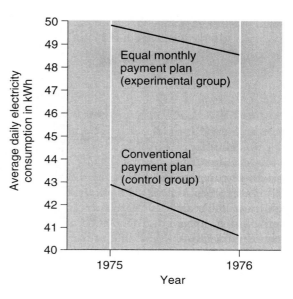

Figure 10.3 Average daily electricity consumption for two payment plans. (Based on data from "Evaluating the Impact of Utility Company Billing Plans on Residential Energy Consumption" by L. J. Becker, V. C. Rabinowitz, and C. Seligman, 1980, *Evaluation and Program Planning, 3,* pp. 159–164.)

posttesting was about the same for both groups. The question now becomes one of interpreting these results. The difference in pretest scores suggests that the two groups were not equivalent at the beginning of the experiment, in which case variables other than the experimental condition may have produced the obtained results. For example, those selecting the equal monthly payment plan used more electricity at the outset and, therefore, may have differed from the control group in a variety of ways.

Cook and Campbell (1979) have pointed out that the rival hypotheses in such a situation tend to be directly related to the results obtained from the experiment. These researchers identified several different experimental outcomes that could occur from the use of a nonequivalent control group design. They then listed the rival hypotheses that could also explain the obtained results. We will first take a look at these outcomes and the rival hypotheses that threaten them and then attempt to relate them to the Becker et al. study.

Increasing treatment effect I outcome
An outcome in which the experimental and the control groups differ at pretesting and only the experimental group's scores change from pre- to posttesting

Outcomes with Rival Hypotheses

Increasing Treatment Effect I Outcome In the **increasing treatment effect I outcome**, illustrated in Figure 10.4, the control group scores reveal no change from pretest to posttest,

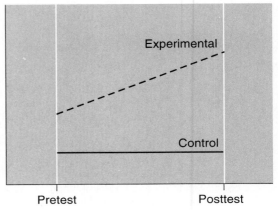

FIGURE 10.4 Increasing treatment effect
I outcome. (From "The Design and
Conduct of Quasi-Experiments and True
Experiments in Field Settings" by T. D.
Cook and D. T. Campbell, in *Handbook of
Industrial and Organizational Psychology*,
edited by M. D. Dunnette. Copyright ©
Rand McNally College Publishing
Company, 1976.)

but the experimental group starts at a higher level and shows
a significant positive change. Such an outcome appears to
suggest that the experimental treatment was effective. However, this outcome could also have occurred as a result of a
selection–maturation effect or a local history effect.

Selection–maturation
effect
The result of selecting
one of two groups in
such a way that its
subjects develop faster
than those in the other
group

A **selection–maturation effect** refers to the fact that one
of the two groups of subjects was selected in such a way that
its subjects were growing or developing faster than the subjects
in the other group. One group may progress faster because its
members are more intelligent or capable than those in the
other group. In the Becker et al. study, an increasing treatment
effect I outcome would have been indicated if the experimental
group had consumed more electricity during pretesting and
had continued to increase consumption between pre- and posttesting, while control group consumption remained stable.
Such an increase could have been caused by the type of payment plan used by the experimental group, but it could also
have been caused by the fact that the salary level of this group
was increasing, so that these subjects were less concerned with
electrical costs. If this were the case, then the posttest increase
could be accounted for by the fact that the selection procedure
happened to place in the experimental group individuals
whose salary levels were increasing more rapidly.

In an attempt to eliminate the potential biasing of this
type of selection–maturation effect, many investigators try to
match subjects. This procedure is supposed to equate subjects
on the matched variables not only at the time of matching but
also during the remainder of the study. If matching is conducted during the pretest, then experimental and control sub-

jects should not differ on the dependent variable measure. If they do not, then it is assumed that they are equated. This equality is supposed to persist over time, so any difference observed during a posttest is attributed to the experimental treatment effect. However, evidence (Campbell and Boruch, 1975; Campbell and Erlebacher, 1970) has revealed that such an assumption could be erroneous because of a statistical regression phenomenon that may occur within the two groups of subjects. This regression phenomenon increases the difference between the two matched groups upon posttesting, apart from any experimental treatment effect. Such a difference could be misinterpreted as being due to a treatment effect or a failure to find a treatment effect, depending on which of the matched groups operated as the experimental group and which operated as the control group.

Assume that we are conducting a study designed to investigate the influence of a Head Start program on children's subsequent school performance. We consider the attitudes of the mothers to be important, so we decide to match on this variable to eliminate its influence. Assume further that the attitude scores obtained from mothers of Head Start children and of non–Head Start children are distributed in the manner shown in Table 10.1. From this table, it is readily apparent that most of the Head Start mothers have lower attitude scores than do the non–Head Start mothers. Therefore, matching involves selecting for the experimental group those Head Start mothers with the highest attitude scores and for the control group those non–Head Start mothers with the lowest attitude scores. In other words, we include only subjects with extreme scores—the subjects most susceptible to the statistical regression phenomenon. This would not be a serious factor if the distributions of scores of the two groups were the same, but they are not. Statistical regression dictates that the Head Start mothers' scores, upon posttesting, will decline and regress toward the mean of their group and that the control subjects' scores will regress or increase toward their group's mean (also illustrated in Table 10.1). Such a regression phenomenon could indicate that the experimental treatment is detrimental when it actually may not have any effect. If the treatment does have a positive effect, this regression effect might lead us to underestimate it.

Another way of attempting to equate subjects by eliminating the selection–maturation bias artifact is to use a variety of statistical regression techniques, such as analysis of covari-

TABLE 10.1 Hypothetical Attitude Scores

Head Start Subjects	Head Start Mothers' Pretest Attitudes	Mothers' Posttest Attitudes	Non–Head Start Subjects	Non–Head Start Mothers' Pretest Attitudes	Mothers' Posttest Attitudes
S_1	5		S_{16}	25	28
S_2	7		S_{17}	27	30
S_3	9		S_{18}	29	32
S_4	11		S_{19}	31	34
S_5	13		S_{20}	33	36
S_6	15		S_{21}	35	
S_7	17		S_{22}	37	
S_8	19		S_{23}	39	
S_9	21		S_{24}	41	
S_{10}	23		S_{25}	43	
S_{11}	25	22	S_{26}	45	
S_{12}	27	24	S_{27}	47	
S_{13}	29	26	S_{28}	49	
S_{14}	31	28	S_{29}	51	
S_{15}	33	30	S_{30}	53	

Matched Subjects

ance and partial correlation. Campbell and Erlebacher (1970) and Campbell and Boruch (1975) have pointed out the fallacy of such an approach, but a discussion of this fallacy is beyond the scope of this text. Suffice it to say that these researchers and others (Cronbach and Furby, 1970; Lord, 1969) have found that such statistical adjustments cannot equate nonequivalent groups unless there is no error in the dependent measures given to the subjects.

Local history effect The result of an extraneous event's influencing either the experimental or the control group, but not both groups

A second rival explanation of the increasing treatment effect I outcome is a **local history effect** (Cook and Campbell, 1975). A general history effect, discussed in Chapter 7, is controlled in the nonequivalent control group design by inclusion of a control group. However, the design is still susceptible to a local history effect, in which some event affects either the experimental or the control group but not both. A local history effect could have operated in the Becker et al. study—if the subjects in the control group had purchased additional insulation for their homes, the control group would have decreased consumption of electricity not because of the type of payment plan but because of the additional insulation. Such a variable would represent a rival hypothesis for any difference observed between the control and the experimental groups.

*** Increasing Treatment and Control Groups Outcome*** In the
increasing treatment and control groups outcome, both the
control group and the experimental group show an increment
in the dependent variable from pre- to posttesting, as is depicted in Figure 10.5. The difference between the increased
growth rates could be the result of an actual treatment effect,
but it could also be due to a type of selection–maturation
interaction. Figure 10.5 indicates that subjects in both groups
are increasing in performance. Note, however, that at the
time of pretesting, the treatment group scored higher on the
dependent variable. This could mean that the subjects in the
experimental treatment group were just naturally increasing
faster on the dependent variable than were the control subjects. The greater difference between groups of subjects at
posttesting might simply reflect the fact that the experimental subjects continued to increase faster on the dependent
variable than did the control subjects. For example, assume
that the dependent variable consisted of a measure of problem-solving ability and that the subjects were six years old at
the time of pretesting and eight years old at the time of
posttesting. Also assume that the experimental subjects were
brighter and therefore increasing in problem-solving ability
more rapidly than were the control subjects. If this were the
case, then we would expect the two groups of subjects to
differ somewhat at pretest time. However, subjects would not
stop increasing in problem-solving ability at age six, and thus
an even greater difference would exist at posttest time, independent of any treatment effect. Where such a differential

*Increasing treatment
and control groups
outcome*
An outcome in which
the experimental and
the control groups differ
at pretesting and both
increase from pre- to
posttesting, but the
experimental group
increases at a faster rate

FIGURE 10.5 Increasing treatment and
control groups outcome. (From "The
Design and Conduct of
Quasi-Experiments and True Experiments
in Field Settings" by T. D. Cook and D. T.
Campbell, in *Handbook of Industrial and
Organizational Psychology,* edited by
M. D. Dunnette. Copyright © Rand
McNally College Publishing Company,
1976.)

growth pattern occurs, we may interpret a greater posttest difference as being the result of a treatment effect when it is really an artifact of a selection–maturation interaction.

Evidence of the existence of a selection–maturation interaction can be seen by looking at the variability of the subjects' scores at pretest and posttest time. Random error dictates that the variability of the scores should be the same on both occasions. A growth factor dictates that the scores should increase in terms of variability, however. Thus an increase in the variability of the scores for the experimental and control groups from pretest to posttest suggests the possibility of the existence of a selection–maturation interaction.

Increasing treatment effect II outcome
An outcome in which the control group performs better than the experimental group at pretesting but only the experimental group improves from pre- to posttesting

Increasing Treatment Effect II Outcome The **increasing treatment effect II outcome,** depicted in Figure 10.6, is an outcome in which the control group and experimental treatment group differ rather extensively at pretest time. However, the experimental group improves over time, presumably because of the experimental treatment, so that the posttest difference is decreased. Such an outcome would be desired when the experimental group was a disadvantaged group and the experimental treatment was designed to overcome the disadvantage. For example, Head Start was initiated to overcome the environmental deprivation experienced by many children in the United States and bring the performance of these disadvantaged individuals up to that of nondisadvantaged children. If a study were conducted to compare the pretest and posttest performances of a group of control individuals (who had not experienced the environmental deprivation) with

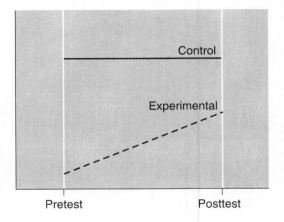

FIGURE 10.6 Increasing treatment effect II outcome. (From "The Design and Conduct of Quasi-Experiments and True Experiments in Field Settings" by T. D. Cook and D. T. Campbell, in *Handbook of Industrial and Organizational Psychology,* edited by M. D. Dunnette. Copyright © Rand McNally College Publishing Company, 1976.)

those of a group of environmentally disadvantaged children who had received the Head Start experimental treatment, we would hope to find the type of effect illustrated in Figure 10.6. However, before we can interpret the increase in performance of the experimental treatment group as being the result of the Head Start experience, several rival hypotheses must be ruled out. The first is a local history effect that affects only one of the two groups of subjects. The second and more likely rival hypothesis is a statistical regression effect—a likely source of confounding, since the subjects in the experimental treatment group are typically selected because of their unusually poor performance or low scores. Consequently, the regression artifact would predict that the scores of this group should increase during posttesting. Statistical regression could, therefore, produce the outcome depicted in Figure 10.6, an outcome that the unwary investigator would interpret as a treatment effect. Therefore, designs that involve administering an experimental treatment to a disadvantaged group should provide a check for the possibility of such a regression artifact.

One indicator of the existence of a regression artifact is the instability of the deprived group's scores in the absence of the experimental treatment. If the deprived group's scores stay consistently low over time, this suggests that the low scores represent the true standing of the individuals. In such cases, a pretest-to-posttest increment would probably represent a true experimental effect, or at least an effect not confounded by the influence of a regression artifact.

Crossover effect
An outcome in which the control group performs better at pretesting but the experimental group performs better at posttesting

Crossover Effect Figure 10.7 depicts the **crossover effect,** an experimental outcome in which the treatment group scores significantly lower than the control group at pretest time but significantly higher at posttest time. This outcome represents the typical interaction effect and is much more readily interpreted than the others discussed because it renders many of the potential rival hypotheses implausible. Statistical regression can be ruled out because it is highly unlikely that the experimental treatment group's lower pretest scores would regress enough to become significantly higher than those of the control group on posttesting. Second, a selection–maturation effect is improbable because it is typically the higher scoring pretest subjects who gain faster. The outcome depicted in Figure 10.7 shows that the subjects scoring lower on the pretest increased their scores more rapidly than did the control group,

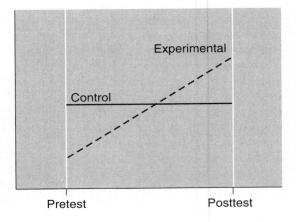

FIGURE 10.7 Crossover effect. (From "The Design and Conduct of Quasi-Experiments and True Experiments in Field Settings" by T. D. Cook and D. T. Campbell, in *Handbook of Industrial and Organizational Psychology*, edited by M. D. Dunnette. Copyright © Rand McNally College Publishing Company, 1976.)

which scored higher on the pretest. This is the opposite of what a selection–maturation outcome would suggest.

TIME-SERIES DESIGN

In such research areas as psychotherapy and education, it is very difficult to find an equivalent group of subjects to serve as a control group. Is the one-group before–after design (discussed in Chapter 9) the only available design in such cases? Is there no means of eliminating some of the rival hypotheses that arise from this design? Fortunately, there is a means for eliminating *some* of these hypotheses, but to do so one must think of mechanisms other than using a control group. "Control is achieved by a network of complementary control strategies, not solely by control-group designs" (Gottman, McFall, and Barnett, 1969, p. 299). These complementary strategies are detailed in the following section.

Interrupted time-series design
A quasi-experimental design in which a treatment effect is assessed by comparing the pattern of pre- and posttest scores of one group of subjects

Interrupted Time-Series Design

The **interrupted time-series design** requires the investigator to take a series of measurements both before and after the introduction of some treatment condition, as depicted in Figure 10.8. The result of the treatment condition is indicated by a

Preresponse measure   *Treatment*   *Postresponse measure*

Y_1 Y_2 Y_3 Y_4 Y_5 X Y_6 Y_7 Y_8 Y_9 Y_{10}

FIGURE 10.8   Interrupted time-series design. (Adapted from *Experimental and Quasi-Experimental Designs for Research* by D. T. Campbell and J. C. Stanley, 1963. Chicago: Rand McNally and Company. Copyright 1963, American Educational Research Association, Washington, D.C.)

discontinuity in the recorded series of response measurements. Consider the study conducted by Lawler and Hackman (1969) in which they tried to identify the benefit derived from employee participation in the development and implementation of an employee incentive plan. Prior research had investigated a variety of payment plans and found that a given plan (say, a bonus plan) may be successful in one instance and not in another, indicating that the success of pay incentive plans is a function of factors other than just the plan itself. Lawler and Hackman hypothesized that a particular pay incentive plan would be more effective if the employees participated in its development, as opposed to having a plan dictated by management. To assess the validity of this theory, Lawler and Hackman had three work groups meet and develop a bonus incentive plan for reducing absenteeism. Absenteeism rates for these work groups were measured before and after the incentive plan was developed. The rates were then converted to a percentage of the number of scheduled hours that the employees actually worked. The average percentage of scheduled hours actually worked for all subjects appears in Figure 10.9. From this figure, you can see that there was a rise in this average percentage and that this rise persisted over the sixteen weeks during which data were collected. All this is a visual interpretation, however. Now it is necessary to ask two questions. First, did a significant change occur following the introduction of the treatment condition? Second, can the observed change be attributed to the treatment condition?

The answer to the first question naturally involves tests of significance, since, as Gottman, McFall, and Barnett (1969, p. 301) have stated, "The data resulting from the best of experimental designs is of little value unless subsequent analyses permit the investigator to test the extent to which obtained differences exceed chance fluctuations." However, before presenting the specific tests of significance, I want to follow the

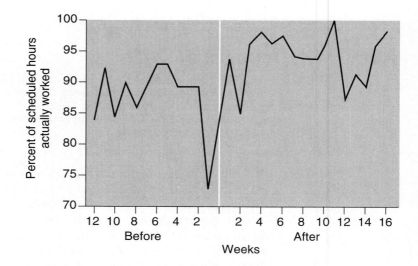

FIGURE 10.9 Mean attendance of the participative groups for the twelve weeks before the incentive plan and the sixteen weeks after the plan. (Attendance is expressed in terms of the percentage of hours scheduled to be worked that were actually worked.) (From "Impact of Employee Participation in the Development of Pay Incentive Plans: A Field Experiment" by E. E. Lawler and J. R. Hackman, 1969, *Journal of Applied Psychology, 53*, pp. 467–471. Copyright 1969 by the American Psychological Association. Reprinted by permission of the author.)

orientation set forth by Campbell and Stanley (1963) and Caporaso (1974) and discuss the possible outcome patterns for time series that would reflect a significant change resulting from an experimental alteration. Let us first take a look at the data that would have been obtained from Lawler and Hackman's 1969 study and a study conducted by Vernon, Bedford, and Wyatt (1924) if they had used only a one-group before–after design. The Vernon et al. study was concerned with investigating the influence of introducing a rest period on the productivity of various kinds of factory workers. These data are presented in Figure 10.10. Note that in *both* studies beautiful data seem to support the hypothesis that the experimental treatment condition produced a beneficial effect. Remember, however, that the one-group before–after design does not include a comparison group, so the increase in performance could have been due to many variables other than the experimental treatment condition.

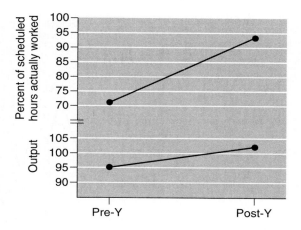

FIGURE 10.10 A one group before–after representation of a portion of the Vernon et al. data and Lawler and Hackman data.

One means of eliminating some of the sources of rival hypotheses is to take a number of pre- and postmeasurements or to conduct an interrupted time-series analysis. When this kind of study is undertaken, we find the data depicted in Figure 10.9 for Lawler and Hackman's study and the data depicted in Figure 10.11 for the Vernon et al. study. The data suggest that the treatment condition investigated by Lawler and Hackman was influential but that the treatment condition investigated by Vernon et al. was not. The pattern of responses obtained by Vernon et al. seems to represent a chance fluctuation rather than a real change in performance.

Visual inspection of a pattern of behavior can be very helpful in determining whether an experimental treatment had a real effect. Caporaso (1974) has presented a number of additional possible patterns of behavior, shown in Figure 10.12, that could be obtained from time-series data. Note that the first three patterns reveal no treatment effect but merely represent a continuation of a previously established pattern of behavior. Lines D, E, F, and G represent *true* changes (I am assuming that they would be statistically significant) in behavior, although line D represents only a temporary shift.

Now let us return to our original question of whether a significant change in behavior followed the introduction of the treatment condition. Such a determination involves tests of significance. The most widely used and, I believe, the most appropriate statistical test is the Bayesian moving average model (Box and Jenkins, 1970; Box and Tiao, 1965; Glass, Tiao, and Maguire, 1971; Glass, Willson, and Gottman, 1975). Basi-

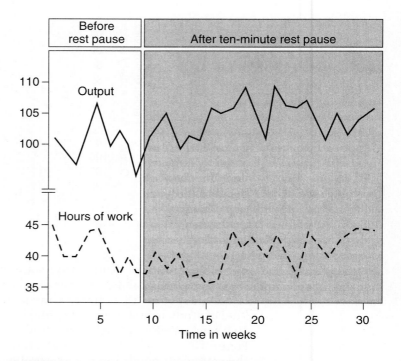

FIGURE 10.11 Effect of a ten-minute rest pause on worker productivity. (Reprinted from *Two Studies of Rest Pauses in Industry* by H. M. Vernon, T. Bedford, and S. Wyatt, 1924. Medical Research Council, Industrial Fatigue Research Board Report No. 25. London: His Majesty's Stationery Office.)

cally, this method consists of determining whether the pattern of postresponse measures differs from the pattern of preresponse measures. To make such an assessment using the moving average model requires many data points. Glass, Willson, and Gottman (1975) recommend that at least fifty data points be obtained. If enough data points cannot be collected to achieve the desired level of sensitivity, Cook and Campbell (1975) advocate using the time-series design. The data can be plotted on a graph and visually inspected to determine whether a discontinuity exists between the pre- and postresponse measures. Naturally, this approach should be used only when one cannot use an appropriate statistical test, and one should remember that the number of preresponse data points obtained must be large enough to identify all the plausible patterns that may exist.

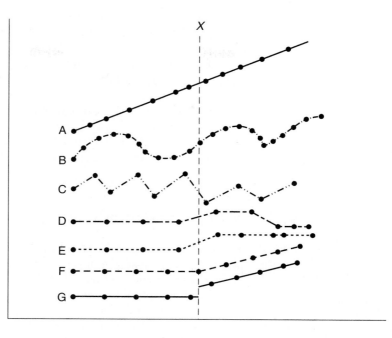

FIGURE 10.12 Possible pattern of behavior of a time-series variable. (From "Quasi-Experimental Approaches to Social Science" by J. A. Caporaso, in *Quasi-Experimental Approaches,* edited by J. A. Caporaso and L. L. Ross, Jr., 1973. Evanston, Ill: Northwestern University Press.)

Lawler and Hackman's analysis of their data revealed a significant difference between the patterns of pre- and postresponse measures. This led them to conclude that a nonrandom change occurred following the introduction of the incentive plan. This brings us to the second question: whether this significant change can be attributed to the employees' participation in the incentive plan. The primary source of weakness in the interrupted time-series design is its failure to control for the effects of history. Considering Lawler and Hackman's study, assume that at about the same time the treatment condition was introduced, some extraneous event occurred that could also have led to an increase in the number of hours worked. Such an extraneous event serves as a rival hypothesis for the significant nonrandom change. The investigator must consider all the other events taking place at about the same time as the experimental event and determine whether they might be rival hypotheses. Actually, Lawler and Hackman included several other control groups in their study to rule out such effects.

The two other potential but unlikely sources of rival hypotheses that could be found in the interrupted time-series design are maturation and instrumentation. The interested reader is referred to Campbell and Stanley (1963, p. 41) for a discussion of the unique situations in which these biases may crop up. Glass, Willson, and Gottman (1975) also present a discussion of sources of invalidity in the time-series experiment.

Multiple Time-Series Design

Multiple time-series design
A time-series design in which a control and an experimental group are included to rule out a history effect rival hypothesis

The **multiple time-series design** is basically an extension of the interrupted time-series design. However, it has the advantage of eliminating the history effect by including an equivalent—or at least comparable—group of subjects that does not receive the treatment condition. As Figure 10.13 shows, in this design one experimental group receives the treatment condition and an equivalent or comparable control group does not. Consequently, the design offers a greater degree of control over sources of rival hypotheses. The history effects, for example, are controlled because they would influence the experimental and control groups equally.

Consider the study conducted by Campbell and Ross (1968) in which they attempted to assess the impact of Connecticut Governor Ribicoff's crackdown on speeding violators in 1955. In one portion of this study, a multiple time-series

	Preresponse measure				Treatment	Postresponse measure			
Experimental group	Y_1	Y_2	Y_3	Y_4	X	Y_5	Y_6	Y_7	Y_8
Control group	Y_1	Y_2	Y_3	Y_4		Y_5	Y_6	Y_7	Y_8

FIGURE 10.13   Multiple time-series design. (Adapted from *Experimental and Quasi-Experimental Designs for Research* by D. T. Campbell and J. C. Stanley, 1963. Chicago: Rand McNally and Company. Copyright 1963, American Educational Research Association, Washington, D.C.)

design was used. The state of Connecticut was naturally used for the experimental group; a number of adjacent states were used for the control group. The number of traffic fatalities was plotted for the years 1951 through 1959, as shown in Figure 10.14. If you look just at the line representing the Connecticut fatality rate, particularly following the year of the crackdown, it seems as though there was a definite decline. If only these data were presented, the design would be of the interrupted time-series type. Remember, however, that history effects may serve as rival hypotheses in such designs. As Campbell and Ross have pointed out, the immediate decline occurring in 1956 could be the result of such effects as less severe winter driving conditions or more safety features on automobiles. These effects may even persist for several years, creating the progressive downward trend indicated.

In order to state conclusively that the downward trend was caused by the crackdown on speeding violators and to eliminate the rival hypothesis of history, it was necessary to use a multiple time-series design that included a comparable control group. As Figure 10.14 illustrates, a comparison group

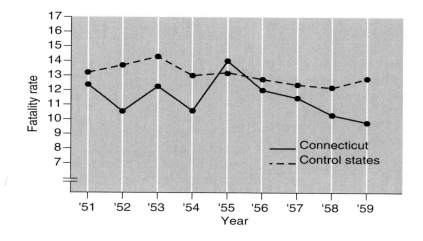

FIGURE 10.14 Connecticut and control states' traffic fatalities for the years 1951–1959. (From "The Connecticut Crackdown on Speeding: Time Series Data in Quasi-Experimental Analysis" by D. T. Campbell and H. L. Ross, 1968, *The Law and Society Review, 3,* p. 44. Reprinted by permission of the Law and Society Association. *The Law and Society Review* is the official publication of the Law and Society Association.)

was incorporated into the study. Campbell and Ross used as their comparison group a pool of adjacent states (New York, New Jersey, Rhode Island, and Massachusetts). This group experienced no progressive decline paralleling the one depicted by the Connecticut data. Graphically, it seems as though the crackdown had a slight effect.

However, it is still necessary to analyze the data statistically to determine if this slight decline is significant. Again, the most appropriate approach seems to involve using the moving-average model discussed by Gottman, McFall, and Barnett (1969) and presented in considerable detail in Glass, Willson, and Gottman (1975).

SUMMARY

This chapter has deviated considerably from the orientation taken in the previous chapters by presenting a number of quasi-experimental designs, which represent approximations of true experimental designs in the sense that they use the experimental mode of analysis in investigating areas that do not allow for complete control of extraneous variables. Quasi-experimental designs are the best type of design available for use in some field studies in which one wants to make causal inferences. Of the different quasi-experimental designs presented, one is a before–after design and two are time-series designs.

The nonequivalent control group design, a before–after design, is the one most frequently used. It is exactly like the before–after true experimental design except that subjects are not randomly assigned to the experimental and control groups, which means that we do not have the necessary assurance that the two groups of subjects are equated. We could attempt to equate subjects on the important variables using matching techniques. However, this still does not assure us that the subjects are totally equated, and it may produce a statistical regression effect. Nonetheless, the design is useful in that it may control for effects such as history and maturation.

With the exception of the multiple time-series design, the time-series designs attempt to eliminate rival hypotheses without the use of a control group. In the interrupted time-series design, a series of measurements is taken on the dependent variable both before and after the introduction of some experimental treatment condition. The effect of that condition is then determined by examining the magnitude of the discontinuity produced by the condition in the series of recorded responses. The primary source of error in this design is the possible history effect. If an equivalent or comparable group of subjects who have not received the experimental treatment condition can be found, the potential history effect can be eliminated by using a multiple time-series design.

STUDY QUESTIONS

1. Define quasi-experimental design.
2. Describe the nonequivalent control group design and what makes this design a quasi-experimental design.
3. How are potential rival hypotheses identified when using the nonequivalent control group design? What are some outcomes that might be confounded by rival hypotheses? Identify the rival hypotheses that might account for the results.
4. What is a time-series design, and how does this design attempt to eliminate the influence of rival hypotheses?
5. What are two types of time-series designs, and how do they differ?

KEY TERMS AND CONCEPTS

Quasi-experimental design

Nonequivalent control group design

Increasing treatment effect I outcome

Selection–maturation effect

Local history effect

Increasing treatment and control groups outcome

Increasing treatment effect II outcome

Crossover effect

Interrupted time-series design

Multiple time-series design

CHAPTER 11

Single-Subject Research Designs

LEARNING OBJECTIVES

1. To understand the strategy used in single-subject designs to rule out the influence of rival hypotheses.
2. To learn the different types of single-subject designs.
3. To be able to identify the situations in which each of the single-subject designs would be appropriate.
4. To gain an understanding of the methodological issues that must be considered in using the single-subject designs.

In 1979 Dr. Winston Shen and Dr. Terence D'Souza reported in the Rocky Mountain Medical Journal *a case of a thirty-one-year-old black male truck driver who was brought into the hospital in an agitated and confused state. He complained that his nerves were "bad" and that he was being attacked by bright shiny flies and bugs controlled by "voodoo." "Get them off me!" he shouted. He yelled that those flying insects were trying to bite him and that someone was pouring gasoline on him.*

Obviously this man was hallucinating; he exhibited all the signs of being emotionally disturbed. During the next twenty-four hours, his mental status remained unchanged. Within another twenty-four hours, however, his hallucinations and disturbed emotional state had disappeared. An interview with the man revealed that he had no personal or family history of mental illness. He denied using street drugs, an assertion that was confirmed by laboratory tests. He further reported that he drank about two cans of beer a day, did not use coffee excessively, had no history of seizures or other neurological deficits, had attended college for two years, and was married with two children. This information provided few clues to the origin of the temporary but severe emotional disturbance that this man had just experienced.

One other aspect of the interview, however, did provide a clue to the possible cause of the brief decline of his emotional stability. During the previous month his job had required him to load and unload crates of soft drinks in a cola factory. One of the benefits of working in such a factory is that there is continuous access to soft drinks. This man had been consuming liberal quantities of cola during this period—about ten cans a day. On the day before his hospitalization, he had consumed twenty to twenty-five cans of cola. Since cola drinks contain about fifty milligrams of caffeine per twelve-ounce can, this man had been consuming quite a bit of caffeine. This provides a reasonable explanation for the transient nature of his emotional disturbance. But this is a post hoc and, therefore, inadequate explanation of his emotional disturbance.

In order to determine if caffeine was the culprit in producing this person's emotional disturbance, we would have to conduct an experiment assessing the effect of caffeine. In trying to do so, we would immediately encounter a difficulty. The effect of caffeine is very idiosyncratic, influenc-

ing different people in diametrically opposed ways. For example, caffeine is typically considered a stimulant, but for some people it seems to act as a sedative, and for others it has little if any apparent effect. Further, some people cannot tolerate large quantities of caffeine, whereas others tolerate massive doses very well. Therefore, if we wanted to conduct an experiment assessing the impact of large quantities of caffeine on emotional disturbance, which individuals would we use as subjects? At present, this question has not been answered, making it very difficult to assess the impact of caffeine on emotional distress. We do know, however, that the person just described experienced a temporary period of emotional distress, and that he had consumed large quantities of caffeine prior to the onset of this distress. Therefore, he may represent a caffeine-sensitive individual. However, in conducting an experiment to determine whether caffeine ingestion was the cause of his emotional distress, we are limited to this one individual. This means that we cannot use either random assignment or inclusion of a control group, the two primary techniques that are typically used to control for the influence of rival hypotheses. How can we control for the influence of rival hypotheses when conducting an experiment on only one subject? The answer is to make use of single-subject designs—designs constructed for use with only one subject and in a manner that controls for the influence of most rival hypotheses.

This chapter will present the most frequently used single-subject designs and demonstrate how each of them enables the investigator to assess the impact of an independent variable while at the same time controlling for the influence of rival hypotheses.

INTRODUCTION

Single-subject research designs
Research designs in which a single subject is used to investigate the influence of a treatment condition

Single-subject research designs are designs that use only one subject to investigate the influence of some experimental treatment condition. Encountering these designs for the first time, most people tend to equate them with case studies, but this is incorrect: single-subject designs experimentally investigate a treatment effect, whereas case studies provide an in-depth description of an individual. Dukes (1965) found that only 30 percent of all single-subject studies are case studies; the great majority are of the experimental variety. A brief look at the history of experimental psychology reveals that psychological research actually began with the intensive study of a single organism. Wundt's use of the method of introspection required a highly trained single subject. Ebbinghaus (1885/1913) conducted his landmark studies on memory using only one subject—himself. Pavlov's (1928) basic findings were the result of experimentation with a single organism (a dog) but were replicated on other organisms. As you can see, single-subject research was alive and well during the early history of psychology. In 1935, however, Sir Ronald Fisher published a book on experimental design that altered the course of psychological research. In it, Fisher laid the foundation for conducting and analyzing multisubject experiments. Psychologists quickly realized that the designs and statistical procedures elaborated by Fisher were very useful and began to adopt them. With the publication of Fisher's (1935) work, psychologists turned from single-subject studies toward multisubject designs. The one notable exception to this tradition was B. F. Skinner (1953), his students, and his colleagues. They developed a general approach that has been labeled the *experimental analysis of behavior.* This method is devoted to experimentation with a single subject (or with only a few subjects) on the premise that the detailed examination of a single organism under rigidly controlled conditions will yield valid conclusions about a given experimental treatment condition. Use of this approach led to the development of a variety of single-subject experimental designs, which form the basis of the single-subject research designs.

The single-subject research designs developed by Skinner (1953) and his colleagues and explicated by Sidman (1960) probably would not have experienced the level of acceptability that they currently enjoy without the growth in popularity of behavior therapy (Hersen and Barlow, 1976). In psychotherapy, the case study has been the primary method of investigation (Bolger, 1965). During the 1940s and 1950s researchers grounded in experimental methodology attacked the case study method on methodological grounds (Hersen and Barlow, 1976). This led some investigators to focus on the percentage of clients who were successfully treated by a given psychotherapy. Eysenck (1952) demonstrated the inadequacy of this method, however, by showing that the percentage of successes achieved by psychotherapy was no greater than that achieved by a spontaneous remission of symptoms. This disturbing evidence led researchers to focus even more on the multisubject design, and they found many difficulties in applying this approach (Bergin and Strupp, 1972). More significantly, they found "that these studies did not prove that psychotherapy worked" (Hersen and Barlow, 1976, p. 12). Such evidence left researchers perplexed, and some (e.g., Hyman and Berger, 1966) wondered if psychotherapy could be evaluated. Other researchers lapsed into naturalistic studies of the therapeutic process; still others engaged in process research, which emphasizes what goes on during therapy and deemphasizes the outcome of therapy. These efforts did little to advance knowledge of psychotherapy. By the 1960s, there was tremendous dissatisfaction with clinical practice and research, prompting the search for other alternatives. Bergin (1966) thought that the multisubject designs failed to demonstrate the effectiveness of therapy because results were averaged. In the studies he reviewed, he noted that some clients improved and some got worse; when results of these two types of clients were averaged, however, the effects canceled each other out, indicating that therapy had no effect. Given such evidence, along with the fact that process research was not beneficial in increasing effectiveness of therapy, some researchers (Bergin and Strupp, 1970) began to advocate returning to the use of experimental case studies, which employ an experimental analogue. Research was making a change back to single-subject research. During the 1960s, however, an appropriate methodology for experimentally investigating the single subject was not apparent. It took the growing popularity of behavior therapy to provide a vehicle for the use of the appropriate methodology. Since behavior therapy involved the application of many of the principles of learning that had been identified in the laboratory, it was but a small step for these applied researchers also to borrow the procedures used to identify these principles. This methodology, successfully used in applied settings, has become accepted for use in identifying the influence of antecedents on individual behavior.

SINGLE-SUBJECT DESIGNS

When planning an experimental study that uses only one subject, it is necessary to use some form of time-series design.

Recall that the time-series design requires that repeated measurements be taken on the dependent variable both before and after the treatment condition is introduced. This is necessary to permit detection of any effect produced by the treatment condition, because it is not possible to include a control group of subjects, such as in the case where we wanted to determine if caffeine was the cause of the emotional disturbance experienced by the truck driver. Here, there is only one subject, and we want to know whether caffeine caused his emotional disturbance. We could administer the caffeine and measure the subject's level of emotional stability, but then we would have no basis for determining whether the caffeine produced the effect, because we would not know how stable the subject was when he was not consuming the caffeine. Without such a comparison, it is impossible to infer any effect of the treatment condition.

What can we use as a basis of comparison in a single-subject design? Since there is only one subject in the study, the comparison responses have to be the subject's own pretreatment responses. In other words, the investigator has to record the subject's responses before and after administration of the independent variable. In the caffeine experiment, we would have to record the subject's level of emotional stability prior to and after consuming caffeine. If we take only one pre- and postresponse measure, we will have a one-group before–after design, which has many disadvantages. To overcome some of those problems (such as maturation), we must obtain multiple pre- and postresponse measures. For example, we could measure the truck driver's level of emotional stability each day over a period of two weeks prior to consuming caffeine and while consuming caffeine. Now we have a time-series design using one subject, which represents descriptive experimentation because it furnishes a continuous record of the organism's responses during the course of the experiment. Using this procedure, we would have a continuous record of the truck driver's level of emotional stability over the course of the entire experiment. This technique is also experimental because it permits us to interject a planned intervention—a treatment condition such as caffeine—into the program. Consequently, it allows us to evaluate the effect of an independent variable.

Although the basic time-series design can be used in single-subject research, we must remember that it is only a quasi-experimental design. Taking repeated pre- and postintervention measures of the dependent variable does allow us to rule out many potential biasing effects, but it does *not* rule out the possibility of a history effect. Risley and Wolf (1972) have pointed out that the ability to detect a treatment effect with the time-series design hinges on the ability to predict the behavior of the subject if the treatment condition had not been administered. When using the time-series design, we collect both pre- and postintervention measures of the dependent variable. In determining whether the treatment or intervention had any effect on behavior, we compare the pre– and post–dependent variable measures to see if there is a change in the level or the slope of the responses. However, in this assessment, the underlying assumption is that the pattern of preresponse measures would have continued if the treatment intervention had not been applied. In other words, the pretreatment responses are used to forecast what the posttreatment responses would have been in the absence of the treatment. If this forecast is inaccurate, then we cannot adequately assess the effects of the treatment intervention, because the pretreatment responses do not serve as a legitimate basis for comparison. The basic time-series design, then, is truly limited in unambiguously identifying the influence of an experimental treatment effect.

A-B-A DESIGN

A-B-A design
A single-subject design in which the response to the treatment condition is compared to baseline responses recorded before and after treatment

In order to improve on the basic time-series design in an attempt to generate unambiguous evidence of the causal effect of a treatment condition, a third phase has been added. This third phase, a withdrawal of the experimental treatment conditions, makes the design an **A-B-A design.** The A-B-A design, depicted in Figure 11.1, represents the most basic of the single-subject research plans. As the name suggests, it has three separate conditions. The A condition is the baseline condition, which is the target behavior as recorded in its freely occurring

A	B	A
Baseline measure	Treatment condition	Baseline measure

FIGURE 11.1   A-B-A design.

Baseline
The target behavior of
the subject in its
naturally occurring state
or prior to presentation
of the treatment
condition

state. In other words, **baseline** refers to a given behavior as observed prior to presentation of any treatment designed to alter this behavior. The baseline behavior thus gives the researcher a frame of reference for assessing the influence of a treatment condition on this behavior. The B condition is the experimental condition, wherein some treatment is deliberately imposed to try to alter the behavior recorded during baseline. Generally, the treatment condition is continued for an interval equivalent to the original baseline period, or until some substantial and stable change occurs in the behaviors being observed (Leitenberg, 1973).

After the treatment condition has been introduced and the desired behavior generated, the A condition is then reintroduced. There is a return to the baseline conditions—the treatment conditions are withdrawn and whatever conditions existed during baseline are reinstated. This second A condition is reinstituted in order to determine if behavior will revert back to its pretreatment level. It is generally assumed that the effects of the treatment are reversible, but this is not always the case. Reversal of the behavior back to its pretreatment level is considered to be a crucial element for demonstrating that the experimental treatment condition, and not some other extraneous variable, produced the behavioral change observed during the B phase of the experiment. If the plan had included only two phases (A and B), as in the typical time-series design, rival hypotheses could have existed. However, if the behavior reverts back to the original baseline level when the treatment conditions are withdrawn, rival hypotheses become less plausible.

Consider the study conducted by Walker and Buckley (1968). These researchers investigated the effect of using positive reinforcement to condition attending behavior in a nine-year-old boy named Phillip. A bright, underachieving child, Phillip was referred to the investigators because he exhibited deviant behavior that interfered with classroom performance. Specifically, Phillip demonstrated extreme distractability, which often kept him from completing academic assignments. The investigators first took a baseline measure of the percentage of time that Phillip spent on his academic assignment. After the percentage of attending time had stabilized, the treatment condition was introduced, which consisted of enabling Phillip to earn points if no distraction occurred during a given time interval. These points could then be exchanged for a model of his choice. When Phillip had completed three successive ten-minute distraction-free sessions, the reinforcement of being able to earn points was withdrawn. Figure 11.2 depicts the results of this experiment. During the first baseline (A) condition, attending behavior was very low. When the treatment contingency (B) of being able to earn points was associated with attending behavior, percentage of attending behavior increased dramatically. When the contingency was withdrawn and baseline conditions were reinstated (A), attending behavior dropped to its pretreatment level.

In this case, the A-B-A design seems to provide a rather dramatic illustration of the influence of the experimental treatment conditions. However, there are several problems with this design (Hersen and Barlow, 1976). The first of these is that the design ends with the baseline condition. From the standpoint of a therapist or other individual who desires to have some behavior changed, this is unacceptable because the

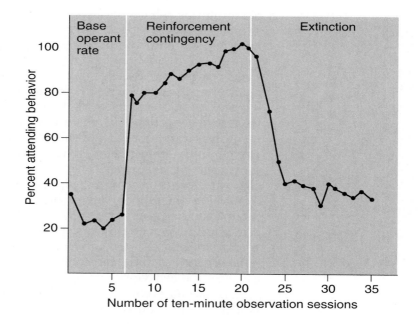

FIGURE 11.2 Percentage of attending behavior in successive time samples during the individual conditioning program. (From "The Use of Positive Reinforcement in Conditioning Attending Behavior" by H. M. Walker and N. K. Buckley, 1968, *Journal of Applied Behavior Analysis, 1*, p. 247. Copyright 1968 by the Society for the Experimental Analysis of Behavior, Inc.)

benefits of the treatment condition are denied. Fortunately, this limitation is easily handled by adding a fourth phase to the A-B-A design in which the treatment condition is reintroduced. We now have an A-B-A-B design, as illustrated in Figure 11.3. The subject thus leaves the experiment with the full benefit of the treatment condition.

Quattrochi-Tubin and Jason (1980) provide a good illustration of the A-B-A-B design. Their study actually used the responses from a single *group* of subjects rather than from a single subject. This design is flexible; it can be used with a single group or with a single subject. Quattrochi-Tubin and Jason investigated a means for getting residents of a nursing home to increase attendance and social interaction in the lounge area instead of remaining in their rooms or passively watching television. (Such increased activity is considered important to the mental and physical well-being of the elderly.) The experiment was divided into four phases, with each phase consisting of four days. The first four days were the baseline phase, during which the experimenters merely recorded, on two different occasions, the number of residents present in the lounge, the number of those present watching television, and the number of those present engaged in social interaction. The second phase was the treatment phase, during which an announcement was made on the public-address system that coffee and cookies were available in the lounge. After four treatment days, the third phase (the baseline conditions) was instituted, which meant that the refreshments were no longer offered. In the fourth phase, refreshments (the treatment condition) were again served. Figure 11.4 depicts the number of elderly residents present in the lounge as well as those engaged in social interaction or watching television during each of the four phases of the experiment. Apparently, attendance and social interaction increased and television watching decreased when refreshments were offered, suggesting that a simple act of incorporating coffee and cookies in a nursing home routine can alter the behavior of its residents.

A second problem of the A-B-A design is not so easily handled. As previously stated, one of the basic requirements of the A-B-A design is that the situation revert to the baseline conditions when the experimental treatment condition is

	A	B	A	B
FIGURE 11.3 A-B-A-B design.	Baseline measure	Treatment condition	Baseline measure	Treatment condition

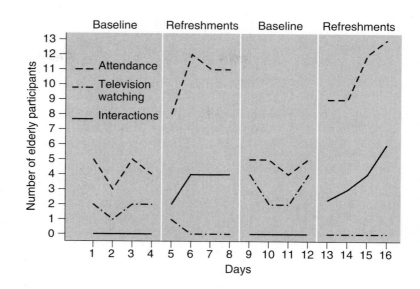

FIGURE 11.4 Attendance, television watching, and social interaction during baseline and refreshment phases in the lounge. (From "Enhancing Social Interactions and Activity among the Elderly through Stimulus Control" by S. Quattrochi-Tubin and L. A. Jason, 1980, *Journal of Applied Behavior Analysis, 13*, pp. 159–163. Copyright 1980 by the Society for the Experimental Analysis of Behavior, Inc.)

withdrawn. This requirement is necessary in order to rule out rival hypotheses such as history, since, if the behavior did not revert to baseline, we would have an A-B, or time-series, design. Quattrochi-Tubin and Jason needed to demonstrate that the attendance, television watching, and social interactions returned to baseline levels once the refreshments were taken away in order to show that the behavior was modified by the experimental treatment. As Gelfand and Hartmann (1968, p. 211) have stated:

> After substantial and apparently reliable behavior modification has taken place, the . . . contingencies should be altered temporarily, for example, reversed, so the problem behavior is once again . . . instated. . . . Correlated changes in the observed response rate provide a convincing demonstration that the target behavior is unmistakably under the therapist's control and not due to adventitious, extratherapeutic factors.

In other words, the reversal is necessary to rule out rival hypotheses such as history.

The problem with the A-B-A design is that a reversal to baseline does not occur with all behavior. Hewett, Taylor, and Artuso (1969) found that removal of a token program increased the target behaviors rather than returning them to baseline. Without the reversal, the experimenter cannot be sure that the change in behavior following introduction of the treatment condition was not caused by some other extratreatment factor. Failure to reverse may be due to a carry-over effect across phases, whereby the treatment condition was maintained so long that a relatively permanent change in behavior took place. In fact, Bijou, Peterson, Harris, Allen, and Johnston (1969) have recommended that short experimental periods be used to facilitate obtaining a reversal effect. This is in line with Leitenberg's (1973, p. 98) statement that "single-case experimental designs are most pertinent to the discovery of short-term effects of therapeutic procedure while they are being carried out." Once the influence of the experimental treatment has been demonstrated, attention can then be placed on its persistence.

Although the argument for shortening the experimental treatment to facilitate reversal (thereby demonstrating cause of the change in behavior) is valid, it applies only to behaviors that will in fact reverse. When the investigator is interested in nontransient effects, none of these arguments is valid because a relatively permanent change is instated.

A last issue concerns a distinction between a reversal and a withdrawal A-B-A design. In discussing the A-B-A design, I have described **withdrawal,** in which the treatment condition is removed during the third (second A) phase of the design. Leitenberg (1973) states that the A-B-A withdrawal design should be distinguished from an A-B-A **reversal design.** The distinction occurs in the third (second A) phase of the A-B-A design. In the withdrawal design, the treatment condition is withdrawn; in the reversal design, the treatment condition is applied to an alternative but incompatible behavior. For example, assume that you were interested in using reinforcement to increase the play behavior of a socially withdrawn $4\frac{1}{2}$-year-old girl, as were Allen, Hart, Buell, Harris, and Wolf (1964). If you followed the procedure used by these investigators, you would record the percentage of time the girl spent interacting with both children and adults during the baseline phase. During treatment (the B phase), praise would be given whenever the girl interacted with other children, and isolated play and interaction with adults would be ignored. During the third phase of

Withdrawal
Removal of the treatment condition

Reversal design
A design in which the treatment condition is applied to an alternative but incompatible behavior so that a reversal in behavior is produced

the experiment (the second A phase), the true reversal would take place. Instead of being withdrawn, the contingent praise would be shifted to interactions with adults so that any time the child interacted with adults she would be praised, and interactions with other children would be ignored. This phase was implemented to see if the social behavior would reverse to adults and away from children as the reinforcement contingencies shifted. Although the A-B-A reversal design can reveal rather dramatic results, it is more cumbersome and thus is used much less frequently than the more adaptable withdrawal design. Therefore, most of the single-subject A-B-A designs that you encounter will be of the withdrawal variety.

INTERACTION DESIGN

A survey of the literature on single-subject designs shows that researchers have not been content to stick to the basic A-B-A design, but instead have extended this basic design in a variety of ways. The most intriguing and valuable extension is used to identify the interactive effect of two or more variables. In discussing multisubject designs, I described interaction as the situation that exists when the influence of one independent variable depends on the specific level of the second independent variable. This definition of interaction was presented because multisubject designs allow us to include several levels of variation for each independent variable being investigated. In a single-subject design, we do not have that degree of flexibility. One of the cardinal rules in single-subject research (Hersen and Barlow, 1976) is that only one variable can be changed from one phase of the research to another. For example, in the A-B-A-B design, we can introduce a specific type or level of reinforcement when changing from the baseline phase to the treatment phase of the experiment. However, only one level of reinforcement can be implemented. Therefore, when I

Interaction effect in single-subject research The combined influence of two or more specific levels of two or more independent variables

discuss an **interaction effect in single-subject research,** I am referring to the combined influence of two or more specific levels of two or more different independent variables. For example, we could investigate the interaction effect of a concrete reinforcement (giving of tokens) and verbal reinforcement (the experimenter saying "good"). It would not be practical to investigate the interaction of different forms of material reinforcement (tokens, points, and candy) with different forms of praise. Therefore, interaction typically refers to the combined influence of two specific variables.

In order to isolate the interactive effect of two variables from the effect that would be achieved by only one of these variables, it is necessary to analyze the influence of each variable separately and in combination. To complicate the issue further, we must do this by changing only one variable at a time. Thus the sequence in which we test for the influence of each variable separately and in combination must be such that the influence of the combination of variables (interaction effect) can be compared with that of each variable separately. Figure 11.5 illustrates this design. In sequence 1, the effect of treatment B is independently investigated, and then the combined influence of treatments B and C is compared to the influence of treatment B alone. In like manner, sequence 2 enables the investigation of the influence of treatment C independently, and then the combined influence of treatments B and C against treatment C. In this way, it is possible to determine if the combined influence of B and C was greater than that of B or C. If it was, then an interactive effect exists. However, if the combined effect was greater than that of one of the treatment variables (C) but not the other (B), then an interactive effect does not exist because the effect can more parsimoniously be attributed to treatment B.

One of the more useful illustrations of a test of an interaction effect is found in the combined studies of Leitenberg, Agras, Thompson, and Wright (1968) and Leitenberg (1973). In the first study, Leitenberg et al. used feedback and praise to overcome a severe knife phobia in a fifty-nine-year-old woman. The dependent variable measure was the amount of time the subject could spend looking at an exposed knife. Following the completion of a trial, the subject was given feedback and/or praise regarding the amount of time spent observing the knife. Praise consisted of verbally reinforcing the subject when she

	Baseline	Single treatment	Baseline	Single treatment	Combined treatment	Single treatment	Combined treatment
Sequence 1	A	B	A	B	BC	B	BC
Sequence 2	A	C	A	C	BC	C	BC

FIGURE 11.5 Single-subject interaction design.

looked at the knife for progressively longer periods of time. Feedback consisted of telling the subject how much time had been spent observing the knife. The specific design of this study is depicted in Figure 11.6. Although it does not correspond exactly to an interaction design, it is close enough to demonstrate the essential components.

The results of this study appear in Figure 11.7. Feedback resulted in an increase in mean viewing time. This increase does not appear to have been altered by the introduction of praise, suggesting that praise had no effect on the knife phobia. During the third phase of the experiment, when praise was withdrawn, the same increase persisted, lending even more support to the notion that praise was ineffective. Therefore, feedback seems to have been the controlling agent. The second half of the study, during which feedback was presented independently and in combination with praise, provides additional support for the notion that feedback is the controlling agent.

Although the apparent conclusion of the Leitenberg et al. (1968) study is that feedback is the sole agent responsible for the reduction in the knife phobia, this is only one of two possible conclusions. As Hersen and Barlow (1976) have pointed out, an alternative interpretation is that praise had an effect but that it was masked by the effect of feedback. Feedback may have been so powerful that it enabled the subject to progress at her optimal rate. If such were the case, then there would have been no room for the effect of praise to manifest itself, which would lead us to conclude erroneously that praise was ineffective when actually it did have some effect. This is one reason that both of the sequences depicted in Figure 11.5 must be incorporated in order to isolate an interaction effect.

Experimental condition

	B Feedback	BC Feedback and praise	B Feedback	A Baseline	B Feedback	BC Feedback and praise
Leitenberg et al. (1968)						
Leitenberg (1973)	Praise	Feedback and praise	Praise	Baseline	Praise	

FIGURE 11.6 Design of two studies used to test the interaction of feedback and praise.

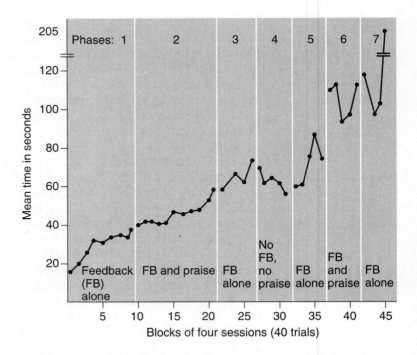

FIGURE 11.7 Time in which a knife was kept exposed by a phobic patient as a function of feedback, feedback plus praise, and no feedback or praise conditions. (From "Feedback in Behavior Modification: An Experimental Analysis in Two Phobic Cases" by H. Leitenberg, W. S. Agras, L. E. Thompson, and D. E. Wright, 1968, *Journal of Applied Behavior Analysis, 1,* p. 136. Copyright 1968 by the Society for the Experimental Analysis of Behavior, Inc.)

In accordance with this requirement, Leitenberg (1973) conducted another experiment on a second knife-phobic patient. In this study, praise was presented independently and then in combination with feedback, as illustrated in Figure 11.6. Otherwise, the procedure of the study was identical to that of the Leitenberg et al. (1968) study. Figure 11.8 depicts the results of the second study. As you can see, the subject made no progress when only praise was administered. When feedback was combined with praise, progress was made. Interestingly, this progress was maintained even when feedback was subsequently discontinued in the third phase of the study. In the fifth and sixth phases of the study, again no progress was made unless feedback was combined with praise.

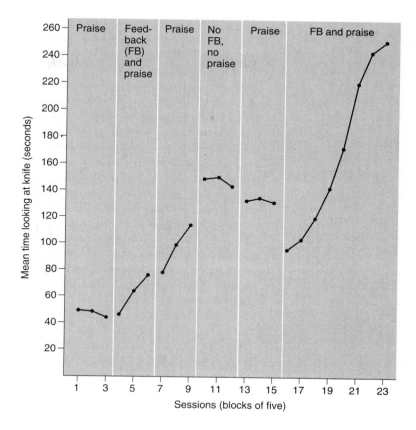

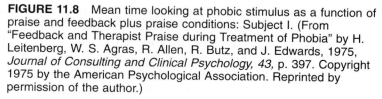

FIGURE 11.8 Mean time looking at phobic stimulus as a function of praise and feedback plus praise conditions: Subject I. (From "Feedback and Therapist Praise during Treatment of Phobia" by H. Leitenberg, W. S. Agras, R. Allen, R. Butz, and J. Edwards, 1975, *Journal of Consulting and Clinical Psychology, 43,* p. 397. Copyright 1975 by the American Psychological Association. Reprinted by permission of the author.)

Taken together, these two studies reveal that feedback alone was the primary agent in helping the patient overcome the knife phobia, because praise alone had no appreciable effect and adding praise to feedback did not produce a marked increase in progress toward overcoming the phobia. These two studies also show the necessity of testing each variable (such as feedback and praise) separately and in combination in order to isolate any interactive effect. Herein lie what may be considered the disadvantages of testing for an interaction effect. First, at least two subjects are typically required since a different subject will have to be tested on each of the two sequences depicted in Figure 11.5. Second, the interaction effect can be demonstrated only under conditions in which each variable alone (for example, feedback) does *not* produce maximum increment in performance on the part of the subject. As pointed out in the Leitenberg et al. study, it was possible that praise was effective in overcoming knife phobia, but the feedback variable was so potent that it enabled the subject to respond at the maximum level, thus precluding any possibility of demonstrating an interactive effect. In such cases, the proper conclusion would be that the two variables being tested were equally effective, and addition of the second variable was not beneficial. Note that a conclusion could not be drawn regarding the possible interactive effect of the two variables, since this effect could not be tested. In this case, the interaction design is quite useful in demonstrating the continued effects of two or more variables.

MULTIPLE-BASELINE DESIGN

One of the primary limiting components of the A-B-A design is its failure to rule out a history effect in situations in which the behavior does not revert back to baseline level when the treatment condition is withdrawn. If you suspect that such a situation may exist, the multiple-baseline design is a logical alternative because it does not entail withdrawing a treatment condition. Therefore, its effectiveness does not hinge upon a reversal of behavior to baseline level.

Multiple-baseline design
A single-subject design in which the treatment condition is successively administered to several subjects or to the same subject in several situations after baseline behaviors have been recorded for different periods of time

In the **multiple-baseline design,** depicted in Figure 11.9, baseline data are collected on several different behaviors for the same individual, on the same behavior for several different individuals, or on the same behavior across several different situations for the same individual. After the baseline data have been collected, the experimental treatment is successively administered to each target behavior. If the behavior exposed to the experimental treatment changes while all others remain at baseline, this provides some evidence for the efficacy of the treatment condition. It becomes increasingly implausible that rival hypotheses would contemporaneously influence each tar-

		T_1	T_2	T_3	T_4
Behaviors,	A	Baseline	Treatment		
people, or	B	Baseline	Baseline	Treatment	
situations	C	Baseline	Baseline	Baseline	Treatment
	D	Baseline	Baseline	Baseline	Baseline

FIGURE 11.9 Multiple-baseline design.

get behavior at the same time as the treatment was administered.

Sulzer-Azaroff and Consuelo de Santamaria (1980) used the multiple-baseline design in a study that investigated the use of feedback and approval or corrective suggestions in reducing the frequency of occurrence of industrial hazards. Again, a single group was used in the study, as opposed to a single subject. The investigators recorded the frequency, type, and location of hazardous conditions found in each of the six departments in the industrial organization in which the study was conducted. This record of hazardous conditions was kept for a three-week baseline period for departments 1 and 2, for a six-week period for departments 4 and 5, and for a nine-week period for departments 3 and 6. At the conclusion of each department's baseline period, the department supervisor was given feedback as to the number and locations of the hazards observed and specific suggestions for improvement. This feedback was given twice a week during the treatment phase. The results of the study, as depicted in Figure 11.10, show that the frequency of hazardous conditions declined in each department as soon as the feedback-and-suggestions (treatment) phase was initiated. Also note that the frequency of the hazardous conditions did not decline in departments 4 and 5, even though the treatment condition had already been administered to departments 1 and 2. Similar conditions existed for departments 3 and 6. Consequently, a change in behavior did not occur until the treatment condition was administered, providing convincing evidence that the feedback and suggestions were the cause of the reduction in the frequency of occurrence of hazardous conditions.

Interdependence of behaviors The influence of one behavior on another

Although the multiple-baseline design avoids the problem of reversibility, it has another basic difficulty. For this design to be effective in evaluating the efficacy of the treatment condition, the target behaviors must not be highly interrelated. This means that there must not be **interdependence of**

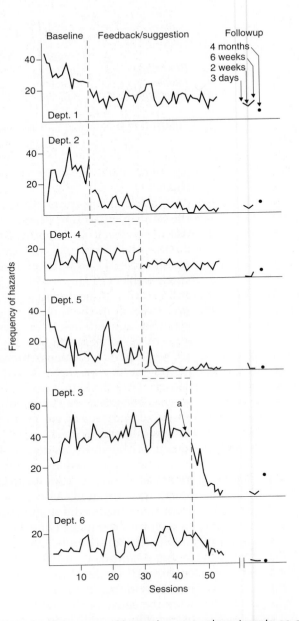

FIGURE 11.10 Frequency of hazards across departments as a function of the introduction of the "feedback package." Data for days following unplanned safety meetings are indicated by an open circle. At point "a" there was a change in supervisors. (From "Industrial Safety Hazard Reduction through Performance Feedback" by B. Sulzer-Azaroff and M. Consuelo de Santamaria, 1980, *Journal of Applied Behavior Analysis, 13,* pp. 287–295. Copyright 1980 by the Society for the Experimental Analysis of Behavior, Inc.)

behaviors being investigated such that a change in one behavior alters the other behaviors. Figure 11.10 indicates that this requirement was satisfied in the Sulzer-Azaroff and Consuelo de Santamaria study, since the reduction in hazardous conditions occurred only when the experimental treatment was administered. However, such independence is not always found. Kazdin (1973), for example, noted that the classroom behaviors of inappropriate motor behavior, inappropriate verbalizations, and inappropriate tasks are interrelated, and a change in one response can result in a change in one of the other responses. In like manner, Broden, Bruce, Mitchell, Carter, and Hall (1970), using a multiple-baseline design across individuals, found that contingent reinforcement changed not only the inattentive behavior of the target subject but also that of an adjacent peer.

The problem of interdependence of behaviors is real and needs to be considered before the multiple-baseline design is selected, because interdependence will destroy much of the power of this design, resting as it does on its ability to demonstrate change whenever the treatment condition is administered to a given behavior. If administering the experimental treatment to one behavior results in a corresponding change in all other behaviors, then when the experimental treatment is administered to the remaining behaviors, it will have less impact and produce less change because the behavior has previously been altered. In such a case, it is not clear what caused the change in behavior. Which behaviors are interrelated is an empirical question. Sometimes data exist on this interdependence, but where none exist the investigators must collect their own.

Kazdin and Kopel (1975) provide several recommendations for cases in which independence cannot be achieved. The first is to select behaviors that are as independent as possible. Since it may be difficult to predict in advance which behaviors are independent, one should consider using different individuals or situations, as their behaviors will probably be more distinct than different behaviors of the same individual.

In attempting to select independent behaviors, investigators often correlate the baseline behaviors. If a low correlation is obtained, they tend to infer that the behaviors are independent. Although this could be an indication of independence, it may suggest a level of independence that does not exist. If the baseline behaviors are quite stable or do not change much, a low correlation could result from this limited fluctuation in behavior. Then we might conclude that the behaviors were independent when they really were not. Even if we have good, valid evidence that baseline behaviors are independent, this does not provide any assurance that they will remain independent after implementation of the treatment condition. For example, implementing a time-out procedure for disruptive behavior with one child in a classroom could affect other children's behavior. Therefore, consideration of independence should take place during the treatment phase as well as during the baseline phase.

A second recommendation made by Kazdin and Kopel for decreasing ambiguity in the multiple-baseline design is to use several baselines (specifically, four or more) to decrease the possibility of dependence across all baselines. Even though some baseline behaviors will be dependent, those that are independent can demonstrate the treatment effect relationship.

A third recommendation is to implement a reversal on one of the baseline behaviors. If this leads to a generalized reversal effect (reversal on other behaviors), then one has added evidence for the effect of the treatment, as well as evidence for a generalized treatment effect.

CHANGING-CRITERION DESIGN

Changing-criterion design
A single-subject design in which a subject's behavior is gradually shaped by changing the criterion for success during successive treatment periods

The **changing-criterion design,** depicted in Figure 11.11, requires an initial baseline measure on a single target behavior. Following this measure, a treatment condition is implemented and continued across a series of intervention phases. During the first intervention or treatment phase, an initial criterion of successful performance is established. If the subject successfully achieves this performance level across several trials, or if the subject achieves a stable criterion level, the criterion level is increased. The experiment moves to the next successive phase, where a new and more difficult criterion level is established while the treatment condition is continued. When behavior reaches this new criterion level and is maintained across trials, the next phase, with its more difficult criterion level, is introduced. In this manner, each successive phase of the experiment requires a step-by-step increase in the criterion measure. "Experimental control is demonstrated through successive replication of change in the target behavior, which

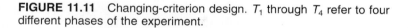

T₁	T₂	T₃	T₄
Baseline	Treatment and initial criterion	Treatment and criterion increment	Treatment and criterion increment

FIGURE 11.11 Changing-criterion design. T_1 through T_4 refer to four different phases of the experiment.

changes with each stepwise change in criterion " (Kratochwill, 1978, p. 66).

Hall and Fox (1977) provide a good illustration of the changing-criterion design in a study of a child named Dennis who refused to complete arithmetic problems. To overcome this resistant behavior, the investigators first obtained a baseline measure of the average number of assigned arithmetic problems (4.25) that he would complete during a forty-five-minute session. Then Dennis was told that a specified number of problems had to be completed correctly during the subsequent session. If he completed them correctly, he could take recess and play with a basketball; if he did not, he would have to miss recess and remain in the room until they were correctly completed. During the first treatment phase, the criterion number of problems to be solved was set at five, which was one more than the mean number completed during the baseline phase. After successfully achieving the criterion performance on three consecutive days, Dennis had to finish an additional problem. The recess and basketball contingencies were maintained. The results of this experiment, shown in Figure 11.12, reveal that Dennis's performance increased as the criterion level increased. When a change in behavior parallels the criterion change so closely, it rather convincingly demonstrates the relative effects of the treatment contingency.

Hartmann and Hall (1976) indicate that successful use of the changing-criterion design requires attention to three factors: the length of the baseline and treatment phases, the magnitude of change in the criterion, and the number of treatment phases or changes in the criterion. With regard to the length of the treatment and baseline phases, Hartmann and Hall state that the treatment phases should be of different lengths; or, if they are of a constant length, then the baseline phases should be longer than the treatment phases. This is necessary to ensure that the step-by-step changes in the subject's behavior are

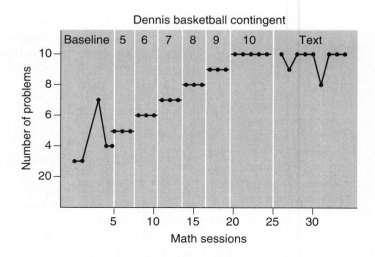

FIGURE 11.12 Number of math problems solved in a changing-criterion design. (From "Changing-Criterion Designs: An Alternative Applied Behavior Analysis Procedure" by R. V. Hall and R. W. Fox, in *New Developments in Behavioral Research: Theory, Method, and Application.* In honor of Sidney W. Bijou, edited by C. C. Etzel, G. M. LeBlanc, and D. M. Baer, 1977. Hillsdale, N.J.: Lawrence Erlbaum Associates. Copyright 1977 by Lawrence Erlbaum Associates. Reprinted by permission of the authors.)

caused by the experimental treatment and not by some history or maturational variable that occurs simultaneously with the criterion change. Additionally, the baseline data should be stable or changing in a direction opposite to that of the treatment condition in order to establish unambiguously that only the treatment condition produced the observed change. With regard to the actual length of each treatment, the rule of thumb is that each treatment phase must be long enough to allow the behavior to change to its new criterion level and then to stabilize. In other words, the new criterion level must be consistently achieved across trials. If the behavior continues to fluctuate between the new and the old criterion level, stability has not been achieved.

The second consideration is the magnitude of the criterion change. Naturally, it must be large enough so that a detectable change can occur. If the behavior is difficult to change, the criterion change should be small enough so that it can be achieved but still large enough to be noticed. If the behavior varies wildly from trial to trial, then the criterion change must be rather large in order to allow the experimenter to detect any change.

Hartmann and Hall (1976) state that two criterion changes may be adequate. This issue is, however, directly dependent on the number of replications that are required to demonstrate convincingly that the behavioral change is the result of the treatment condition. For this reason, Kratochwill (1978) recommends at least four criterion changes. When the subject's behavior is quite variable, Hall and Fox (1977) suggest including a reversal in one of the treatment phases. This reversal could consist of reverting back to baseline or to a former criterion level. Such a reversal would provide additional evidence of the influence of the treatment condition.

The changing-criterion design has not been used extensively, and all its applications and limitations have yet to be identified (Kratochwill, 1978). It seems to be a useful design in studies that require shaping of behavior over a period of time (Hall and Fox, 1977) or in cases in which step-by-step increases in accuracy, frequency, duration, or magnitude are the therapeutic goals (Hartmann and Hall, 1976), as may be the case in learning to write or read.

METHODOLOGICAL CONSIDERATIONS IN USING SINGLE-SUBJECT DESIGNS

The preceding discussion of single-subject research designs by no means represents an exhaustive survey, but presents the most basic and commonly used designs. Regardless of which design is used, there are several common issues that one must consider when attempting to conduct a single-subject study.

Baseline

Baseline has been defined as the target behaviors in their freely occurring state. Repeatedly, investigators (such as Gelfand and Hartmann, 1968) have emphasized the importance of the baseline data in single-subject research. A prime concern is obtaining a **stable baseline,** because the baseline data serve as the standard against which change induced by the experimental treatment condition is assessed. The essential question is,

Stable baseline
A set of responses characterized by the absence of any trend and by little variability

When has a stable baseline been achieved? There is no final answer. As Sidman (1960, p. 258) states:

> There is, again, no rule to follow for the criterion will depend upon the phenomenon being investigated and upon the level of experimental control that can be maintained. . . . By following behavior over an extended period of time, with no change in the experimental conditions, it is possible to make an estimate of the degree of stability that can eventually be maintained; a criterion can then be selected on the basis of these observations.

The crucial question is, What criteria should be selected? Kazdin (1978) suggests that a stable baseline is characterized by the absence of trend and by only a slight degree of variability in the data. An absence of trend means that the baseline data should not demonstrate an increase or a decrease over time. But what is "a slight degree of variability"? Kazdin does not clarify this issue, but Sidman (1960) believes that the overall mean response of baseline behavior should be stable within a 5 percent range.

Sometimes it is impossible to eliminate a baseline trend. If the trend occurring during the baseline phase is opposite that which is expected during the intervention phase when the experimental treatment condition is administered, the experiment demonstrates that the treatment condition is powerful enough not only to produce an effect but also to reverse a previous trend. If the baseline change is in the same direction as is expected from the intervention, it is difficult to draw an unambiguous conclusion regarding the influence of the treatment condition. In such a case, it is best to wait for the baseline to stabilize before introducing the treatment condition. If this cannot be done, one can resort to an alternating-treatments design in which the two treatments are designed to change the trend in opposite directions.

Excessive variability is also a concern. Variability will, in most cases, be more of a problem with human subjects than with animals, because a greater degree of control can generally be exercised over the animals. McCullough, Cornell, McDaniel, and Mueller (1974), for example, found that the number of irrelevant comments made by high school students during a fifty-minute class period ranged from 17 to 104 during an eight-day period. Although this may be somewhat atypical, it does illustrate the extent to which baseline data can fluctuate with humans. When extreme fluctuations or unsystematic variations exist in the baseline data, one should check all components of the study and try to identify and control the sources of the variability. Sometimes the fluctuation can be traced to sources that are important to the validity of the experiment, such as unreliability in scoring subject behavior. When the sources cannot be identified or controlled, one can artificially reduce the variability by averaging data points across consecutive days or sessions. This averaging substantially reduces variability and allows the effect of the treatment condition to be accurately assessed. However, it does distort the day-to-day pattern of performance.

There is one additional problem to be considered in obtaining baseline frequencies on humans: the potential reactive effect of the assessment on the behavior under study (Webb, Campbell, Schwartz, and Sechrest, 1966). The fact that baseline data are being taken may itself have an effect on the behavior. This was vividly demonstrated in the classic Hawthorne studies and has also been demonstrated by McFall (1970) and Gottman and McFall (1972), who showed that monitoring of one's own behavior can have a significant influence on that behavior. If one monitors frequency of smoking, one increases the number of cigarettes smoked, whereas if one monitors the frequency of not smoking, one smokes less.

Changing One Variable at a Time

A cardinal rule in single-subject research is that only one variable can be changed from one phase of the experiment to the next (Hersen and Barlow, 1976). Only when this rule is adhered to can the variable that produced a change in behavior be isolated. Assume that you want to test the effect of reinforcement on increasing the number of social responses emitted by a chronic schizophrenic. In an attempt to employ an A-B-A design, you first measure baseline performance by recording the number of social responses. Following baseline, you give the schizophrenic a token (which can be redeemed for cigarettes) and say "good" after each social response. At this point you are violating the rule of one variable, because two types of reinforcement are being administered. If the number of social responses increases, you will not know which type of reinforcement is responsible for changing the behavior. In fact, it may not be either reinforcer independently but the combined (interactive) influence that is the catalyst. To isolate the separate and combined influences of the two reinforcers, you would need an interaction design.

Length of Phases

An issue that must be given consideration when a single-subject study is being designed is the length of each phase of the study. Although there are few guidelines to follow, most experimenters advocate continuing each phase until some semblance of stability has been achieved. Johnson (1972) believes that in the study of punishment each phase should be continued until a lack of trend and a constant range of variability are realized. Although this is the ideal, in many clinical studies it is not feasible. Additionally, following this suggestion leads to unequal phases, which Hersen and Barlow (1976) consider to be undesirable. According to these investigators, unequal phases (particularly when the treatment phase is extended in time to demonstrate a treatment effect) increase the possibility of a confounding influence of history or maturation. For example, if the baseline phase consisted of recording responses for seven days and the treatment phase lasted fourteen days, we would have to entertain the possibility of a history or maturation variable affecting the data if a behavioral change did not take place until about the seventh day of the treatment phase. Because of such potential confounding influences, Hersen and Barlow suggest using an equal number of data points for each phase of the study.

There are two other issues that relate directly to the length of phases: carry-over effects and cyclic variations (Hersen and Barlow, 1976). Carry-over effects in single-subject A-B-A-B designs usually appear in the second baseline phase of the study as a failure to reverse to original baseline level. When such effects do occur or are suspected, many single-subject researchers (for example, Bijou et al., 1969) advocate using short treatment condition phases (B phases). These effects become particularly problematic in a single-subject drug study.

Hersen and Barlow (1976) consider cyclic variations a neglected issue in the applied single-subject literature. It is of paramount concern when subjects are influenced by cyclic factors, such as the menstrual cycle in females. Where the data may be influenced by such cyclical factors, it is advisable to extend the measurement period during each phase to incorporate the cyclic variation in both baseline and treatment phases of the study. If this is not possible, then the results must be replicated across subjects that are at different stages of the cyclic variation. If identical results are achieved across subjects regardless of the stage of the cyclic variation, then meaningful conclusions can still be derived from the data.

CRITERIA FOR EVALUATING CHANGE

The single-subject designs discussed in this chapter attempt to rule out the influence of extraneous variables by strategies such as replicating the intervention effect over time, which is quite different from the control techniques employed by multisubject experimental designs. Similarly, single-subject designs use different criteria for evaluating treatment effects than do multisubject designs. The two criteria that are usually used in single-subject research are an experimental criterion and a therapeutic criterion (Kazdin, 1978).

Experimental Criterion

Experimental criterion
In single-subject research, repeated demonstration that a behavioral change occurs when the treatment is introduced

The **experimental criterion** requires a comparison of pre- and postintervention behavior. In making this comparison, most experimenters using a single-subject design do not employ statistical analyses, which is definitely a source of controversy, as illustrated in Exhibit 11.1. Instead of using statistical analysis, these researchers rely on replicating the treatment effect over time.

When it can be demonstrated that behavior repeatedly changes as the treatment conditions change, the experimental criterion has been fulfilled. In actual practice, the experimental criterion is considered to be met if the behavior of the subject during the intervention phase does not overlap with his or her behavior during the baseline phase, or if the trend of the behavior during baseline and intervention phases differs.

Therapeutic Criterion

Therapeutic criterion
Demonstration that the treatment condition has eliminated a disorder or has improved everyday functioning

The **therapeutic criterion** refers to the clinical significance or value of the treatment effect for the subject. Does the treatment effect eliminate some disorder for the subject or does it

EXHIBIT 11.1   Analysis of Data Obtained from Single-Subject Designs

In the past, when single-subject research designs were conducted predominantly by Skinner, his colleagues, and his students, statistical analysis of single-subject data was shunned. It was deemed to be unnecessary because the studies were conducted on infrahumans and sufficient experimental control of extraneous variables could be established to enable the experimental effect to be determined by visual inspection of the data.

As single-subject designs have become more popular, some people have insisted on the need for statistical analysis of the data. This point of view is by no means universal, however.

The arguments against the use of statistical analysis are as follows.

1. Statistical analysis of the data provides evidence of a treatment effect only by demonstrating if the effect is statistically significant. It offers no evidence regarding the treatment's clinical effectiveness. For example, even though a treatment condition that was applied to reduce irrational thought patterns in schizophrenic individuals produces a statistically significant decline in such thought patterns, the patient may not have improved enough to operate effectively outside of an institutional setting.

2. Statistical tests hide the performance of the individual subject because they lump subjects together and focus only on average scores. Consequently, a treatment condition that benefited only a few individuals might not achieve statistical significance and would therefore be considered ineffective when in fact it was beneficial for some individuals.

There are two basic arguments that support the use of statistical analysis.

1. Visual inspection of the data obtained from single-subject designs will not provide an accurate interpretation when a stable baseline cannot be established. When data are not statistically analyzed, investigators must use the trend and the variability of the data to reach a conclusion as to whether the treatment condition produced an effect. If the baseline data and the treatment data have different trends or different levels of performance, then a decision is typically made that the treatment condition produced an effect, particularly if there is a stable baseline. However, if there is a great deal of variability in the data, it is difficult to interpret the data without statistical analysis. Statistical analysis can analyze extremely variable data more objectively than can individuals.

2. Visual inspection of the data leads to unreliable interpretation of the treatment effects. For example, Gottman and Glass (1978) found that the thirteen judges given data from a previously published study disagreed on whether the treatment effect was significant. Seven said a treatment effect existed, and six said it did not.

The proponents and opponents of statistical analysis each have valid points to make. However, doctrinaire positions that unequivocally advocate one strategy to the exclusion of the other would seem to do more harm than good. When a stable baseline and limited variability can be achieved, statistical analysis probably adds little to the interpretation of the data. When they cannot, statistical analysis should be used in addition to visual analysis.

enhance the subject's everyday functioning? This criterion is much more difficult to demonstrate than is the experimental criterion. For example, a self-destructive child may demonstrate a 50 percent reduction in self-destructive acts following treatment but still engage in fifty instances of such behavior every hour. Even though the experimental criterion has been satisfied, the child is still far from reaching a normal level of behavior.

Social validation
Determination that the treatment condition has significantly changed the subject's functioning

In an attempt to resolve this problem, researchers have included a procedure known as social validation in some experiments. **Social validation** of a treatment effect consists of determining if the treatment effect has produced an important change in the way the client can function in everyday life. (For example, after treatment, can a claustrophobic ride in an elevator?) This validation is accomplished by either a social comparison method or a subjective evaluation method.

Social comparison method
A social validation method in which the subject is compared with nondeviant peers

Subjective evaluation method
A social validation method in which others' views of the subject are assessed to see if those others perceive a change in behavior

The **social comparison method** involves comparing the behavior of the client before and after treatment with the behavior of his or her nondeviant peers. If the subject's behavior is no longer distinguishable from that of the nondeviant peers, then the therapeutic criterion has been satisfied. The **subjective evaluation method** involves assessing whether the treatment has led to qualitative differences in how others view the subject. Individuals who normally interact with the subject and are in a position to assess the subject's behavior may be asked to provide a global evaluation of the client's functioning on an assessment instrument, such as a rating scale or a behavioral checklist. If this evaluation indicates that the client is functioning more effectively, then the therapeutic criterion is considered to have been satisfied. Each of these methods has its limitations, but both provide additional information regarding the therapeutic effectiveness of the experimental treatment condition.

RIVAL HYPOTHESES

When discussing and reading literature on single-subject designs (for example, Leitenberg, 1973), one gets the distinct impression that these designs can effectively identify causal relationships. However, it seems wise to heed Paul's (1969) claim that only multisubject designs are capable of establishing causal relationships:

This is the case because the important classes of variables for behavior modification research are so closely intertwined that the only way a given variable can be "systematically manipulated" alone somewhere in the design is through the factorial representation of the variables of interest in combination with appropriate controls (p. 51).

Paul does admit that the reversal and multiple-baseline designs provide the strongest evidence of causal relationships that can be attained from single-subject designs.

What types of rival hypotheses exist in the single-subject designs presented? The issues of nonreversible changes and interdependence of behavior have already been discussed and the discussions need not be repeated. A number of studies (for example, Packard, 1970) have shown that instructions alone can change behavior. If different instructions are given for the baseline and experimental treatment phases, it is difficult to determine whether the effect was due to the treatment, the instructions, or some combination of the two. The best we can do is to maintain constant instructions across the treatment phases while introducing, withdrawing, and then reintroducing the therapeutic treatment condition (Hersen, Gullick, Matherne, and Harbert, 1972). Experimenter expectancies are another source of error in single-subject designs. In most studies, the researcher is acutely aware of the time periods devoted to baseline and to the experimental treatment, which may lead to differential reactions on his or her part. These differential reactions may lead the subject's behavior to change in the desired direction. A last possible biasing effect has to do with sequencing. (For an extended discussion of such effects in actual research, the reader is referred to Poulton and Freeman, 1966.) Since the same subject must perform in all phases of the experiment, order effects and carry-over effects may exist. It is difficult to separate the effects of the particular sequence of conditions from the effect of the treatment condition. If a change in behavior occurs, it could be the result of the sequence effect, the treatment effect, or some combination of the two.

USING MULTISUBJECT OR SINGLE-SUBJECT DESIGNS

Since both single-subject and multisubject designs have been presented, you might wonder which is better or which should be used in a given study. In multisubject research, the basic strategy is to assign a group of subjects randomly to various treatment conditions. The independent variable is then manipulated, and statistical tests are used to determine if there is a significant mean difference in the responses of the subjects in the various treatment groups. In single-subject research, on the other hand, we attempt to assess the effects of a given independent variable by comparing the subject's performance during presentation of the independent variable with performance when the independent variable is not present. Control over a behavior is demonstrated if the behavior can be altered at will by altering the experimental operations.

Traditionally, psychological research has conformed to a multisubject strategy, and most psychologists have accepted this as the research strategy to use. However, as Dukes (1965) has pointed out, single-subject research has always been with us, even though its impact, with a few notable exceptions, has not been great. With the publication of Sidman's book in 1960 and the proliferation of behavior therapy research, single-subject experimentation has come into its own. Perhaps the most frequently cited advantage of the single-subject over the multisubject research design is that the former bypasses the variability due to intersubject differences found in multisubject designs. As Kazdin (1973) has noted, this is desirable because intersubject variability is a function of the research design and not a feature of the behavior of the individual subject. Also, group averages often misrepresent individual behavior.

Years ago, Cronbach (1957) advocated the integration of the experimental and correlational approaches to research in psychology. Experimental multisubject research was dedicated to identifying nomothetic laws of behavior. One of the big obstacles in the path of experimental psychology was individual difference or error variance, which needed to be reduced by any possible device. Cronbach (1957, p. 674) proceeded to identify various ways in which this could be accomplished. One device he advocated was the correlation approach: including the aptitudes of the subjects in an experiment to determine how they interacted with the experimental treatments. In other words, he advocated a science of aptitude by treatment interactions. In a more recent article, Cronbach (1975) reported that this science was flourishing, but at the same time he stated that it was no longer sufficient. In order to reduce the error variance and to explain and predict behavior, dimensions of the situation and the person had to be taken into account.

Another approach that Cronbach (1957) said could be used to eliminate error variance or individual difference was to follow B. F. Skinner's lead and use only one subject. This is the approach strongly advocated by Sidman (1960). There is a definite advantage to studying the individual organism. Variables affecting its behavior can be more highly controlled, so the effect of the treatment condition in isolation can be seen. The results of the study can be generalized only to another identical organism in the same controlled setting, however. Cronbach (1975) has documented the fact that behavior is an interactive function of the situation, the subject, and the experimental treatment.

It is impossible to state that either multisubject or single-subject research is the preferable mode. Rather, it seems as though the two techniques should be integrated, just like the correlational and experimental approaches. This is particularly true in light of the fact that the results obtained also seem to be dependent on the research design used. For an extended discussion of this issue, see Grice (1966) and D'Amato (1970, pp. 29–30). The question is, How should the two be integrated? A number of individuals (Kazdin, 1973; Leitenberg, 1973; Paul, 1969; Shine, 1975) have suggested that the single-subject approach may be the best means for starting an investigation because of its economy in research time and costs. The single-subject approach could be used as an initial probing process to investigate promising experimental treatment conditions and determine if they are functionally related to behavior. However, the single-subject experiment should serve only as a mapping device and not as a final indication of causality because of the possibility of confounding effects from extraneous variables that cannot be controlled. It is important to realize that the fact that a promising hypothesis is not supported with this initial probe does not discredit it. There is often a great deal of variability in the behavior of different individuals, and an experimental treatment condition that does not work on one individual may be effective on another. Therefore, if the hypothesis does not receive support from the individual on which it was tested, it should be tested on other individuals before it is discarded.

Paul (1969), Kazdin (1973), and Shine (1975) believe that the experimental treatment should then be investigated from a multisubject approach. Switching to the multisubject approach allows us to control for competing rival hypotheses and to examine the degree of generality of the findings ("Is the experimental treatment condition effective when administered to others?"). If a significant effect is found with the multisubject approach, generality is established.

Use of the multisubject approach does, however, bring up the objections voiced by single-subject researchers. Multisubject research focuses on mean differences among groups of subjects, which are not representative of individual performance. Seldom, if ever, does an experimental treatment change the behavior of all subjects in the same way. This individual difference is considered to be error in the system. The single-subject approach should again be used to attempt to identify what is causing the individual variation in response to the experimental treatment—Shine's (1975) detection and identification function. Each subject's data should be analyzed to determine the magnitude of effect that the experimental treatment had. Once the subjects have been clustered according to the effectiveness of the experimental treatment on their behavior, the experimenter must find what is common to the subjects that are clustered but differs across clusters of subjects. Once the variables have been identified, they must be verified in a multisubject research design.

The process that has just been advocated consists of a continuing interaction between the multi- and single-subject approaches. It seems that this interactive approach will ultimately lead to more accurate prediction and explanation of behavior.

SUMMARY

In conducting an experimental research study that uses only one subject, you must reorient your thinking, because extraneous variables cannot be controlled by using a randomization control technique, nor can they be handled by the inclusion of a control group. To begin to rule out the possible confounding effect of extraneous variables, you must take some form of a time-series approach. This means that multiple pre- and post-measures on the dependent variable must be made in order to exclude potential rival hypotheses such as maturation. The most commonly used single-subject design is the A-B-A type, which requires the investigator to take baseline measures before and after the experimental treatment effect has been introduced. The experimental treatment effect is demonstrated by a change in behavior when the treatment condition is introduced and a reversal of the behavior to its pretreatment level when the experimental treatment condition is withdrawn. The success of this design depends on the reversal.

Many extensions of the basic A-B-A design have been made; the most valuable one attempts to assess the combined or interactive effect of two or more variables. The influence of each variable is assessed separately and in combination. Additionally, the influence of the combination of variables, or the interaction of the two or more variables, must be compared with that of each variable separately. This means that at least two subjects must be used in the study.

A third type of single-subject design is the multiple-baseline design. This design avoids the necessity for reversibility required in the A-B-A design by calling for successive administration of the experimental treatment condition to different subjects. The influence of the treatment condition is revealed if a change in behavior occurs simultaneously with the introduction of the treatment condition. Although the multiple-baseline design avoids the problem of reversibility, it requires that the behaviors under study be independent.

The changing-criterion design is useful in studies that require a shaping of behavior over a period of time. This plan requires that, following baseline, a treatment condition be implemented and continued across a series of intervention phases. For each intervention phase the criterion that must be met in order to advance to the next intervention phase is progressively more difficult. In this way, behavior can gradually be shaped to a given criterion level.

In addition to a basic knowledge of the single-subject designs, you should also have a knowledge of some of the methodological considerations required to appropriately implement the plans. These include the following:

1. *Baseline.* A stable baseline must be obtained, although some variation will always be found in the freely occurring target behaviors.
2. *Changing one variable at a time.* A cardinal rule in single-subject research is that only one variable can be changed from one phase of the experiment to another.

3. *Length of phases.* Although there is some disagreement, the rule seems to be that the length of the phases should be kept equal.
4. *Criteria for evaluating change.* An experimental or a therapeutic criterion must be used to evaluate the results of a single-subject design to determine if the experimental treatment condition produced the desired effect.
5. *Rival hypotheses.* Alternative theories must be considered, including the effect of variables such as instructions, experimenter expectancies, and sequencing effects.

Since both single-subject and multisubject designs are available, the researcher has to decide which type of design to use in a given study. The best approach seems to be an integration of the two approaches; the single-subject approach is an efficient means of identifying possible causal relations that can be validated by the multisubject approach.

STUDY QUESTIONS

1. Outline the historical development of the acceptance of single-subject designs.
2. What underlying assumption must exist for single-subject designs to be useful in psychological experimentation?
3. Identify and describe the various single-subject designs and explain when each should be used.
4. What disadvantages exist with each of the various single-subject designs?
5. What methodological issues must be considered when using a single-subject design?
6. When has a baseline measure achieved stability?
7. What is the difference between the experimental and the therapeutic criterion for evaluating change?
8. Should the data from single-subject designs be subjected to statistical analysis?
9. What rival hypotheses must be considered when single-subject designs are used?

10. When should you use multisubject designs, and when should you use single-subject designs?

KEY TERMS AND CONCEPTS

Single-subject research design

A-B-A design

Baseline

Withdrawal

Reversal design

Interaction design

Interaction effect in single-subject research

Multiple-baseline design

Interdependence of behavior

Changing-criterion design

Stable baseline

Experimental criterion

Therapeutic criterion

Social validation

Social comparison method

Subjective evaluation method

CHAPTER 12

Data Collection

LEARNING OBJECTIVES

1. To learn the type of decisions that must be made after the research design has been established but before data collection begins.
2. To understand why it is necessary to conduct a pilot study prior to data collection.
3. To gain an appreciation of the necessity of debriefing.

Depression is a mental disorder identified primarily as a dysphoric mood, characterized as depressed, sad, blue, hopeless, low, down in the dumps, or irritable. In 1978 the President's Commission on Mental Health estimated that approximately 20 percent of the population will experience, in their lifetime, depression that is serious enough to warrant clinical intervention. At any point in time, approximately 6 to 7 percent of the population will experience depression. Such data indicate that depression is a major public health concern, with the potential to create extreme human suffering, disability, and even suicide. A variety of techniques have been investigated in an attempt to treat this disorder. Psychotherapy and medication are the most commonly prescribed treatments. However, other less typical treatments such as jogging have also been investigated.

Because of the seriousness of depression, research has focused on trying to identify the treatments that are most effective. Currently, this research indicates that antidepressant medication and one or two forms of psychotherapy are the most effective treatment modalities, although none is effective with all individuals and some individuals do not seem to be helped by any of the treatments. Therefore, the current emphasis in depression research has changed somewhat, from a focus on developing therapeutic treatments to a focus on attempting to identify the types of patients who respond best to a given treatment and to identifying the characteristics of the therapist that result in the most effective treatment. In fact, the National Institute of Mental Health (NIMH) has completed a six-year, $10 million study that investigated just this issue.

In conducting this study, the researchers naturally had to design their study and identify their independent and dependent variables. The independent variable consisted of three different treatments that had previously been identified as the most effective in treating depressed patients. In selecting the dependent variables, the investigators assessed a variety of variables in addition to depression, such as the research participants' interpersonal styles and personalities. To control for the influence of extraneous variables, the investigators included control techniques such as random assignment of subjects to treatment conditions. After these basic design and control decisions have been made, however, there are still many choices the researcher must make before beginning actual data collection. For example, if you were going to conduct this study, you would have to decide on your source of subjects. Where will you get a sample of depressed subjects? Once you identify a source, you must decide whether to use all the depressed subjects or a subsample of them. Remember that depression can range from mild to very severe. Additionally, some of the subjects will currently be taking some type of medication, and others will have previously been in therapy. Will you include those individuals in your sample, or will you limit your sample to depressed individuals who have never received any type of therapy? These are just some of the decisions you will have to make concerning the subject sample.

INTRODUCTION

Once a study has been designed, the researcher may feel that most of the decisions required for completion of the experiment have been made. In fact, however, a great deal of decision making lies ahead. For example, the researcher must determine what types of subjects are to be used in the study, where they are to be obtained, and how many should be used. The researcher must decide if human subjects are to be used and, if they are, what instructions should be given. As you can see, the design provides only the framework of the study. Once established, this outline must be implemented. This chapter is oriented toward answering the questions that may arise in implementing a research design. These questions will, of necessity, be answered in a general way, since each study has its own unique characteristics, but this discussion should provide a basic framework that will assist you in conducting your own research study.

SUBJECTS

Psychologists investigate the behavior of organisms, and there are a wealth of organisms that can potentially serve as subjects. What determines which organism will be used in a given study? In some cases, the question asked dictates the type of organism used. If, for example, a study is to investigate imprinting ability, then one must select a species, such as ducks, that demonstrates this ability. In most studies, however, the primary determining factor is precedent: most investigators use subjects that have been used in previous studies. As Sidowski and Lockard (1966, pp. 7–8) state:

> Most of the common laboratory animals are mammals; man, several species of monkey, numerous rodents, a few carnivores, and one cetacean, the porpoise. Other than mammals, teleost fishes and one species of bird, the pigeon, have mainly represented the other classes of chordates; amphibians and reptiles have been rare. The 21 phyla below the chordates have been underrepresented. . . .[1]

Other than humans, precedent has established the albino variant of the brown rat as the standard laboratory research animal. The concentrated use of the albino rat in infrahuman research has not gone without criticism. Beach (1950) and Lockard (1968) have eloquently criticized the fact that psychologists have focused too much attention on the use of this particular animal. As Lockard has argued, rather than using precedent as the primary guide for selecting a particular organism as a subject, one should look at the research problem and select the type of animal that is best for its study.

Once a decision has been made regarding the type of organism to be used, the next questions are where to get the subjects and how many subjects to use. Researchers who use rats typically select from one of three strains: the Long-Evans hooded, the Sprague-Dawley albino, and the Wistar albino. The researcher must decide on the strain, sex, age, and supplier of the albino rats, since each of these variables can influence the results of the study (see Sidowski and Lockard, 1966).

Once the albino rats have been selected, ordered, and received, they must be maintained in the animal laboratory. The Animal Welfare Act of 1966 regulates the care, handling, treatment, and transportation of animals used in research, with the exception of a few such as birds and rats. However, if an institution receives Public Health Service funds for research, the care of even rats and birds is specified. The Na-

---

[1] From *Experimental Methods and Instrumentation in Psychology* by J. B. Sidowski and R. B. Lockard. Copyright © 1966 by McGraw-Hill, Inc. Used with permission of McGraw-Hill Book Co.

tional Academy of Sciences Institute for Laboratory Animal Resources (ILAR) developed a *Guide for the Care and Use of Laboratory Animals.* The purpose of this guide was to assist scientific institutions in using and caring for laboratory animals in professionally appropriate ways. The recommendations in this publication are the ones suggested by National Institutes of Health policy. They are also the standards used by the American Association for the Accreditation of Laboratory Animal Care (AAALAC) for its accreditation of institutions. Therefore, the guidelines suggested in this manual are the ones that all researchers should seriously consider using even if their or their institution's research does not have Public Health Service funding.

Researchers selecting humans as their subjects experience varying degrees of ease in finding participants. In most university settings, the psychology department has a subject pool consisting of introductory psychology students. These students are motivated to participate in a research study because it is a course requirement, they are offered an improved grade, or they are offered this activity as an alternative to some other requirement such as writing a term paper. Disregarding the ethical issue of such coerced participation and the possible bias that it may produce (see Cox and Sipprelle, 1971), subject pools provide a readily available supply of subjects for the researcher. For many types of research studies, however, introductory psychology students are not appropriate. A child psychologist who wishes to study kindergarten children usually will try to solicit the cooperation of a local kindergarten. Similarly, to investigate incarcerated criminals, one must seek the cooperation of prison officials as well as the criminals.

When one has to draw subjects from sources other than a departmental subject pool, a new set of problems arises because many individuals other than the subject become involved. Assume that a researcher is going to conduct a study using kindergarten children. The first task is to find a kindergarten that will allow her or him to collect the data needed for the study. In soliciting the cooperation of the individual in charge, the researcher must be as tactful and diplomatic as possible, because many people are not receptive to psychological research. If the person in charge agrees to allow the researcher to collect the data, the next task is to obtain the parents' permission to allow their children to participate. This frequently involves having parents sign permission slips that explain the nature of the research and the tasks required of their children. Where an agency is involved, such as an institution for mentally retarded persons, one might be required to submit a research proposal for the agency's research committee to review.

The NIMH study discussed in the beginning of this chapter was conducted at three different universities. The investigators had to make arrangements with each university not only to allow them to conduct the study at that site but also to permit them to solicit as subjects the depressed individuals who were seeking treatment from that university clinic.

After identifying the subject population, the researcher must select individual subjects from that group. Ideally, this should be done randomly. In a study investigating kindergarten children, a sample should be randomly selected from the population of kindergarten children. However, this is often impractical—not only in terms of cost and time but also in terms of the availability of the subjects. Not all kindergartens or parents will allow their children to participate in a psychological study. Therefore, human subjects are generally selected on the basis of convenience and availability. The kindergarten children used in a study will likely be those who live closest to the university and who cooperate with the investigator.

Because of this restriction in subject selection, the researcher may have a built-in bias in the data. For example, the children whose parents allow them to participate may perform differently than would children whose parents restrict their participation. Rosenthal and Rosnow (1975) have summarized research exposing the differences in the responses of volunteer and nonvolunteer subjects. If random selection is not possible, the next best solution is to assign subjects randomly to treatment conditions. In this way, the investigator is at least assured that no systematic bias exists among the various groups of available subjects. Because of the inability to select subjects randomly, the investigator **must** report the nature of subject selection and assignment, in addition to the characteristics of the subjects, to enable other investigators to replicate the experiment and assess the compatibility of the results. In the NIMH study, for example, the depressed individuals who were used as research participants had to meet a variety of inclusion and exclusion criteria after they had volunteered to participate in the study. The inclusion criteria included meeting the research diagnostic criteria for a current episode of a Major Depressive Disorder and having a score of at least 14 on the

Hamilton Rating Scale for depression. Exclusion criteria included the presence of other psychiatric disorders such as panic disorders, alcoholism, or drug use. Consequently, the subjects for this study were not selected randomly but were chosen on the basis of whether they volunteered and whether they met certain criteria.

It is similarly inappropriate to assume that any sample of albino rats is representative of the population of all rats. As Sidowski and Lockard (1966, pp. 8–9) point out,

> Freshly received animals are not uniform products from an automatic production line, nor are they a random sample from the world's population of rats. Animal suppliers differ greatly in such environmental practices as the ambient temperature, light–dark cycle, type of food, cage size and animal density, and the physical arrangements of food and water devices. . . . To further complicate the picture, two shipments from the same firm may not be equivalent. Most companies use tiers of cages, with some high and some almost on the floor. The high animals may be in as much as ten times the illumination as the low ones because of ceiling and light fixtures. Vertical gradients of temperature are also common, with the high animals warmer. . . . Since shipments of animals tend to be drawn by the supplier from the same cage, a given shipment is not a random sample but rather an overly homogeneous subset not representative of the range of conditions within the colony.[2]

Before leaving this section on subject selection I want to alert you briefly to one additional source of bias. This is the tendency of most studies to make use of male subjects. Most animal studies are conducted on male rats (Keller, 1984). The most reasonable explanation for this practice is that female rats have a four-day estrus cycle which complicates experiments, but Keller (1984) does not believe that this explanation is justification for the preponderant use of male rats. A similar situation seems to exist with regard to human research. Males seem to be overrepresented in psychological research (Unger & Crawford, 1992), and in some entire areas of study, such as achievement motivation, findings are based totally on males (McClelland, (1953). This apparent bias with regard to choice

---

[2]From *Experimental Methods and Instrumentation in Psychology* by J. B. Sidowski and R. B. Lockard. Copyright © 1966 by McGraw-Hill, Inc. Used with permission of McGraw-Hill Book Co.

of subjects does not lead to inaccurate information. Rather, it leads to a psychology of predominately male behavior rather than human behavior. To the extent that females respond differently than males, data obtained only from males would be an inaccurate representation of the behavior of females. This is a problem of generalization, which will be discussed further in Chapter 14.

SAMPLE SIZE

After you have decided which type of organisms will be used in the research study and have obtained access to a sample of such subjects, you must determine how many subjects are needed to test the hypothesis adequately. This decision must be based on issues such as the design of the study and the variability of the data. The relationship between the design of the study and sample size can be seen clearly by contrasting a single-subject and a multisubject design. Obviously, a single-subject design requires a sample size of one, so sample size is not an issue. In multisubject designs, however, the sample size is important because the number of subjects used can theoretically vary from two to infinity. We usually want more than two subjects, but it is impractical and unnecessary to use too many subjects. Unfortunately, few guidelines exist for deciding how large the sample size must be. The primary guide used by most researchers is precedent, which may be just as inappropriate in sample size selection as in subject selection (Beach, 1950; Lockard, 1968). The issue surrounding sample size in multisubject designs really boils down to the number of subjects needed in order to detect an effect caused by the independent variable, if such an effect really exists. As the number of subjects within a study increases, the ability of our statistical tests to detect a difference increases; that is, the power of the statistical test increases. This is why some investigators, when asked how many subjects should be used, state that the larger the sample size, the better the study. As the sample size increases, however, the cost in terms of both time and money also increases. From an economic standpoint we would like a relatively small sample. Researchers must balance the competing desires of detecting an effect and reducing cost. They must select a sample size that is within their cost constraints but still provides the ability to detect an independent variable effect.

I am aware of only two sources that provide some specific indication of the number of subjects needed in a multisubject research study. The National Education Association (NEA) Research Division ("Small Sample Techniques," 1960) has developed a formula for determining the sample size needed for a sample to be representative of a given population. Based on this formula, Krejcie and Morgan (1970) have computed the sample size required for populations of up to 1,000,000 when the .05 confidence level is desired. The results of these calculations appear in Figure 12.1. One difficulty that arises when one attempts to use these results, however, is that it is seldom possible or practical to select subjects randomly from a given population. Although this is a procedural difficulty and not one inherent in the formula, it still leaves us with the problem of determining sample size for a study that cannot randomly select from the population.

Cowles (1974) is definitive in his suggestion for the required number of subjects. Based on such considerations as the power of the statistical test, the significance level (.05), and the strength of the relationship between the independent and dependent variables, Cowles suggests that thirty-five subjects be used for most preliminary studies. If an analysis-of-variance design with several levels of the independent variable is being used, then fifteen subjects per cell is recommended. This is only one approach, however.

If you do not want to make use of this recommendation, it is possible to obtain a more exact indication of the number of subjects required for your particular study. Computation of the sample size, however, requires consideration of the actual size of the effect created by the independent variable, the significance level selected by the investigator, and the power of the statistical test (the probability that it will yield statistically significant results). These issues are beyond the scope of this text. If you are interested in these issues, Cohen (1988) provides the most complete presentation I have seen. Streiner (1990) presents a succinct overview of the procedure to be used in identifying sample size. Designers of the NIMH study used such a statistical procedure to determine that 240 subjects were needed for the study.

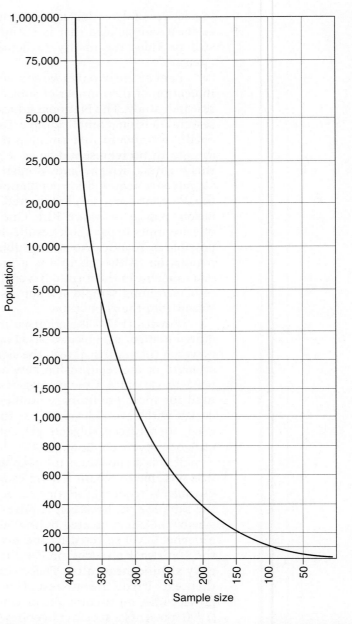

FIGURE 12.1 Relationship between sample size and total population. (From "Determining Sample Size for Research Activities" by R. V. Krejcie and D. W. Morgan, 1970, *Educational and Psychological Measurement, 30,* p. 609. Copyright 1970 by Frederick Kuder. Reprinted by permission.)

APPARATUS

In addition to securing subjects, the investigator must identify the manner in which the independent variable will be presented and the way in which the dependent variable will be measured. In some studies the presentation and manipulation of the independent variable requires the active participation of the investigator, and the measurement of the dependent variable involves the administration of a variety of psychological assessment instruments. For example, Nezu (1986) investigated the effectiveness of two different types of therapy in treating depression. These treatment modalities required active intervention on the part of the experimenter, which meant that the investigator was actively participating in the manipulation of the independent variable. To assess the effectiveness of the various treatment modalities, Nezu administered several different depression inventories. Consequently, psychological assessment instruments were used as dependent variable measures. In other studies, a specific type of apparatus must be used to arrive at a precise presentation of the independent variable and to record the dependent variable. For example, assume that you were conducting a study in which the independent variable involved presenting words on a screen for different periods of time. You could try to control manually the length of time during which the words were presented, but because it is virtually impossible for a human to consistently present words for a very specific duration of time, a tachistoscope is typically used. Similarly, if the dependent variable is the recorded heart rate, we could use a stethoscope and count the number of times per minute a subject's heart beats. It is, however, much more accurate and far simpler to use an electronic means for measuring this kind of dependent variable. The use of such automatic recording devices also reduces the likelihood of making a recording error as a function of experimenter expectancies or some type of observer bias.

In the past the apparatus typically used by the researcher were electromechanical devices with strange-sounding names like memory drum, operant chamber, relay rack, and snap leads. Although these devices served their purpose well, each was designed for a specific function. Memory drums, for example, were designed to present words or nonsense syllables to a subject at a specified rate, and they could perform only this function. If your research needs went beyond what could be done by the memory drum, you had to acquire another piece of equipment that would perform this extended function. With the advent of the microcomputer, researchers acquired an extremely flexible apparatus. The microcomputer can be programmed to present as many different independent variables and record as many different types of responses as creativity will allow. Additionally, the researcher is not tied to one specific computer. Rather, the role of the computer in stimulus presentation and recording of responses is preserved in the computer program, and this program is typically saved on a disk, which enables the researcher to reconfigure any compatible computer at a moment's notice. Finally, the computer has enabled researchers to investigate areas previously inaccessible to behavioral scientists.

To illustrate this advantage, consider the research of Breitling, Guenther, and Rondot (1986). These investigators attempted to study the electroencephalographic (EEG) topography of normal subjects during a series of hand movements, using a technique of brain electrical mapping. This technique involves monitoring the electrical activity of the brain from a variety of different sites. Breitling et al. monitored sixteen different brain sites. The brain electrical activity recorded from each of the sixteen different areas of the brain was then analyzed and converted to a graph representing the brain wave patterns recorded by each electrode. Finally, these brain wave patterns were used to produce a map of the electrical activity of the brain. As you can see, brain electrical mapping requires careful recording of extremely low-voltage electrical signals from the brain. This by itself requires a sensitive piece of equipment. Then these electrical signals must be analyzed and converted to an electrical map of the brain, which calls for a detailed and sophisticated analysis that can be accomplished within a reasonable amount of time only by a computer. In fact, electrical brain mapping is a technique that has been developed in the last few years with the advent of smaller, more powerful and portable computers. Figure 12.2 illustrates the apparatus used by Breitling et al. (1986) and reveals that these investigators used numerous pieces of equipment to conduct their experiment and that the computer played a vital role.

Since the apparatus for a given study can serve a variety of purposes, the investigator must consider the particular study being conducted and determine the type of apparatus that is most appropriate. One journal, *Behavioral Research Methods, Instruments, and Computers*, is devoted specifically to apparatus and instrumentation. If you have difficulty in identifying an instrument or a computer program that will perform a certain func-

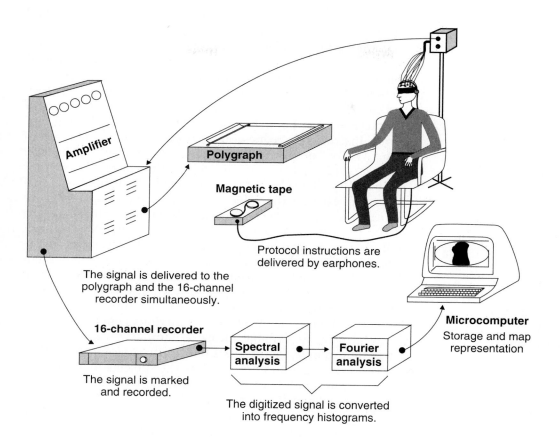

Amplifier

Polygraph

Magnetic tape

Protocol instructions are
delivered by earphones.

The signal is delivered to the
polygraph and the 16-channel
recorder simultaneously.

16-channel recorder

The signal is marked
and recorded.

**Spectral
analysis**

**Fourier
analysis**

Microcomputer

Storage and map
representation

The digitized signal is converted
into frequency histograms.

FIGURE 12.2 Apparatus used to provide an electrical map of the
brain. (Reprinted from "Motor Responses Measured by Brain Electrical
Activity Mapping" by D. Breitling, W. Gunther, and P. Rondot, 1986,
Behavioral Neuroscience, 100, pp. 104–116.)

tion, you might find it helpful to consult this journal and the
previous research conducted in your area of investigation.

INSTRUCTIONS

The investigator who conducts an experiment using human
subjects must prepare a set of instructions. This brings up such
questions as "What should be included in the instructions?"

and "How should they be presented?" Sidowski and Lockard (1966) state that instructions serve the purposes of defining the task, directing attention, developing a set, and perhaps motivating the subject. Instructions are important, and considerable care must be exercised in their formulation. They must include a clear description of the purpose, or disguised purpose, and the task that the subjects are to perform. Certain types of instructions may be ineffectual in producing the desired outcome. Instructions requesting that the subjects "pay attention," "relax," or "ignore distractions" are probably ineffective because subjects are constrained by other factors that limit their ability to adhere to the commands. Instructions sometimes request that the subjects perform several operations at the same time. If this is not possible, then they will choose one of the possible operations to perform, and the experimenter will not know which choice was made. For example, if the subjects receive the instruction to work quickly and accurately, they may concentrate on accuracy at the expense of speed because both speed and accuracy cannot be achieved simultaneously. This means that the experimenter will not know which component of the instructions contributed most to the dependent variable measure. Similarly, vague instructions (for example, instructions telling the subjects to imagine, guess, or visualize something) allow the subjects to place their own interpretations on the task. It is best to avoid such instructions whenever possible.

As you can see, instructions should be clear, unambiguous, and specific, but at the same time they should not be too complex because of the possibility of a memory overload (Sutcliffe, 1972). Beginning researchers often think that directions should be extremely terse and succinct. Although this style is good for writing the research report, in writing instructions one runs the risk that the subjects will not grasp important points. Instructions should be very simple, down to earth, and at times even redundant. This will provide added assurance that the subjects understand all the instructions.

SCHEDULING OF SUBJECTS

Scheduling of subjects' participation in the experiment involves consideration not only of when the researcher has time available but also of the type of subjects being used. With rats, for example, there is the problem of the lighting cycle. As Sidowski and Lockard (1966, p. 10) have noted:

> Rats and other nocturnal animals are most active in the dark phase of the lighting cycle and do most of their eating and drinking then. From the animal's point of view, the light portion of the day is for sleeping and inactivity but may be interrupted by an experimenter who requires him to run or bar-press for food. It is unfortunate that the amount of lighting and the timing of the cycle are usually arranged for the benefit of the caretaker and not the animals or the experimenter.[3]

This problem is most commonly handled by leaving the light on in the animal laboratory twenty-four hours a day. In this way, the animals will not form a schedule based on the light/dark phase of the lighting cycle, and thus the lighting cycle should not represent a source of confounding.

When one is scheduling human subjects, there is a completely different set of issues to consider. First, the experiment must be scheduled at a time when the experimenter and the subjects are all available. Some subjects will undoubtedly fail to show up, so it is often advisable to allow for limited rescheduling. Some subjects who do not show up at the designated time will not want to be rescheduled, which creates a problem with the randomization control technique. In such instances, the researcher may want to use replacement subjects, in which case more subjects must be selected than the experiment calls for and then replacement subjects must be scheduled to substitute for those who drop out.

PROCEDURE

After the scheduling of subjects has been completed, you must specify the procedure to be used in data collection. The events to take place in the experiment must be arranged so that they flow smoothly. Awareness of what is to take place is not sufficient; the investigator must carefully think through the whole

---

[3]From *Experimental Methods and Instrumentation in Psychology* by J. B. Sidowski and R. B. Lockard. Copyright © 1966 by McGraw-Hill, Inc. Used with permission of McGraw-Hill Book Co.

experiment and specify the sequence in which each activity is to take place, laying down the exact procedure to be followed during data collection. For animal research, this means not only specifying the conditions of the laboratory environment and how the animals are going to be handled in the laboratory, but also specifying how they are to be maintained in their maintenance quarters and how they are to be transferred to the laboratory. These are very important considerations, since such variables can influence the subjects' behavior in the laboratory. For an extended discussion of transient and environmental factors that can influence animals' behavior, see Sidowski and Lockard (1966, pp. 10–14).

With human subjects, the researcher must specify what the subjects are to do, how they are to be greeted, and the type of nonverbal behavior (looking at the subject, smiling, using a particular tone of voice in reading instructions, etc.) as well as verbal behavior in which the experimenter is to engage. Friedman (1967) has shown the wide variety of ways in which the same experimenter may react to different subjects, both verbally and nonverbally. Every effort should be made to eliminate these variations. In the NIMH study, attempts were made to maximize the possibility that the therapists would treat the depressed patients in a consistent manner by having them go through an intensive training session before actually administering therapy to the depressed subjects. The therapists were also checked periodically during the study to ensure that they maintained a consistent approach to conducting therapy.

Pilot study
An experiment that is conducted on a few subjects prior to the actual collection of data

Once each of these phases has been specified, the investigator must then conduct a pilot study. A **pilot study** is a run-through of the experiment with a small number of subjects. It is in fact a pretest of the experiment and should be conducted as conscientiously as if data were actually being collected. The pilot study can provide a great deal of information. If the instructions are not clear, this will show up either in the debriefing session or by virtue of the fact that the subjects do not know what to do after the instructions have been read.

The pilot study can also indicate whether the independent variable manipulation produced the intended effect. Debriefing can help to determine if fear, surprise, or some other state was actually generated. If none of the pilot subjects reports the particular emotion under study, then their help can be solicited in assessing why it was not generated, after which changes can be made until the intended state is induced. In a similar manner, the sensitivity of the dependent variable can be checked. Pretesting may suggest that the dependent variable is too crude to reflect the effect of the manipulation and that a change in a certain direction would make it more appropriate.

The pilot study also gives the researcher experience with the procedure. The first time the experimenter runs a subject, he or she is not yet wholly familiar with the sequence and therefore probably does not make a smooth transition from one part of the study to another. With practice, one develops a fluency in carrying out these steps, which is necessary if constancy is to be maintained in the study. Also, when running pilot subjects, the experimenter tests the procedure. Too much time may be allowed for certain parts and not enough for others, the deception (if used) may be inadequate, and so on. If there are problems, the experimenter can identify them before any data are collected, and the procedure can be altered at this time.

There are many subtle factors that can influence the experiment, and the pilot phase is the time to identify them. Pilot testing involves checking all parts of the experiment to determine if they are working appropriately. If a malfunction is isolated, it can be corrected without any damage to the experiment. If a malfunction is not spotted until after the data have been collected, it *may* have had an influence on the results of the study.

INSTITUTIONAL APPROVAL

After you have designed your study and have decided on such aspects as the nature of the subjects, where you will obtain the subjects, and any instructions they will receive, you must obtain approval from one or two institutional committees, before you can actually carry out the study. If you are conducting a study that uses animals as subjects, you must receive approval from the Institutional Animal Care and Use Committee (IACUC). If you are conducting a study that uses humans as subjects, you must receive approval from the Institutional Review Board (IRB). In either case you must prepare a research proposal that details all aspects of the research design, including the type of subjects you propose to use and the procedures that will be employed in conducting the study. This detail is necessary because these two committees review the research proposal to determine if it is ethically acceptable.

The IACUC reviews studies to determine if animals are used in appropriate ways. Specifically, the IACUC reviews studies to determine such things as whether the procedures employed avoid or minimize pain and discomfort to the animals, whether sedatives or analgesics are used in situations requiring more than momentary or slight pain, whether activities involving surgery include appropriate pre- and postoperative care, and whether methods of euthanasia are in accordance with accepted procedures. If the study procedures conform to acceptable practices, the IACUC will approve the study and you can then proceed with data collection. However, if it does not approve the study, the committee will detail the questionable components and the investigator can revise the study in an attempt to overcome the objections. Of course, the investigator can also refuse to compromise and not conduct the study.

The IRB reviews studies involving human subjects to determine if the procedures used are appropriate. The primary concern of the IRB is the welfare of human subjects. Specifically, the IRB will review proposals to ensure that subjects provide informed consent for participation in the study and that the procedures used in the study do not harm the participants. This committee has particularly difficult decisions to make when a procedure involves the potential for harm. Some procedures, such as administering an experimental drug, have the potential for harming subjects. In such instances the IRB must seriously consider the potential benefits that may accrue from the study relative to the risks to the subject. Thus the IRB frequently faces the ethical questions discussed briefly in Chapter 5. Sometimes the Board's decision is that the risks to the human participants are too great to permit the study; in other instances the decision is that the potential benefits are so great that the risks to the human participants are deemed to be acceptable. Unfortunately, the ultimate decision seems to be partially dependent on the composition of the IRB, as Kimmel (1991) has revealed that males and research-oriented individuals who worked in basic areas were more likely to approve research proposals than were women and individuals who worked in service-oriented contexts and were employed in applied areas.

Although there may be differences among IRB members with regard to the way ethical questions are resolved, the Board's decision is final and the investigator must abide by it. If the IRB refuses to approve the study, the investigator must either redesign the study to overcome the objections of the IRB, supply additional information that will possibly overcome the objections of the IRB, or not conduct the study.

Receiving approval from the IRB or the IACUC is one of the hurdles that investigators must overcome in order to conduct their proposed studies. Conducting research without such approval can cause investigators and their institutions to be severely reprimanded and jeopardize the possibility of receiving Public Health Service funding for future research projects.

DATA COLLECTION

Once you have laid out the procedure, obtained institutional approval, tested the various phases of the experimental procedure with the pilot study, and eliminated the bugs, you are ready to run subjects and collect data. The primary rule to follow in this phase of the experiment is to adhere as closely as possible to the procedure that has been laid out. A great deal of work has gone into developing this procedure, and if it is not followed exactly, you run the risk of introducing contaminants into the experiment. If this should happen, you will not have the well-controlled study you worked so hard to develop, and you may not attain an answer to your research question.

INFORMED CONSENT

When the subjects arrive at the experimental site, the first task of the experimenter is to obtain their consent to participate in the study—unless, of course, the study falls under the exempt category. The exempt category includes (but is not limited to) activities such as interviews, survey procedures, or observation of public behavior in which the subject is anonymous and any disclosure of the subject's response outside the context of the research would not have a detrimental effect on the subject. To determine if your study is exempt from the requirement to obtain informed consent, you should contact your institution's Institutional Review Board. You should also be aware of the fact that just because you think your study meets the criteria for exemption does not mean that you can circumvent the IRB. If you think your study is exempt, you must complete your institution's exemption protocol form and receive confirmation from the IRB that it is indeed exempt.

If your research proposal does not fall into the exempt category, you must inform the research participants of all aspects of the study that may influence their decision to participate. This information is typically provided in written form. Ideally, an informed consent statement should be written in simple, first-person, layperson's language. If a potential subject does not read or speak the language in which the form is written, the terms must be explained in detail. If the research participant is a minor over the age of seven, he or she must give assent. When minors are the research participants, a separate form written to their level of understanding must be provided.

The informed consent statement should be prepared so that it includes the following elements.

1. What the study is about, where it will be conducted, the duration of the study, and when the research participant will be expected to participate should be specified.
2. The statement should list what procedures will be followed and whether any of them are experimental. In the description of the procedures the attendant discomforts and risks should be spelled out.
3. Any benefits to be derived from participation in the study and any alternative procedures that may be beneficial to the subject should be identified.
4. If the research participant will receive any monetary compensation, this should be detailed, including the schedule of payments and the effect (if any) on the payment schedule in the event the subject withdraws from the study. If course credit is to be given, the statement should provide an explanation of how much credit will be received and whether the credit will still be given if the research participant withdraws from the study.

5. If the study involves responding to a questionnaire or survey, subjects should be informed that they can refuse to answer, without penalty, any questions that make them uncomfortable.
6. Studies that investigate sensitive topics such as depression, substance abuse, or child abuse should provide information on where assistance for these problems can be obtained, such as from counselors, treatment centers, and hospitals.
7. The subjects must be told that they can withdraw from the study at any time without penalty.
8. The subjects must be informed as to how the records and data obtained will be kept confidential.

As you can see, an informed consent statement is quite involved and attempts to provide research participants with complete information about the study so that they can make an intelligent and informed choice as to whether they want to participate. Exhibit 12.1 gives an illustration of an informed consent prepared for a study investigating the biological versus psychological bases of fatigue in depression. Only after informed consent has been obtained can you proceed to collect the data necessary for answering the research question.

DEBRIEFING, OR POSTEXPERIMENTAL INTERVIEW

Postexperimental interview
An interview with the subject following completion of the experiment during which all aspects of the experiment are explained and the subject is allowed to comment on the study

Once the data have been collected, there is the tendency to think that the job has been completed and the only remaining requirement (other than data analysis) is to thank the subjects for their participation and send them on their way. However, the experiment does not—or should not—end with the completion of data collection. In most studies, following data collection there should be a **postexperimental interview** with the subjects that allows them to comment freely on any part of the experiment. This interview is very important for several reasons. In general, the interview can provide information regarding the subjects' thinking or strategies used during the experiment, which can help explain their behavior. Orne (1962) used this interview to assess why subjects would persist

Exhibit 12.1   Example of an Informed Consent Statement

You are invited to participate in a study that is investigating one of the potential causes of fatigue and low energy levels. Persistent fatigue is experienced by somewhere between 21 and 41 percent of the U.S. population and is particularly prevalent among individuals suffering from depression. Although lower than normal energy levels are very common, in most cases fatigue is assumed to represent a symptom of stress or depression. Fatigue itself has, therefore, attracted little attention from the scientific community.

Procedure   If you volunteer to participate in this study, you will be asked to complete several psychological inventories and a rather lengthy interview, which will provide us with information regarding your present psychological state. You will then be asked to consume two glasses of Koolaid and complete several psychological tests every thirty minutes for the next two hours. After completing this task you will be given a diet to follow for the next two weeks. After adhering to this diet for two weeks, you will be asked once again to drink two glasses of Koolaid and complete several psychological tests every thirty minutes for the next two hours.

We anticipate that approximately sixty people will participate in the study and that the study will take a year to complete. It will be conducted in the psychophysiological laboratory located in the psychology building.

Discomforts and Risks from Participating in the Study   We will try to make the study as comfortable as possible. However, you may experience some symptoms such as nervousness or shakiness the first few days while following the diet. These should disappear within several days. It is also possible, if you are experiencing depression or fatigue, that the dietary intervention may not eliminate these feelings.

Alternatives   You may decline to participate in the study. If you are experiencing depression, you could, as an alternative, receive drug therapy or psychotherapy as a treatment for your depression. If you are experiencing fatigue, you could make an appointment with your physician to determine if there is a physical basis for the fatigue.

Expected Benefits   Based on the results of prior research we have found that some individuals experience a decline in their feelings of fatigue and depression as a result of the dietary intervention we are using. If you are experiencing fatigue or depression, you may find that these feelings decline as a result of participating in the study. If you are a psychology student, you will receive credit (specify amount of credit) for participating in this study.

Confidentiality of the Results   The results of the study will be kept strictly confidential and will be maintained in a locked file cabinet. At no time will we release the results of the study to anyone other than individuals working on the project without written consent.

Freedom to Withdraw   You may, at any time you desire, withdraw from the study and discontinue participating. In the event you do so, you will be given credit for the number of hours you have participated. Also, we will furnish you with alternative suggestions of sources where you can receive treatment for your depression should you be experiencing depression.

_____   _____
Date   Signature of Participant

at a boring, repetitive task for hours. Martin, in the course of conducting learning studies with extremely bright subjects, found that these subjects could learn a list of nonsense syllables in one trial.[4] Upon seeing such a performance, he essentially asked them, "How did you do that?" They relayed a specific strategy for having accomplished this task, which led to another study (Martin, Boersma, and Cox, 1965) investigating strategies of learning.

Tesch (1977) has identified three specific functions of debriefing. First, debriefings have an ethical function. In many studies, research participants are deceived about the true purpose of an experiment. Ethics dictate that we must undo such deceptions, and the debriefing session is the place to accomplish this. Other experiments generate some negative affect in the subjects or in some other way create physical or emotional stress. (For example, electric shock creates physical pain and failure at a task can create problems with self-esteem.) The researcher must attempt to return the subjects to their preexperimental state by eliminating any stress that the experiment has generated. Second, debriefings have an educational function. The typical rationale used to justify requiring the participation of introductory psychology students in experiments is that they learn something about psychology and about psychological research. The third function of debriefing is methodological. Debriefings are frequently used to provide evidence regarding the effectiveness of the independent variable manipulation or of the deception. They are also used to probe the extent and accuracy of subjects' suspicions and to give the experimenter an opportunity to convince the subjects not to reveal the experiment to others.

Sieber (1983b) has added a fourth function. She states that subjects should, from their participation in the study, derive a sense of satisfaction from the knowledge that they have contributed to science and to society. This perceived satisfaction should come from the debriefing procedure.

Given these functions of debriefing, how do we proceed? Two approaches have been used. Some investigators use a questionnaire approach, in which subjects are handed a postexperimental survey form to complete. Others use a face-to-face interview, which seems to be the best approach because it is not as restrictive as the questionnaire.

If you want to probe for any suspicions that the subjects may have had about the experiment, this is the first order of

---

[4]C. J. Martin, 1975, personal communication.

business. Aronson and Carlsmith (1968) believe that the researcher should begin by asking the subjects if they have any questions. If so, the questions should be answered as completely and truthfully as possible. If not, the experimenter should ask the subjects if all phases of the experiment—both the procedure and the purpose—were clear. Next, depending on the study being conducted, it may be appropriate to ask a subject to "comment on how the experiment struck him, why he responded as he did, how he felt at the time, etc. Then he should be asked specifically whether there was any aspect of the procedure that he found odd, confusing, or disturbing" (p. 71).[5]

If the experiment contained a deception and the subjects suspected that it did, they are almost certain to have revealed this fact by this time. If no suspicions have been revealed, the researcher can ask the subjects if they thought there was more to the experiment than was immediately apparent. Such a question cues the subjects that there must have been. Most subjects will therefore say yes, so this should be followed with a question about what the subjects thought was involved and how this may have affected their behavior. Such questioning will give the investigator additional insight into whether the subjects had the experiment figured out and also will provide a perfect point for the experimenter to lead into an explanation of the purpose of the study. The experimenter could continue "the debriefing process by saying something like this: 'You are on the right track, we *were* interested in some problems that we didn't discuss with you in advance. One of our major concerns in this study is . . .' " (Aronson and Carlsmith, 1968, p. 71).[6] The debriefing should then be continued in the manner suggested by Mills (1976). If the study involved deception, the reasons that deception was necessary should be included. The purpose of the study should then be explained in detail, as well as the

---

[5]Reprinted by special permission from "Experimentation in Social Psychology" by E. Aronson and J. M. Carlsmith, in *The Handbook of Social Psychology*, 2nd edition, volume 2, edited by G. Lindzey and E. Aronson, 1968, p. 71. Reading, MA: Addison-Wesley.

[6]Reprinted by special permission from "Experimentation in Social Psychology" by E. Aronson and J. M. Carlsmith, in *The Handbook of Social Psychology*, 2nd edition, volume 2, edited by G. Lindzey and E. Aronson, 1968, p. 71. Reading, MA: Addison-Wesley.

specific procedures for investigating the research question. This means explaining the independent and dependent variables and how they were manipulated and measured. As you can see, the debriefing requires explaining the entire experiment to the subjects.

The last part of the debriefing session should be geared to convincing the subjects not to discuss any components of the experiment with others, for obvious reasons. This can be accomplished by asking the subjects not to describe the experiment to others until after the date of completion of the data collection, pointing out that communicating the results to others may invalidate the study. If the study were revealed prematurely, the experimenter would not know that the results were invalid and the subjects would probably not tell (Altemeyer, 1971), so the experimenter would be reporting inaccurate data to the scientific community. Aronson (1966) has found that we can have reasonable confidence that the subjects will not tell others; but Altemeyer (1971) has shown that if subjects do find out, they will probably not tell the experimenter.

At this point you might wonder whether this debriefing procedure accomplishes the functions it is supposed to accomplish. The ethical function will be accomplished quite well if these procedures are followed. The educational function is fulfilled less completely in debriefing. Most investigators seem to think, or rationalize, that the educational function is served if the subjects participate in the experiment and are told of its purpose and procedures during debriefing. Tesch (1977) believes that this function would be better served if the researcher also required the subject to write a laboratory experience report, which would relate the experimental experience to course material. However, data indicate that subjects perceive psychological experiments to be most deficient in educational value, although they view debriefing in general to be quite effective (Smith and Richardson, 1983). It is possible that our psychological experiments are not as educational as might be hoped, and that even a good debriefing procedure cannot adequately enhance their educational value. The methodological function seems to be served quite well, since the validity of the experiment is often dependent on it. The investigator sometimes does extensive pilot study work to ensure that, for example, the manipulation checks actually verify the manipulations.

SUMMARY

Following completion of the study design, the investigator must make a number of additional decisions before beginning to collect data. The investigator must first decide on the type of organism to be used in the study. Although precedent has been the primary determining factor guiding the selection of a particular organism, the research problem should be the main determinant. The organism that is best for investigating the research problem should be used when possible.

Once the question of type of organism has been resolved, the researcher needs to determine where these organisms can be attained. Infrahumans, particularly rats, are available from a number of commercial sources. Most human subjects used in psychological experimentation come from departmental subject pools, which usually consist of introductory psychology students. If the study calls for subjects other than those represented in the subject pools, the investigator must locate an available source and make the necessary arrangements.

In addition to identifying the source subjects, the experimenter needs to determine how many subjects should be used. For studies in which subjects can be randomly selected from the population, the NEA Research Division has published a formula that can be used for this purpose. However, most experimental studies do not lend themselves to use of this formula. Based on a variety of statistical issues, one recommendation is that at least fifteen subjects per cell be used.

Instructions must also be prepared for studies using human subjects. The instructions should include a clear description of the purpose (or disguised purpose) of the task required of the subject.

Next, the investigator must specify the procedure to be used in data collection—the exact sequence in which all phases of the experiment are to be carried out, from the moment the investigator comes in contact with the subjects until that contact terminates. It is helpful to conduct a pilot study to iron out unforeseen difficulties. Once the procedure has been streamlined, it is necessary to obtain institutional approval before the study can be conducted. After approval is received, you can begin to collect the data.

When the subject arrives at the experimental site, the first task of the experimenter is to obtain the research participant's consent to participate in the study. This means that the subject must be informed of all aspects of the study that may affect his or her willingness to participate. Only after this information has been conveyed and the subject agrees to participate can you proceed with the study and collect the data that will answer the research question.

Immediately following data collection, the experimenter must conduct a postexperimental interview, or debriefing session, with the subjects. During this interview, the experimenter attempts to detect any suspicions that the subjects may have had. Additionally, the experimenter explains to the subjects the reasons for any deceptions that may have been used as well as the entire experimental procedure and purpose.

STUDY QUESTIONS

1. Discuss the important issues to consider with regard to the subjects used in psychological experimentation.
2. What purpose does the apparatus serve in psychological experimentation, and what are the advantages served by microcomputers?
3. What is a pilot study, and what purpose does it serve?
4. What purpose does debriefing serve, and how should it be conducted?

KEY TERMS AND CONCEPTS

Pilot study   IRB
IACUC   Postexperimental interview

CHAPTER 13

Hypothesis Testing

LEARNING OBJECTIVES

1. To gain an appreciation of the relationship between research design and statistical analysis.
2. To understand the nature of hypothesis testing.
3. To learn to select the appropriate statistical test for analyzing data.
4. To understand the types of errors that may occur in hypothesis testing.

Lenard Hebert, a forty-year-old recovering drug addict, started using drugs during his tour as a Marine in Vietnam in 1967. This began a twenty-five-year history of doing whatever drug was most popular. When he returned home in the 1970s, he "did strictly reefer. A good militant did natural, herbal things. Then I went disco. I snorted cocaine for ten years. It was chic because it was so expensive." When crack appeared on the drug scene, Hebert switched from cocaine and turned his apartment into a crack den. This ultimately led to his downfall. Just before he arrived at the Phoenix House, a residential drug treatment center, he was sleeping in abandoned cars and shelters for the homeless (Hurley, 1989).

Hebert's story resembles that of thousands of other Americans who, since the early 1960s, have been engaged in a tremendous social experiment with drugs. During the early 1960s, LSD was the most abused drug. Emphasis shifted to marijuana in the mid-1960s, to heroin from 1969 to 1971, and to cocaine in the late 1970s and the 1980s. In the 1990s, a smokable, fast-acting form of methamphetamine known as "ice" may well be the drug of choice.

Given the devastating effects of drugs on society, there has been an ongoing attempt to find effective treatments as well as to identify factors that predispose individuals to drug abuse. Although it might seem that everyone would be equally predisposed to drug abuse, there is actually a lot of variability. If the predisposing factors could be identified, extra efforts could be directed toward helping susceptible individuals to overcome their tendency to abuse drugs. The best way to combat drug abuse would be through effective prevention.

One researcher who has focused attention on prevention is Susan Schenk. Dr. Schenk's efforts in this area began rather serendipitously while she was a graduate student at Concordia University in Montreal. She had designed a drug study that required the use of about forty rats. Because she did not have the funds to purchase the rats, she did what all resourceful graduate students do: she scrounged. Another graduate student had just completed a study on the effects of social interactions on rats from the time of weaning to about nine weeks of age. Since he no longer had any use for the rats, he let Schenk use them in her drug study. Much to her surprise, these rats did not demonstrate the expected drug effect. Schenk speculated that this was because the rats had been provided with an enriched social environment (burrows, tubes, extensive handling by the experimenter, access to other rats, etc.) rather than being housed alone in the typical sterile environment of a rectangular steel cage. She decided to conduct a study investigating the role of environmental factors in drug abuse.

To investigate the role of environment in drug abuse, Schenk, Lacelle, Gorman, and Amit (1987) housed rats either in isolation or four to a cage for a period of six weeks. They then inserted catheters into their jugular veins and trained the rats to press a lever that would give them an infusion of 1.0 mg/kg of cocaine. After the rats had learned to press the lever to get cocaine, both groups were observed on three different occasions to determine the number of times they pressed the lever during a three-hour test period. On one occasion, each lever press delivered a 1.0 mg/kg infusion of cocaine; on an other occasion, a 0.5 mg/kg dose; and on the third occasion, a 0.1 mg/kg dose.

Schenk then analyzed the data to determine whether the environmental factors she manipulated affected the extent to which the rats self-administered each of the three cocaine dosages. Making such a determination requires the use of statistics. Although most students studying research methods have had a basic statistics course, they usually find it difficult to apply statistical principles to a research study. In this chapter I will take a very applied approach, showing how to use statistics to make sense of collected data. You will see that statistics and research design go hand in hand and that statistical analysis must be considered when a study is designed.

INTRODUCTION

The question in Schenk's cocaine study was whether environmental conditions predispose an organism to drug abuse. To answer this question, the researchers reared rats in either a sterile or a socially enriched environmental condition and then assessed the extent to which the rats self-administered three different dosages of cocaine. The amount of self-administration was measured by the number of times the rats pressed a lever that infused a measured amount of cocaine into their jugular veins.

Suppose that you hypothesized that rats would increase their abuse of cocaine as their access to it increased. To test this hypothesis, you trained two groups of rats to press a lever that would give them either 0.5 mg/kg (Group 1) or 1.0 mg/kg (Group 2) of cocaine. The hypothetical results of your experiment are displayed in Table 13.1. Once such data have been collected, they must be statistically analyzed to obtain an answer to the research question and to determine whether the stated hypothesis has been supported. Therefore, whenever you reach this stage in the research process, you are engaged in hypothesis testing.

TESTING THE HYPOTHESIS

Hypothesis testing is a decision-making process. To conduct an experiment, you formulate a scientific hypothesis or make

TABLE 13.1 Number of Lever Presses Made by Rats Receiving One of Two Cocaine Dosages

Group 1 (0.5 mg/kg dose)	Group 2 (1.0 mg/kg dose)
35	25
25	19
29	22
27	21
24	20
27	21
26	22
25	25
27	29
29	21
26	22
20	20
$\Sigma X_1 = 320$	$\Sigma X_2 = 267$
$\Sigma X_1^2 = 8{,}672$	$\Sigma X_2^2 = 6{,}027$
$\overline{X}_1 = 26.67$	$\overline{X}_2 = 22.25$
$N = 12$	$N = 12$

a prediction of the relationship among the variables being investigated. You then design a study and collect data to test the validity of the stated hypothesis. After the data have been collected, you must examine them to determine whether there is support for the scientific hypothesis. You will recall from Chapter 4 that you must actually deal with two hypotheses: the scientific hypothesis and the null hypothesis. The scientific hypothesis is a statement of the predicted relationship among the variables being investigated, and the null hypothesis is a statement of no relationship among the variables being tested. Any time we test a hypothesis statistically, we test the null hypothesis, because it is impossible to test the scientific hypothesis directly. Evidence for the existence of the scientific hypothesis is always obtained indirectly through a rejection of the null hypothesis. If you can reject the null hypothesis (that there is no relationship among the variables being investigated), then a relationship must exist among the variables.

As Figure 13.1 illustrates, a test of the null hypothesis actually involves determining whether you are dealing with one or two populations. If you do not reject the null hypothesis, you are stating that only one population of subjects exists. If you reject the null hypothesis, you are stating that two populations exist. The null hypothesis in the cocaine study conducted by Schenk et al. was that environmental conditions did *not* have any effect on the extent to which the rats abused drugs. As Figure 13.1(a) shows, a finding that the two groups of rats did not differ significantly in their tendency to abuse cocaine would indicate that both groups came from a common population with a similar tendency to abuse drugs. In this case the null hypothesis would not be rejected, providing no support for the scientific hypothesis. As Figure 13.1(b) illustrates, however, a finding that the two groups of rats differed in their tendency to abuse drugs would indicate that they came from two different populations. In this case the null hypothesis would be rejected, providing support for the scientific hypothesis.

Once you have collected data on your research subjects, you must make a decision to reject or not reject the null hypothesis. Making such a decision requires an analysis of the data collected on the dependent variable. For the cocaine study, you would analyze the number of lever presses made by the two groups of rats, because this was the response measure

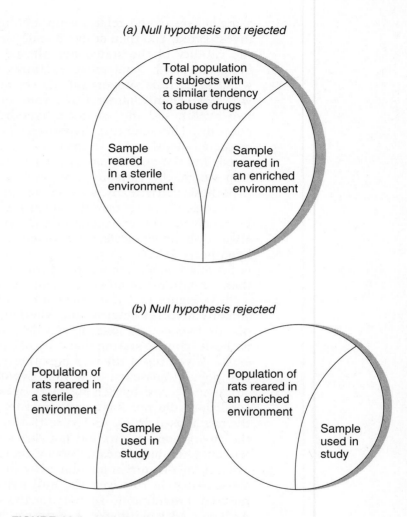

FIGURE 13.1 Hypothesis testing.

used to determine whether the two groups of rats had a differential tendency to abuse cocaine. One type of analysis might involve computing the mean number of lever presses made by the rats receiving the 0.5 mg/kg cocaine dosage and comparing that to the mean number of lever presses made by the rats receiving the 1.0 mg/kg dosage. Such an analysis would tell you which group of rats self-administered cocaine most frequently.

THE MEAN

The **mean** is the arithmetic average of a group of numbers. It is computed by dividing the sum of all the scores by the number of scores in the group:

$$\text{Mean} = \frac{\text{sum of scores}}{\text{number of scores}} \quad \text{or} \quad \overline{X} = \frac{\Sigma X}{N}$$

where

$\overline{X}$ = mean
ΣX = sum of scores
N = number of scores

If we compute the mean number of lever presses for the two groups of rats based on the data in Table 13.1, we have

$$\overline{X}_1 = 26.67 \quad \text{and} \quad \overline{X}_2 = 22.25$$

where Group 1's mean is denoted by $\overline{X}_1$ and Group 2's mean is denoted by $\overline{X}_2$. This clearly shows that the group of rats receiving the 0.5 mg/kg dose pressed the lever more often (and thus self-administered cocaine more often) than did the rats receiving the 1.0 mg/kg dose. Such data would seem to indicate that the concentration of the cocaine dose makes a difference, in which case the null hypothesis should be rejected. Remember, though, that the groups' mean scores reflect only the central tendencies of the two groups. We want to determine whether the experimental treatment conditions produced a *real* effect. In other words, we must determine whether the difference between the groups' mean scores is so large that it is unlikely to be due to chance. This is a difficult task, because even a large group mean difference could occur by chance. No two groups of subjects are alike, so the rats in the two groups would be expected to respond somewhat differently even if they received the same dose of cocaine. To determine whether the difference between the group mean scores is due to chance or to the independent variable, we need some indication of the variability of the subjects' scores in each group.

STANDARD DEVIATION

The two primary and interrelated measures of variability used in psychological research are *variance* and **standard deviation**.

They both provide an index of the extent to which the scores in a group vary about their mean. However, variance is the average of the sum of the squared deviations of the scores about their mean, whereas standard deviation is the square root of the average of the sum of squared deviations of the scores about their mean. Consequently, standard deviation is the square root of variance. A small variance or standard deviation indicates that the scores cluster closely about the mean, whereas a large variance or standard deviation indicates that the scores deviate considerably from the mean.

The variance of a group of scores is computed most economically by using the following formula:

$$s^2 = \frac{\Sigma X^2 - \dfrac{(\Sigma X)^2}{N}}{N-1}$$

where
s^2 = variance
X = individual scores
N = number of scores in the group

The standard deviation of a group of scores is computed most economically by using the following formula:

$$s = \sqrt{\frac{\Sigma X^2 - \dfrac{(\Sigma X)^2}{N}}{N-1}}$$

where
s = standard deviation
X = individual scores
N = number of scores in the group

The formulas for variance and standard deviation show clearly that standard deviation is the square root of variance. The discussion that follows will focus on standard deviation because standard deviation is the measure of variability that is most frequently reported. However, as you will see at a later point in this chapter, variance is the measure that is used in some more elaborate statistical tests.

Using the data presented in Table 13.1, we can compute the standard deviation of the number of lever presses made by the two groups of rats as follows:

$$s_1 = \sqrt{\dfrac{8{,}672 - \dfrac{(320)^2}{12}}{12 - 1}} = 3.55$$

$$s_2 = \sqrt{\dfrac{6{,}027 - \dfrac{(267)^2}{12}}{12 - 1}} = 2.80$$

The standard deviation for Group 1 is 3.55, and the standard deviation for Group 2 is 2.80. Thus more variability exists among the lever presses made by the rats receiving the 0.5 mg/kg dose. It is important to note that the standard deviation of the scores in the 0.5 mg/kg group, 3.55, approximates the difference between the means of the two groups, 4.42. The size of the standard deviation is important because it gives us some idea of whether the group mean differences are real (that is, due to manipulation of the independent variable) or are due to chance.

Figure 13.2 presents a graph of the distribution of lever presses made by rats receiving the two cocaine dosages. The two distributions overlap, suggesting that the mean difference between the two cocaine treatment conditions may be due to chance and that the subjects may have come from one popula-

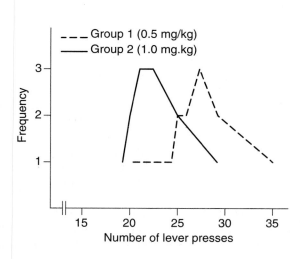

FIGURE 13.2 Distribution of the number of lever presses made by two groups of rats.

tion with the same tendency to abuse cocaine. It is also possible, however, that the group mean difference represents a real difference due to the experimental treatment condition. The point is that we don't know just by looking at the data—we are still uncertain as to whether to reject the null hypothesis.

How can we assess whether the observed difference associated with the experimental treatment condition is real or just a chance difference due to variability in subject performance? The necessary information could be gained by repeating the experiment with different subjects and seeing whether the same results occurred again. This would give evidence of the reliability of the obtained findings. The more often the study was repeated with similar results, the more faith we would have that the experimental treatment conditions produced the results, for if the difference were due to chance, it should average out to zero over many replications of the experiment. Sometimes the rats receiving the 0.5 mg/kg dose would press the lever more often than would the rats receiving the 1.0 mg/kg dose, and sometimes the rats receiving the 1.0 mg/kg dose would press the lever more often than would the rats receiving the 0.5 mg/kg dose. The absence of a reliable finding would suggest that a given difference between the two groups of subjects was the result of chance.

Although it is possible to determine whether an obtained difference between scores is real by repeating the study many times on different groups of subjects, this is not a very economical approach. It is more economical to conduct the experiment on a sample of subjects and then use some mechanism to infer from the data obtained whether the observed difference was due to chance or to a real difference—one that mirrored differences in the population. Fortunately there is a mechanism that allows us to make an inference to the population from data collected on a sample of subjects. This mechanism is referred to as inferential statistics.

Inferential statistics is a set of statistical tests that enable us, with some degree of error, to infer the characteristics of a population from a sample. This is accomplished by determining the probability of an event's happening by chance. Through inferential statistical analysis, we can estimate the amount of difference that could be expected between the group mean scores by chance and then compare this value to what was actually found. If the actual difference is much greater than what would be expected by chance, we say that the difference is a real one.

SELECTION OF A STATISTICAL TEST

At this point in the research process you must analyze your data statistically to determine whether the difference in the group mean scores is so large that it cannot reasonably be attributed to chance. You must decide which statistical test to use in analyzing your data—a decision dictated by the design of your experiment. Figure 13.3 illustrates the intimate relationship between statistics and research design. Specific statistical tests are best suited to the analysis of data collected using specific types of designs. This is why you must consider the statistical analysis of your data while you are designing the study. It is possible to design a study that is not amenable to any type of statistical analysis or that is amenable only to a type of statistical analysis that does not provide an answer to the research question.

To provide some practical experience in selecting a statistical test, let us use Figure 13.3 to choose the appropriate statistical test for the two-group cocaine study. In this study rats were administered one of two different cocaine doses, so the study had a between-subjects design. There was one independent variable, with two levels consisting of the two different concentrations of cocaine. A different group of rats was administered each cocaine dose, so there were two groups of subjects. Therefore, according to Figure 13.3, the appropriate statistical test for analyzing the data is an independent samples *t*-test.

INDEPENDENT SAMPLES *t*-TEST

Independent samples
t-test
A statistical test for
analyzing data collected
from a two-group
between-subjects design

The **independent samples** *t*-test should be familiar to those of you who have had a statistics course. It is a statistical test for analyzing the data obtained from two different groups of subjects to determine whether the group mean difference score is so large that they could not reasonably be attributed to chance. The formula for the *t*-test is

$$t = \frac{\overline{X}_1 - \overline{X}_2}{\sqrt{\dfrac{\Sigma X_1^2 - \dfrac{(\Sigma X_1)^2}{N} + \Sigma X_2^2 - \dfrac{(\Sigma X_2)^2}{N}}{N_1 + N_2 - 2} \left(\dfrac{1}{N_1} + \dfrac{1}{N_2} \right)}}$$

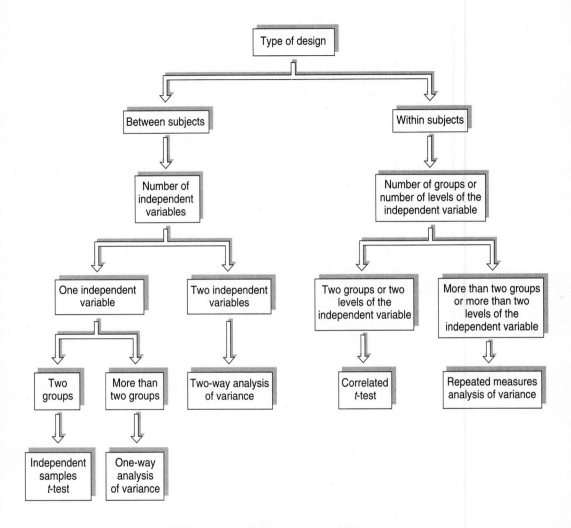

FIGURE 13.3 Decision tree for selecting the appropriate statistical test.

Although it is not readily apparent, the t value obtained from this formula represents a ratio of the difference between the group mean scores to the average variability within the two groups:

$$t = \frac{\text{group mean difference}}{\text{average within-group variability}}$$

Therefore, the greater the t value, the greater the between-group mean difference compared to the average within-group variability and the greater the probability that the group differences are real and not due to chance. Recall that earlier we computed the mean number of lever presses for the two groups of rats and found that the difference between these mean scores was 4.42. Then we computed the standard deviation of each group's scores and found that the extent of their variation about the mean suggested that the difference between the group means might be due to chance and not to the treatment conditions. However, this assessment was completely impressionistic and not based on any rigorous quantitative assessment. The *t*-test makes a similar assessment in a precise quantitative manner by comparing the difference between the group means to the natural variability that exists between the scores of the individual rats.

If we analyze the lever press data in Table 13.1 by means of the *t*-test, we obtain the following result.

$$t = \frac{26.67 - 22.25}{\sqrt{\dfrac{8{,}672 - \dfrac{(320)^2}{12} + 6{,}027 - \dfrac{(267)^2}{12}}{12 + 12 - 2} \left(\dfrac{1}{12} + \dfrac{1}{12}\right)}}$$

$$= \frac{4.42}{1.305}$$

$$= 3.39$$

The value of *t* is 3.39, which means that the between-group mean difference is 3.39 times greater than the average within-group variability. Now we must determine whether this difference is large enough for us to say that it is significant, leading to a rejection of the null hypothesis and therefore providing support for the scientific hypothesis.

SIGNIFICANCE LEVEL

Is there a guideline that determines how large a difference must be to be considered real? Only rarely can we be absolutely sure that the obtained difference between group means is not due to chance. Even very large differences could occur by chance, although the probability would be very low. Figure

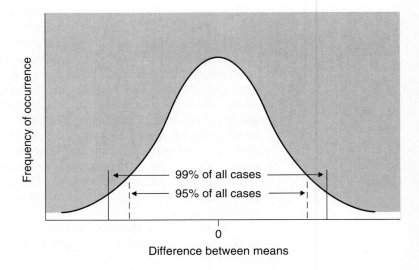

FIGURE 13.4 Sample distribution of mean difference scores.

13.4 illustrates the distribution of group mean differences that would be expected to occur by chance. Note that the mean of this distribution is equal to zero, which means that the average difference between such groups, if the null hypothesis is true, is zero. Note also that although the tails of the distribution approach the baseline, they never touch it. The fact that a very large difference *could* occur by chance means that, except in rare cases with a restricted and finite population, we can never be completely sure that a difference is real and not due to chance. We can, however, determine the probability that a given difference is the result of chance. Ninety-five percent of all chance mean difference scores will fall between the dashed lines in Figure 13.4; 99 percent of all chance mean difference scores will fall between the solid lines. If the difference we find between the group mean scores in an experiment is so large that it falls outside the dashed lines, the likelihood that it occurred by chance is only 5 in 100. Similarly, if the group difference obtained in an experiment is so large that it falls outside the solid lines, the likelihood that it occurred by chance is only 1 in 100.

How certain do we have to be before we will say that the obtained difference between groups is real? The most common practice is to state a significance level that must be reached. A **significance level** is a statement of the probability that an observed difference is a chance difference. The most commonly used significance levels are .05 and .01. If, before calculating your statistical tests, you decide that the .05 significance level is to be used, this means that you will accept as a real difference only one that is so large that it could have occurred by chance only 5 times in 100. If the .01 significance level is selected, then the difference can be expected to occur only 1 time in 100 by chance.

Now let us look at the results of the statistical analysis of the data obtained from the cocaine experiment. The *t*-test provided a *t* value of 3.39. Assuming that we decided to use the .05 significance level, we need to determine whether this *t* value is so great that it would have occurred by chance only 5 times in 100. To determine whether the obtained *t* value is large enough to be significant, we must consult a table of critical values of *t*, using the appropriate number of degrees of freedom for the study. You may recall from your statistics class that the *t*-test has $N - 2$ degrees of freedom, where N is the total number of scores. The cocaine study has 24 scores, so there are $24 - 2$, or 22, degrees of freedom. The critical values of *t* are listed in almost any statistics book. Consulting such a table reveals that the critical value of *t* for the cocaine study is 2.07. The *t* value we computed for the cocaine study is 3.39, which is greater than the critical value of *t* required for significance at the .05 level. This provides the information needed to make a decision regarding the null hypothesis. Because the obtained value of *t* is greater than the critical value of *t*, the likelihood that the difference in the mean scores of the two groups of rats occurred by chance is *less* than 5 in 100. Thus the null hypothesis of no difference between the two groups can be rejected. Now that we have concluded that a difference exists between the two groups in self-administration of cocaine, it is important to identify the dose condition that produced the greatest amount of self-administration. To make this determination, we must look at the mean number of lever presses made by the rats in the two groups. Table 13.1 shows that Group 1 pressed the lever more often and hence demonstrated the greater tendency to self-administer cocaine.

Now let us look at a method for analyzing data from a design that uses more than two groups.

ANALYSIS OF VARIANCE

Analysis of variance
A general statistical
procedure for analyzing
data obtained from a
between-subjects design
with one or more levels
of the independent
variable and one or more
independent variables

Analysis of variance is an extension of the *t*-test. It is a general statistical procedure appropriate for analyzing data generated from a research design that uses more than two levels of one independent variable and/or more than one independent variable.

One-Way Analysis of Variance

One-way analysis of
variance
A statistical test for
analyzing data collected
with a simple
randomized design

The simplest form of analysis of variance, frequently abbreviated as ANOVA, **is one-way analysis of variance.** The one-way ANOVA is the statistical test applied to data collected on the basis of a simple randomized subjects design. Assume that, based on the results of the two-group experiment discussed above, you hypothesize that rats will increase their rate of self-administration of cocaine as access to cocaine decreases. This hypothesis sounds logical because a rat receiving a 0.5 mg/kg dose for each lever press would have to press the lever twice as often as would a rat receiving a 1.0 mg/kg dose to receive the same amount of cocaine.

To test this scientific hypothesis, you allow three groups of rats to press a lever to receive one of the following doses of cocaine: 0.1 mg/kg (Group 1), 0.5 mg/kg (Group 2), or 1.0 mg/kg (Group 3). Table 13.2 shows the hypothetical data obtained from this study.

Referring to Figure 13.3, you can see which type of analysis is required to analyze the data obtained from this three-group experiment. The data were collected on the basis of a between-subjects design because there were three different groups of subjects. There was only one independent variable, and it had more than two groups of subjects. Therefore the appropriate statistical analysis is the one-way ANOVA.

Total variation
A measure of the overall
variability of the scores
obtained from a given
experimental design

Analysis of variance is a statistical procedure that allows us to estimate the probability that the observed difference between the means of three groups is the result of chance factors. The process is to partition the total variation in the data into two independent sources of variation. The **total variation** in any data set represents a measure of the overall variation of the scores obtained from all subjects within the experiment. This total variation is calculated by summing the

TABLE 13.2 Number of Lever Presses Made by Rats Receiving One of Three Cocaine Dosages

Group 1 (0.1 mg/kg dose)	Group 2 (0.5 mg/kg dose)	Group 3 (1.0 mg/kg dose)
35	126	67
41	64	40
4	19	32
8	33	10
29	24	45
27	75	22
12	48	41
0	30	20
1	30	24
17	29	40
30	25	16
19	44	15
1	40	21
$\Sigma X_{0.1} = 224$	$\Sigma X_{0.5} = 587$	$\Sigma X_{1.0} = 393$
$\Sigma X_{0.1}^2 = 6{,}252$	$\Sigma X_{0.5}^2 = 36{,}729$	$\Sigma X_{1.0}^2 = 14{,}901$
$\overline{X}_{0.1}^2 = 17.23$	$\overline{X}_{0.5} = 45.15$	$\overline{X}_{1.0} = 30.23$
$N = 13$	$N = 13$	$N = 13$

$$\Sigma X_{\text{tot}} = 1{,}204$$
$$\Sigma X_{\text{tot}}^2 = 57{,}882$$
$$N_{\text{tot}} = 39$$

squared deviations of all the scores from the mean of all the scores:

$$\text{Total variation} = \Sigma X^2 - \frac{(\Sigma X)^2}{N}$$

where

X = individual scores
N = total number of scores

(Students who are familiar with statistics should recognize this variation as the total sum of squares.) Using this formula to calculate the total variation of the data in Table 13.2, we have

$$\text{Total variation} = \Sigma X^2 - \frac{(\Sigma X)^2}{N}$$

$$= 57{,}882 - \frac{(1{,}204)^2}{39}$$

$$= 57{,}882 - 37{,}169.64$$

$$= 20{,}712.36$$

Between-groups variation
Variability arising from
the effect of the
independent variable and
chance factors

This total variation can be divided into two estimates of independent variation: one that reflects the variation *within* the various treatment groups and another that reflects the variation *between* the various treatment groups. The **between-groups variation** is actually a measure of two different sources of variability in the data. One source is the effect that may be produced by the independent variable. Consider the data from the cocaine study, shown in Table 13.2. In this study each of the three groups of subjects received a different amount of cocaine. Each of the three amounts of cocaine was hypothesized to have a different effect on the rats. In other words, the three groups of rats should have pressed the lever a different number of times depending on their group assignment. Therefore, some of the variation in response between the three different groups could be due to the effect of the independent variable, typically referred to as the treatment effect.

The other source of variation in response between the three different groups is the effect of chance factors. If the independent variable had absolutely no effect on the extent to which the rats pressed the lever, we would still expect the rats in each of the three groups to respond somewhat differently, simply because of chance factors. Thus between-groups variation is composed of variation arising from the independent variable, or the treatment effect, and that arising from chance factors:

Between-groups variation

$$= \text{treatment effect} + \text{chance variation}$$

The between-groups variation is calculated as follows:

Between-groups variation

$$= \frac{(\Sigma X_1)^2}{N_1} + \frac{(\Sigma X_2)^2}{N_2} + \cdots + \frac{(\Sigma X_n)^2}{N_n} - \frac{(\Sigma X_{\text{tot}})^2}{N_{\text{tot}}}$$

where

$(\Sigma X_1)^2$ = sum of the scores in Group 1 squared
$(\Sigma X_2)^2$ = sum of the scores in Group 2 squared
$(\Sigma X_{\text{tot}})^2$ = sum of all the scores squared
N_1 = number of scores in Group 1
N_2 = number of scores in Group 2
N_{tot} = number of scores in the entire study

(Students familiar with statistics should recognize the between-groups variation as the between sum of squares.) If this formula is applied to the data in Table 13.2, we have the following:

Between-groups variation

$$= \frac{(\Sigma X_1)^2}{N_1} + \frac{(\Sigma X_2)^2}{N_2} + \frac{(\Sigma X_3)^2}{N_3} - \frac{(\Sigma X_{tot})^2}{N_{tot}}$$

$$= \frac{(224)^2}{13} + \frac{(587)^2}{13} + \frac{(393)^2}{13} - \frac{(1,204)^2}{39}$$

$$= 42,245.69 - 37,169.64$$

$$= 5,076.05$$

*Within-groups variation
A measure of variability
induced by chance
factors estimated from
the variability of the
responses of the subjects
within each
experimental group*

The **within-groups variation** is a measure of the variation of the responses of the subjects within the experimental treatment groups. The within-groups variation is therefore a measure of chance variability, because all subjects within each experimental group should be treated the same way or be exposed to the same treatment condition. For example, in the cocaine study the rats in each group were reared under identical environmental conditions and received the same dose of cocaine for each lever press. Therefore, the variation in the response of the rats within each experimental treatment group has to be the result of chance factors:

Within-groups variation = chance variability

The within-groups variation is calculated as follows:

$$\text{Within-groups variation} = \Sigma X^2 - \frac{\Sigma X_{gp}^2}{N_{gp}}$$

where

ΣX^2 = sum of all the scores in the study after each score has been squared

ΣX^2_{gp} = sum of the scores in the treatment groups after the scores in each treatment group have been totaled and then squared

N_{gp} = number of scores in each group (this assumes an equal number of scores in each group)

(Students familiar with statistics should recognize the within-groups variation as the within sum of squares.) If this formula is applied to the data in Table 13.2, we have the following:

Within-groups variation

$$= (35)^2 + (41)^2 + (4)^2 + \cdots + (15)^2 + (21)^2 - \frac{(224)^2 + (587)^2 + (393)^2}{13}$$

$$= 57,882 - 42,245.69$$

$$= 15,636.31$$

Now that we have computed the between- and within-groups variation, we can see that the total variation of the scores obtained in the experiment is equal to the sum of the between-groups variation and the within-groups variation:

Total variation = between-groups variation + within-groups variation

20,712.36 = 5,076.05 + 15,636.31

20,712.36 = 20,712.36

Once the between- and within-groups variation has been calculated, it is necessary to divide each of these measures by its appropriate degrees of freedom to generate an estimate of between- and within-groups variance. In computing ANOVA this variance is referred to as **mean square,** or MS. The number of degrees of freedom for the between-groups measure is the number of groups minus 1, or $3 - 1$ for the cocaine study, since three different groups of rats were used. Therefore, the between-groups variance, or mean square, is

Mean square
A variance estimate obtained by dividing a sum of squares by its appropriate degrees of freedom

$$MS_{bet} = \frac{\text{between-groups variation}}{\text{degrees of freedom}}$$

$$= \frac{5,076.05}{2}$$

$$= 2,538.02$$

The number of degrees of freedom for the within-groups variance is the total number of scores in the experiment minus the number of groups, or $39 - 3 = 36$. Therefore, the within-groups variance, or mean square, is computed as follows:

$$MS_{within} = \frac{\text{within-groups variation}}{\text{degrees of freedom}}$$

$$= \frac{15,636.31}{36}$$

$$= 434.34$$

Now that the between- and within-groups mean squares have been calculated, it is possible to compute the one-way analysis of variance. As the name implies, analysis of variance involves the analysis of different variances. This analysis, commonly referred to as the *F*-test, is performed by forming a ratio of the between-groups variance, or mean square, to the within-groups variance, or mean square. This ratio is referred to as the *F*-ratio:

$$F\text{-ratio} = \frac{MS_{bet}}{MS_{within}}$$

Because of the relationship that exists between these two measures of variance, this ratio provides a test of the independent variable. Remember that the between-groups variance is a measure of both the experimental treatment effect and chance variability, and the within-groups variance is a measure of just chance variability. Therefore,

$$\frac{MS_{bet}}{MS_{within}} = \frac{\text{treatment effect} + \text{chance variability}}{\text{chance variability}}$$

Both the numerator and the denominator of the right-hand side of this equation contain a measure of chance variability, and these two measures cancel each other out. Thus only the treatment effect is left. If there is no treatment effect, then both variance measures will be measuring only chance variability and the ratio will be 1.

$$\frac{0 + \text{chance variability}}{\text{chance variability}} = 1$$

As the ratio of between- to within-groups variance gets larger, the probability increases that the difference between the groups is due to the independent variable and not to chance. Similarly, as the ratio approaches 1, the probability increases that chance explains the difference between the groups.

Let us now compute the *F*-ratio for the data in Table 13.2.

$$F\text{-ratio} = \frac{MS_{bet}}{MS_{within}}$$

$$= \frac{2,581.1}{434.34}$$

$$= 5.94$$

The value of the *F*-ratio is 5.94, indicating that the between-groups variance is six times greater than the within-groups variance. The fact that the *F*-ratio is much greater than 1 suggests that the between-groups differences were not due to chance.

Analysis of Variance Summary Table

The preceding computation of the one-way ANOVA for the experiment on self-administration of cocaine may appear to involve numerous and difficult calculations. Actually, this is the simplest of the ANOVA designs, so more complex designs may appear even more intimidating with all the calculations they require. Fortunately, once you have mastered the basics of ANOVA, there is seldom a need to perform the mathematical calculations by hand. There are many statistical packages that make use of computers to perform the calculations. You need only a summary of the results of these calculations to make a decision regarding the outcome of an experiment. The results of any analysis of variance computations are typically presented in an *analysis of variance summary table*, such as the one in Table 13.3. This table summarizes calculations for the cocaine experiment.

You can see that the summary table not only provides a summary of the ANOVA calculations but also provides a summary of the information relevant to the *F*-ratio. The table is arranged so that the columns describe the various quantities obtained in the analysis. There are separate columns for the sums of squares (SS), degrees of freedom (df), mean squares (MS), and *F*-ratio. The rows of the table define the sources of variation in the data: between groups, within groups, and total variation. Additionally, the table clarifies the relationships between these quantities. From the table it can be seen that the total variation is partitioned into between- and within-groups variation. These two sums *must* add up to the total variation. Similarly, the between and within degrees of freedom, df, must

TABLE 13.3 ANOVA Summary Table for a One-Way Analysis of Variance

Source of Variation	SS	df	MS	F
Between groups	5,076.05	2	2,538.02	5.94
Within groups	15,636.31	36	434.34	
Total	20,712.36	38		

add up to the total df. The table also reveals that the mean square, MS, is obtained by dividing the sums of squares, SS, by the degrees of freedom, df. That is, the mean square between is obtained by dividing the between sums of squares by the between df, and the mean square within is obtained by dividing the within sums of squares by the within df. Finally, Table 13.3 reveals that the *F*-ratio is obtained by dividing the between mean square by the within mean square; the resulting value appears in the column labeled *F*.

Interpreting the Results of the Analysis of Variance

Once the *F*-ratio has been computed, we must determine the actual probability that the group mean differences were due to chance. If we are operating at the .05 significance level, we must determine whether the *F*-ratio is so large that the results would occur by chance only 5 times in 100. To make this assessment, we must, as we did with the results of the *t*-test, consult a table of critical values of *F*, using the appropriate degrees of freedom. These degrees of freedom can easily be obtained from an analysis of variance summary table. Table 13.3, for example, reveals that there are 2 degrees of freedom for the between-groups variation and 36 degrees of freedom for the within-groups variation. These are the two values for degrees of freedom that must be used when a table of critical values is consulted to determine the critical value of *F* required for significance. For these degrees of freedom, a table of critical values of *F* reveals that an *F*-ratio of 3.26 is required for significance at the .05 level. Because the *F*-ratio we computed from the cocaine data exceeds this critical value, we reject the null hypothesis and accept the scientific hypothesis that the different cocaine dosages differentially affected the number of lever presses made by the rats in the various groups.

Now that we have concluded that the dosages differentially affected the response rate, we must return to the data and identify the dosage that produced the greatest effect. The *F*-ratio told us only that there was a significant difference between the three groups. It did not tell us which group pressed the lever most frequently and thus had the greatest tendency to abuse cocaine. To obtain such information we must use descriptive statistics, because we want to describe the response rate of each group of rats. The best way to do so is to compute the mean number of lever presses for each group. From Table 13.2 we know that the group receiving the 0.5 mg/kg dose of cocaine had the greatest number of responses (mean = 45.15) and the group receiving the 0.1 mg/kg dose had the smallest number of responses (mean = 17.23). Now we know not only that the groups differed significantly but also which group self-administered, or abused, cocaine the most.

It would be appropriate to conduct another statistical test to determine which of the groups differed significantly from one another. The F-test and ANOVA tell us only that there is a significant difference among the three groups, not which groups are significantly different. It may be that the 0.1 mg/kg dose group produced significantly fewer responses than did the other two groups but that these two groups did not produce significantly different numbers of responses. Such information is needed to completely interpret the data from such an experiment. Acquiring this information requires knowledge of post hoc tests, however, and they are beyond the scope of this text.

Two-Way Analysis of Variance

Two-way analysis of variance
A statistical test for analyzing data collected with a factorial design

The **two-way ANOVA** is a statistical test that is applied to data collected from a factorial design. You will recall from Chapter 9 that a factorial design is one in which two or more independent variables are studied simultaneously to determine their independent and interactive effects on the dependent variable. The two-way ANOVA is the simplest form of test to apply to a factorial design; it is used to analyze data from studies that investigate the simultaneous and interactive effects of two independent variables. For example, assume that you want to conduct an experiment that investigates whether consuming caffeine at an early age sensitizes individuals to cocaine abuse.

To test this hypothesis you randomly assign one group of young rats to receive a dose of 1 gm of caffeine per 5 kg of body weight in drinking water and another group to receive quinine-laced drinking water each day for a period of two months. Both of these substances are bitter, so the taste of the water should be similar for both groups. This manipulation will create one group of rats that is preexposed to caffeine and another that is not. Then you randomly assign the subjects in each of these two groups to receive one of the following doses of cocaine: 0.1 mg/kg, 0.5 mg/kg, or 1.0 mg/kg. This provides two independent variables—caffeine pretreatment and cocaine dose—and gives six different groups of subjects, as illustrated in Table 13.4.

Table 13.4 shows hypothetical data on number of lever presses obtained from this study, where each lever press provides the rats with a predetermined 0.1, 0.5, or 1.0 mg/kg dose of cocaine. Figure 13.3 can be used to determine the type of analysis required with the data obtained from this six-group study. The data were collected using a between-subjects design because there were six different groups of subjects. There were two independent variables, so the appropriate statistical test is a two-way analysis of variance.

Like the one-way analysis of variance, the two-way ANOVA is accomplished by partitioning the total variability in the data into the components that reflect the sources of variation in the experiment. The total variation in the one-way ANOVA was partitioned into between and within variation, where the between variation was the variation that included the effect of the independent variable and the within variation was the measure of chance variability.

TABLE 13.4 Number of Lever Presses Made by Rats Receiving One of Three Cocaine Doses and Preexposed or Not Preexposed to Caffeine

	Group 1 (A1) (0.1 mg/kg dose)	Group 2 (A2) (0.5 mg/kg dose)	Group 3 (A3) (1.0 mg/kg dose)	
Caffeine preexposure (B1)	35 41 14 18 29 27	64 19 33 24 75 48	67 40 52 60 45 72	$\Sigma B1 = 763$ $\overline{X}_{B1} = 42.39$ $N_{B1} = 18$
Non-exposure to caffeine (B2)	12 0 1 17 30 19	30 30 29 25 44 40	41 20 24 40 16 15	$\Sigma B2 = 433$ $\overline{X}_{B2} = 24.06$ $N_{B2} = 18$
	$\Sigma A1B1 = 164$ $\Sigma A1B2 = 79$ $\Sigma A1 = 243$ $\overline{X}_{A1} = 20.25$ $N_{A1} = 12$ $\Sigma X_{tot} = 1{,}196$	$\Sigma A2B1 = 263$ $\Sigma A2B2 = 198$ $\Sigma A2 = 461$ $\overline{X}_{A2} = 38.42$ $N_{A2} = 12$ $\Sigma X_{tot}^2 = 51{,}884$	$\Sigma A3B1 = 336$ $\Sigma A3B2 = 156$ $\Sigma A3 = 492$ $\overline{X}_{A3} = 41.00$ $N_{A3} = 12$ $N_{tot} = 36$	

For the two-way ANOVA, the total variation is partitioned into four different components.

1. A component that measures the variation for the independent variable of cocaine dose, or factor A;
2. A component that measures the variation for the independent variable of caffeine pretreatment, or factor B;
3. A component that measures the interaction between factors A and B, or between cocaine dose and caffeine pretreatment; and
4. A measure of chance variability.

The total variation in the data is calculated by summing the squared deviations of all scores from the mean of all scores. The raw score formula is

$$\text{Total variation, or } SS_{tot} = \Sigma X^2 - \frac{(\Sigma X)^2}{N}$$

Using this formula to calculate the total variation of the data in Table 13.4, we have

$$SS_{tot} = \Sigma X^2 - \frac{(\Sigma X)^2}{N}$$

$$= 51,884 - \frac{(1,196)^2}{36}$$

$$= 51,884 - 39,733.78$$

$$= 12,150.22$$

The variation for the independent variable of cocaine dose, factor A, is calculated as follows:

$$\text{Factor A variation, or } SS_A = \frac{(\Sigma X_{A1})^2}{n_{A1}} + \frac{(\Sigma X_{A2})^2}{n_{A2}} + \frac{(\Sigma X_{A3})^2}{n_{A3}} - \frac{(\Sigma X_{tot})^2}{N_{tot}}$$

Using this formula to calculate the variation for factor A, the cocaine dose independent variable, we have

$$SS_A = \frac{(243)^2}{12} + \frac{(461)^2}{12} + \frac{(492)^2}{12} - \frac{(1,196)^2}{36}$$

$$= 42,802 - 39,733.78$$

$$= 3,068.22$$

The variation for the caffeine pretreatment independent variable, factor B, is calculated as follows:

$$\text{Factor B variation, or SS}_B = \frac{(\Sigma X_{B1})^2}{n_{B1}} + \frac{(\Sigma X_{B2})^2}{n_{B2}} - \frac{(\Sigma X_{tot})^2}{N_{tot}}$$

If this formula is applied to the data in Table 13.4, we have the following:

$$SS_B = \frac{(763)^2}{18} + \frac{(433)^2}{18} - \frac{(1,196)^2}{36}$$

$$= 42,758.78 - 39,733.78$$

$$= 3,025$$

The variation for the interaction between the cocaine dose and caffeine pretreatment is calculated as follows:

Interaction effect, or $SS_{A \times B}$

$$= \frac{(\Sigma X_{A1B1})^2}{n_{A1B1}} + \frac{(\Sigma X_{A2B1})^2}{n_{A2B1}} + \frac{(\Sigma X_{A3B1})^2}{n_{A3B1}} + \frac{(\Sigma X_{A1B2})^2}{n_{A1B2}}$$

$$+ \frac{(\Sigma X_{A2B2})^2}{n_{A2B2}} + \frac{(\Sigma X_{A3B2})^2}{n_{A3B2}} - \frac{(\Sigma X_{tot})^2}{N_{tot}} - SS_A - SS_B$$

$$= \frac{(164)^2}{6} + \frac{(263)^2}{6} + \frac{(336)^2}{6} + \frac{(79)^2}{6} + \frac{(198)^2}{6}$$

$$+ \frac{(156)^2}{6} - \frac{(1196)^2}{36} - 3,068.22 - 3,025$$

$$= 46,457.0 - 39,733.78 - 3,068.22 - 3,025$$

$$= 630$$

The chance variation in the data, also known as error, can be calculated using the following formula:

Chance variation, or SS_{error}

$$= \Sigma X^2 - \left[\frac{(\Sigma X_{A1B1})^2}{n_{A1B1}} + \frac{(\Sigma X_{A2B1})^2}{n_{A2B1}} + \frac{(\Sigma X_{A3B1})^2}{n_{A3B1}} + \frac{(\Sigma X_{A1B2})^2}{n_{A1B2}} \right.$$

$$\left. + \frac{(\Sigma X_{A2B2})^2}{n_{A2B2}} + \frac{(\Sigma X_{A3B2})^2}{n_{A3B2}} \right]$$

However, this variation is more commonly calculated by subtraction, as follows:

$$SS_{error} = SS_{tot} - SS_A - SS_B - SS_{A \times B}$$
$$= 12,150.22 - 3,068.22 - 3,025 - 630$$
$$= 5,427$$

The next step in computing the two-way ANOVA is to calculate the mean square for each source of variation. This is accomplished by dividing each source of variation by its appropriate degrees of freedom. The formulas for computing the degrees of freedom, df, for each source of variation are as follows:

$$df_A = A - 1, \text{ or } 3 - 1 = 2$$
$$df_B = B - 1, \text{ or } 2 - 1 = 1$$
$$df_{A \times B} = (A - 1)(B - 1) = (3 - 1)(2 - 1) = 3$$
$$df_{error} = N - AB, \text{ or } 36 - (3)(2) = 30$$

Once the degrees of freedom have been calculated, you can compute the mean square for each effect, as follows:

$$MS_A = \frac{SS_A}{df_A} = \frac{3,068.22}{2} = 1,534.11$$

$$MS_B = \frac{SS_B}{df_B} = \frac{3,025}{1} = 3,025$$

$$MS_{A \times B} = \frac{SS_{A \times B}}{df_{A \times B}} = \frac{630}{2} = 215$$

$$MS_{error} = \frac{SS_{error}}{df_{error}} = \frac{5,427}{30} = 180.9$$

After the mean square values have been computed, it is possible to compute the F-ratio for each of the three effects of interest: the two main effects of cocaine dose and caffeine pretreatment and the interaction of these two effects. These F-ratios are computed by dividing the mean square of each effect by the MS_{error}. The F-ratio for the cocaine main effect is

$$F = \frac{MS_A}{MS_{error}} = \frac{1,534.11}{180.9} = 8.48$$

The F-ratio for the caffeine pretreatment main effect is

$$F = \frac{MS_B}{MS_{error}} = \frac{3,025}{180.9} = 16.72$$

The *F*-ratio for the interaction effect is

$$F = \frac{MS_{A \times B}}{MS_{error}} = \frac{215}{180.9} = 1.19$$

Table 13.5 is an ANOVA summary table for these analyses. To determine whether the main effects and the interaction effect are significant you must compare the computed *F*-value with a table of critical values of *F* (found in any statistics textbook) using the appropriate degrees of freedom. You will remember from your statistics course or from the previous section on one-way ANOVA that you must enter the *F*-table using the degrees of freedom corresponding to the mean square values in the numerator and denominator of the *F*-ratio. For example, the *F*-ratio for the cocaine main effect is obtained by computing the ratio of the cocaine mean square to the error mean square. There are 2 degrees of freedom for the cocaine mean square and 30 degrees of freedom for the error mean square. This means that you would enter a table of critical values of *F* with 2 and 30 degrees of freedom. Using the appropriate df in a table of critical values reveals that an *F*-ratio of 3.32 is needed for the cocaine main effect to be considered significant, an *F*-ratio of 4.17 is needed for the caffeine pretreatment to be considered significant, and an *F*-ratio of 2.42 is needed for the interaction effect to be considered significant. If these critical values are compared with the *F*-ratios shown in the ANOVA summary in Table 13.5, it can be seen that the *F*-ratios for the cocaine and caffeine main effects exceed the critical values, whereas the *F*-ratio for the interaction effect

TABLE 13.5 ANOVA Summary Table for a Two-Way Analysis of Variance

Source of Variation	SS	df	MS	F
Cocaine	3,068.22	2	1,534.11	8.48
Caffeine pretreatment	3,025.00	1	3,025.00	16.72
Cocaine × caffeine pretreatment	630.00	2	315.00	1.19
Total	12,150.22	30		

does not. This means that you would reject the null hypothesis for the cocaine and caffeine main effects but fail to reject the null hypothesis for the interaction effect. Therefore, you would conclude that both cocaine and caffeine differentially affect lever pressing but caffeine pretreatment does not differentially influence the effect of cocaine (the interaction).

To determine how cocaine dose and caffeine pretreatment affect self-administration of cocaine, as measured by lever pressing, it is necessary to look at the mean scores for each treatment group. The mean scores for the three cocaine dose groups reveal that the 0.1 mg/kg dose group had the lowest number of lever presses (mean = 20.25) whereas the 1.0 mg/kg group had the greatest number of lever presses (mean = 41), indicating that lever pressing increases as cocaine dose increases. The group that was pretreated with caffeine had the greatest number of lever presses (mean = 42.39), whereas the nonpretreated group had fewer responses (mean = 24.06). The experiment, therefore, reveals that as cocaine dose increases, the extent to which rats self-administer, or abuse, cocaine also increases. It also reveals that if rats are pretreated with caffeine they will increase lever pressing, or that caffeine increases the extent to which rats self-administer cocaine.

Rejecting Versus Failing to Reject the Null Hypothesis

The foregoing should give some appreciation of the necessity of performing statistical tests on the data. If the statistical tests reveal that a significant difference (one that has reached the specified significance level) exists between the scores of the various groups, then the null hypothesis is rejected and the scientific hypothesis is accepted as real. If the obtained difference does not reach the specified significance level, then the experimenter *fails* to reject the null hypothesis. The expression *fails to reject* is used because it is very difficult to obtain evidence supportive of a null or no-difference conclusion. At first glance, it seems that if the null hypothesis cannot be rejected, it should logically be accepted. To see why the null hypothesis cannot be accepted if the significance level is not attained, consider the following experiment. Nation, Bourgeois, Clark, and Hare (1983) studied the effects of chronic cobalt exposure on the behavior of adult rats. The experimental group of rats was fed laboratory chow laced with cobalt chloride, and their lever-pressing speed was compared with that of a control group of rats who were fed standard laboratory chow. The results of one component of this study revealed that there was not a significant difference between lever-pressing responses of the control and experimental groups at the .05 significance level. Consequently, the null hypothesis could not be rejected. However, although the experimental and control groups did not differ in lever-pressing response rates at the .05 level, a difference in response rates did exist. In order to accept the null hypothesis, we must be able to state that the observed variance represents a chance difference, and to assume that any observed difference is entirely due to chance is very hazardous. In the Nation et al. experiment, it is possible that the cobalt did have an effect but that the effect was too weak to be detected at the .05 significance level. In such a situation, some of the observed difference in response would be due to the treatment condition, which means that it would be inappropriate to accept a no-difference or null conclusion. Thus the expression *fails to reject* is used in place of *accept*.

Although it is logically impossible to accept the null hypothesis of no difference, "practical concerns demand that we sometimes have to provisionally act as though the null hypothesis were true" (Cook, Gruder, Hennigan, and Flay, 1979). As Greenwald (1975) has pointed out, there is a pervasive anti-null-hypothesis prejudice, which can lead to a variety of behavioral symptoms, such as "continuing research on a problem when results have been close to rejection of the null hypothesis ('near significant'), while abandoning the problem if rejection of the null hypothesis is not close" (Greenwald, 1975, p. 3). Such undesirable behavioral manifestations have led Greenwald to conclude that we should do research in which any outcome, including a null hypothesis outcome, is possible. This means, however, that the research must be conducted in such a manner as to allow for the tentative acceptance of the null hypothesis.

Harcum (1990) takes a similar position. While recognizing that it is very risky to consider accepting the null hypothesis or even making a null prediction, Harcum also recognizes the useful role for null predictions. Null predictions can help establish the boundary conditions for a phenomenon. For example, a null hypothesis is useful in determining when a drug dose is so low that it has no demonstrable effect. Null predictions can also help in deciding which variables are not worth further study.

Although null hypothesis predictions can be valuable in scientific research, they should be approached with caution. It is easy to support a null prediction just by using a weak research design that does not eliminate confounding variables or one that does not present a strong independent variable. In an attempt to avoid such problems, Harcum (1990) has proposed several criteria that must be met to reach a null conclusion. These are as follows:

1. No data that present a problem for the null conclusion may exist. In other words, there may be no studies that refute the null conclusion. For example, the results of studies regarding the effect of carbohydrates on mood of normal individuals are contradictory. Some studies (e.g., Spring, Chiodo, Harden, Bourgeois, & Lutherer, 1989) have demonstrated a mood-altering effect of carbohydrates, whereas other studies (e.g., Christensen & Redig, 1993) have not. In such an instance it would be inappropriate for Christensen and Redig to reach a null hypothesis conclusion that carbohydrates did not affect mood. Rather than reaching a null conclusion, these investigators concluded that the effect of carbohydrates on mood was ephemeral and that a strong design with a large number of subjects was required to demonstrate a mood-altering effect of carbohydrates.

2. The probability of the null hypothesis's being false must be large. In other words, if the statistical test you used to analyze the data indicated a significance level of .06, you would be on very shaky ground in attempting to reach a null conclusion. However, a significance level of, say, .45 would provide more assurance of a null conclusion's being correct.

3. The research design must be sufficiently powerful. This is a key element in the ability to predict and reach a null conclusion. If the research design does not control for the influence of rival hypotheses or confounding variables or is conducted inappropriately, then a null result will occur, not because of the absence of an effect, but because of a lack of precision in the research design. However, a null result gains credence if confounding variables are controlled and the research is conducted appropriately.

Even when criteria such as these are met, acceptance of a null result must be approached with caution. Additional guidelines for research that may lead to support for the tentative acceptance of the null hypothesis may be found in Greenwald (1975) and Cook, Gruder, Hennigan, and Flay (1979).

POTENTIAL ERRORS IN THE STATISTICAL DECISION-MAKING PROCESS

In the preceding section, primary concern was given to determining when one should accept a difference as being significant. The commonly accepted significance levels of .05 and .01 mean that a wrong decision would occur only 5 times or 1 time in 100, respectively. In setting this stringent level, we are being very conservative, making sure that the odds of being correct are definitely in our favor. It is somewhat like going to a horse race with the intention of maximizing the possibility of winning—you increase your chances of winning if you bet on a horse that has only a 5 percent chance of losing rather than on one that has a 50 percent chance of losing. In like manner, scientists have set the odds so that they are quite sure of making the correct decision. Note, however, that even with this stringent significance level, scientists will be wrong a given percentage of the time: 5 percent of the time if they operate at the .05 significance level, and 1 percent of the time if they operate at the .01 significance level. In other words, if we conduct the same experiment one hundred times, we can expect that five of these times we will, by chance alone, obtain a mean difference large enough to allow us to reject the null hypothesis at the .05 significance level. This means that we could be wrong when we reject the null hypothesis, whether we are operating at the .05 or the .01 significance level.

Type I error
False rejection of the null hypothesis

When we falsely reject the null hypothesis, we commit a **Type I error.** Type I error is controlled by the significance level that is set. If the .05 significance level is set, the probability of being wrong and committing a Type I error is 5 in 100. It may appear that there is an easy solution to this type of problem: simply set a more stringent significance level, such as .0001. There are two difficulties with this approach. First, it is not possible to eliminate Type I error, since, by definition, there will always be some possibility of obtaining a chance finding as large as the observed difference. Second, and more important, another type of error tends to be inversely related to Type I error. As the probability of falsely rejecting the null hypothesis (Type I error) decreases, the probability of failing to reject the null hypothesis when it is false, called **Type II error,** tends to increase. Note the phrase "tends to increase"—there is no *direct* relationship between Type I and Type II errors. However, the risk of committing a Type I error generally increases as the

Type II error
Failure to reject the null hypothesis when it is false.

TABLE 13.6   Possible Outcomes in Hypothesis Testing

		True Situation	
		Null Hypothesis True	*Null Hypothesis False*
Decision	Do not reject null hypothesis	Decision correct	Type II error
	Reject null hypothesis	Type I error	Decision correct

risk of committing a Type II error decreases, and the risk of committing a Type II error generally increases as the risk of committing a Type I error decreases. Therefore, we face a dilemma. We must make a decision based on the results of our experiment, but we always run the risk of making an error. Table 13.6 illustrates this dilemma. If you do not reject the null hypothesis and it is true, you have made a correct decision; if it is false, you have committed a Type II error. If you reject the null hypothesis and it is false, you have made a correct decision; if it is true, you have committed a Type I error.

As scientists, we must weigh the hazards of committing each of the two types of errors and determine which mistake would be more detrimental. In most instances, it is assumed that a Type I error is worse, which is why the rather stringent .05 or .01 significance level is used as the decision point for rejection of the null hypothesis. However, there are times when committing a Type II error would be very detrimental. For example, if a drug were needed to combat a deadly epidemic, it would be important to make sure that an effective drug was identified. Avoiding a Type II error in such an instance would be considered more important, so the significance level would be raised to, say, the .15 level, to maximize the chances of identifying a drug that would combat the disease.

SUMMARY

Following collection of the data, the investigator must make a decision as to whether to reject the null hypothesis. If the mean response has been computed for each group of subjects, we know which group made the greatest number of responses. However, some differences in the mean responses of groups are to be expected on the basis of chance variability in responding. Computing the variance or standard deviation of the responses in each group provides an index of variability of the data and establishes some basis for assessing whether group mean differences are due to chance or are real differences resulting from manipulation of the independent variable. However, if we rely solely on a measure of variability, our decisions will lack precision and will be based primarily on judgment.

The most appropriate procedure for testing the null hypothesis is to analyze the data using one of the available statistical tests. The statistical test appropriate for analyzing the data collected is dictated by the design of the study. Every experimental study involves one or more independent variables, several levels of which are administered either to a single group of subjects or to different groups of subjects. These factors dictate the type of statistical analysis to be applied to the data. If two levels of one independent variable are used in a study and each level of the independent variable is administered to a different group of subjects, then an independent samples *t*-test is the appropriate statistical test. If more than two levels of one independent variable are used in a study and each level of the independent variable is administered to a different group of subjects, then a one-way analysis of variance is the appropriate statistical test.

No matter what statistical test is selected, the results must be used to make a decision as to whether the obtained group difference is large enough to be considered real or is more likely to be due to chance factors. This decision is based on the significance level set; the significance levels typically used in psychological research are .05 and .01. If the results of the statistical test reveal that the group mean differences are so large that they could occur by chance only 5 times or less in 100, then, if we accepted the .05 significance level, we would reject the null hypothesis and accept the scientific hypothesis.

Note, however, that there will still be an error either 1 percent or 5 percent of the time. This is referred to as a Type I error—the probability of accepting a hypothesis that is false. The Type I error could be decreased by setting a more stringent level for acceptance of the hypothesis, but then the probability of rejecting the hypothesis when in fact it is true—a Type II error—tends to increase. There needs to be a balance between these two types of error, and the typical balance is obtained by using the .05 and .01 significance levels.

STUDY QUESTIONS

1. Why is hypothesis testing a decision-making process, and what kinds of decisions are you required to make?
2. What is the mean, and what can it be used for in the analysis of data?
3. What is the standard deviation, and what is it used for?
4. What type of design would require the use of an independent samples t-test? a correlated samples t-test? a one-way analysis of variance?
5. When a one-way analysis of variance is conducted, the total variability is broken down into two types of variability. What are these two types of variability, and what are they used for?
6. When a two-way analysis of variance is conducted, total variability is broken down into what different types of variability? What are these types of variability used for?

KEY TERMS AND CONCEPTS

Mean

Variance

Standard deviation

Independent samples t-test

Significance level

One-way analysis of variance

Total variation

Between-groups variation

Within-groups variation

Mean square

F-ratio

Two-way analysis of variance

Type I error

Type II error

CHAPTER 14

External Validity

LEARNING OBJECTIVES

1. To understand the significance of external validity.
2. To understand which characteristics of an experiment can threaten external validity.
3. To understand the caution that should be exercised in assessing the external validity of an experiment.
4. To understand the relationship between external and internal validity.

The giant Canada goose, a magnificent specimen weighing up to twenty-six pounds, with a wing span of as much as seven feet, was believed to be extinct for about three decades. In 1962, however, a flock of giant Canada geese was identified in Rochester, Minnesota. The next year the Fergus Falls, Minnesota, Fish and Game Club purchased six mated pairs; over the two years that followed, fifty more young birds were purchased. By 1982 that small flock had grown to a population of five thousand.

With this increase in the goose population, there have inevitably been some complaints. If all the giant Canada goose did was stand around on green lawns looking regal, or float quietly on ponds in the park, no one would object. But the goose is an eating and defecating machine. People love Canada geese until they have to step around goose droppings on the golf course, clean the backyard before letting their children out to play, or even chase geese out of their swimming pool. Farmers complain that the geese damage their grain fields by grazing on tender young wheat, oats, and barley.

In an attempt to control the burgeoning goose population, people have moved the geese to zoos and other towns. In each instance, however, the result is the same: the goose population rapidly expands and becomes a nuisance. Nevertheless, public opinion still leans heavily toward protection. Attempts to control the growing population have included performing vasectomies on the male geese, breaking goose eggs, training dogs to drive geese out of golf courses and parks, and trying to stop the birds from foraging in areas where they are unwanted by, for example, fertilizing grass in locations other than parks and golf courses, since well-fertilized grass attracts the geese. Another solution is to use some of the results of behavioral research.

A number of years ago John Garcia and his colleagues (Garcia and Koelling, 1967) identified a phenomenon that has become known as conditioned taste aversion. Conditioned taste aversion describes a situation in which an animal avoids a food that it ate about the same time as it got sick, whether or not the food was what made the animal sick. For example, Garcia demonstrated that a rat would avoid a food if it got sick from being exposed to radiation at the same time as it was eating that particular food. Conover (1985) reasoned that the phenomenon of conditioned taste aversion might offer a way to control the giant

Canada goose population in places where the birds foraged. He proposed spraying the grass on golf courses and parks with a chemical repellent that would make the geese ill when they digested the grass they had eaten. To determine if this was a viable solution to the problem, Conover set out to demonstrate, using the procedure outlined by Garcia for producing conditioned taste aversion, that the geese would actively avoid an area that had been sprayed with this repellent. In this study Conover (1985) found that spraying the repellent on grass was effective in repelling the geese for up to two months. Here, then, is an instance in which the results of laboratory behavioral research have been experimentally demonstrated to generalize to a real-life situation and to members of another species. In most of our experimental work, we strive for this type of generalizability. In some instances we are interested only in the response of the subjects participating in a particular experiment (Mook, 1983), but usually we would like to be able to generalize the results of our study to individuals responding in different settings and at different times. There are, however, a number of characteristics of experiments that limit a researcher's ability to generalize the results. This chapter is devoted to discussing these limiting characteristics.

INTRODUCTION

The immediate purpose of any well-conducted experiment is to assess the relationship between the independent and the dependent variables, but the long-range goal is to understand the basic underlying laws of behavior. In Chapter 1 it was stated that one of the basic assumptions of any science is that there are uniformities in nature. Science attempts to isolate these uniformities in order to explain the world in which we live. Psychology is the branch of science that endeavors to understand the behavior of organisms by isolating the lawful relationships that exist in nature. This purpose of psychology transcends the question of internal validity and addresses the question of **external validity**—the extent to which the results of an experiment can be generalized to and across different persons, settings, and times. Assessment of external validity is an inferential process because it involves making broad state-

External validity
The extent to which the results of an experiment can be applied to and across different persons, settings, and times

ments based only on limited information. For example, stating that a study conducted on twenty college students in a psychology laboratory is externally valid implies that the results obtained from this experiment would also be true for all college students responding in any of a variety of settings and at different times. Such inference is a necessary component of the scientific process, because all members of a defined population can seldom be studied in all settings at all times. In order to generalize the results of a study, we must identify a target population of people, settings, and times and then randomly select individuals from these populations so that the sample will be representative of the defined population. For a variety of reasons (cost, time, accessibility), most studies do not randomly sample the specified population. Failure to randomly sample the population means that a study may contain characteristics that threaten its external validity. Such threats fall into three broad categories: lack of population validity, ecological validity, and time validity (Bracht and Glass, 1968; Willson, 1981). Becoming aware of some of the factors in these three categories that limit the generalizability of the results of a study will help you to design studies that circumvent these difficulties.

POPULATION VALIDITY

Population validity
The extent to which the results of a study can be generalized to the larger population

Target population
The larger population to which the results are to be generalized

Experimentally accessible population
The population of subjects available to the investigator

Population validity refers to the ability to generalize from the sample on which the study was conducted to the larger population of individuals in which one is interested. Bracht and Glass (1968) have reiterated Kempthorne's (1961) distinction between the target population and the experimentally accessible population. The **target population** is the larger population (such as all college students) to whom the experimental results are generalized, and the **experimentally accessible population** is the one that is available to the researcher (say, the college students at the university at which the investigator is employed). Two inferential steps are involved in generalizing from the results of the study to the larger population, as illustrated in Figure 14.1. First, we have to generalize from the sample to the experimentally accessible population. This step can be easily accomplished if the investigator *randomly* selects the sample from the experimentally accessible population. If the

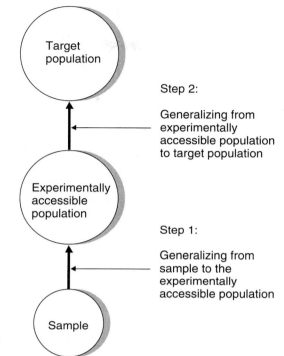

FIGURE 14.1 Two-step inferential process involved in generalizing from the sample to the target population.

Step 2:

Generalizing from experimentally accessible population to target population

Step 1:

Generalizing from sample to the experimentally accessible population

sample is randomly selected, it should be representative, which means that the characteristics of the experimentally accessible population can be inferred from the sample. If you conduct an experiment on a sample of fifty subjects randomly selected from a given university, you can say that the obtained results are characteristic of students at that university.

The second step in the generalization process requires moving from the experimentally accessible population to the target population. This ultimate generalization seldom can be made with any degree of confidence, because only rarely is the experimentally accessible population representative of the target population. For example, assume you are conducting a study using college students as the target population. You would want to be able to say that the results of the study will hold for all college students. To be able to make such a statement, you would have to select randomly from the target population, which is rarely possible. Therefore, you probably will have to settle for randomly selecting from the nonrepresentative, experimentally accessible population.

It is difficult to select randomly even from the experimentally accessible population. This is why experimenters select subjects based on availability or precedent. The two categories of organisms that have most often been used in psychological studies are the albino rat and the college sophomore taking an introductory psychology course. A number of individuals have attacked the use of these two categories of organisms. Beach (1950, 1960), Boice (1973), Ehrlich (1974), Eysenck (1967), Kavanau (1964, 1967), Richter (1959), and Smith (1969) have all questioned the degree to which the research results produced from laboratory rats can be generalized to other animals. Also, as Sidowski and Lockard (1966) point out, the rats that are used in a given study are definitely not a random sample of the experimentally accessible population.

Even more attention has been directed toward research conducted on human subjects, much of which has been aimed at identifying laws governing human behavior. Implicit in this objective is the notion that the results from the sample of subjects on which the research was conducted will generalize to all humans—the target population. There is increasing suspicion among behavioral researchers that the subjects in research studies are not representative of humans in general. As far back as 1946, McNemar (1946, p. 333) issued a warning regarding the biased nature of the human subject research pool. He stated that "the existing science of human behavior is largely the science of the behavior of sophomores." Table 14.1 summarizes the results of surveys that have computed the percentage of studies using college students. You can see from this table that, at least for the journals surveyed, the percentage of psychological studies using college students increased from 1949 to 1964 and then stabilized at about 75 percent through 1973. A more recent survey (Sears, 1986) indicates that this trend not only has continued but may have become even more entrenched.

The potential lack of generalizability when college students are used as the predominant source of subjects in psychological experiments is illustrated by Oakes (1972). He showed that he could not replicate, using a non–college student population, a reinforcement effect that had reliably been demonstrated on college students. Steele and Southwick (1985) found that college students behaved less emotionally and impulsively in laboratory studies than did the general population. Such differences should not be extremely surprising, because college students differ in a number of ways from the general population. Sears (1986) has pointed out that college students

TABLE 14.1 Percentage of Studies Using College Students as Subjects

Author	Source	Year(s)	Percentage Using College Students
Christie (1965)	*Journal of Abnormal and Social Psychology*	1949	20
Christie (1965)	*Journal of Abnormal and Social Psychology*	1959	49
Smart (1966)	*Journal of Abnormal and Social Psychology*	1962–1964	73
Smart (1966)	*Journal of Experimental Psychology*	1963–1964	86
Schultz (1969)	*Journal of Personality and Social Psychology*	1966–1967	70
Schultz (1969)	Journal of Experimental Psychology	1966–1967	84
Carlson (1971)	*Journal of Personality and Social Psychology* and *Journal of Personality*	1968	66
Higbee and Wells (1972)	*Journal of Personality and Social Psychology*	1969	76
Levenson, Gray, and Ingram (1976)	*Journal of Personality and Social Psychology*	1973	72
Levenson, Gray, and Ingram (1976)	*Journal of Personality*	1973	74
Sears (1986)	*Journal of Personality and Social Psychology, Personality and Social Psychology Bulletin,* and *Journal of Experimental Social Psychology*	1980	82

are likely to have less crystallized attitudes, a less formulated sense of self, stronger cognitive skills, stronger tendencies to comply with authority, and more unstable peer group relationships than their noncollege counterparts. Researchers focusing on college students run the risk of presenting a rather narrow and inaccurate portrait of human nature. Oakes (1972), however, has made an interesting point in defense of using the college student in research:

> The point I am suggesting is that research with college students is just as valid as research drawing on any other subject population. A behavioral phenomenon reliably exhibited is a genuine phenomenon, no matter what population is sampled in the research in which it is demonstrated. For any behavioral phenomenon, it may well be that members of another population that one could sample might have certain behavioral characteristics that would preclude the phenomenon being demonstrated with that population. Such a finding would suggest a

restriction of the generality of the phenomenon, but it would not make it any less genuine. No matter what population a researcher samples, whether it be psychology students, real-people volunteers, public school students, or whatever, there are probably some behavioral phenomena that would be manifested differently in that population due to an interaction effect of the particular characteristics of that subject population.

This suggests, then, that the generalizability of the results of behavioral research is not a function of the population sampled, but rather that the external validity of the research depends on the interaction of subject characteristics and the particular behavioral phenomenon with which one is concerned. For some behavioral phenomena, probably those closer to the reflex level of response, there may be no interaction with subject characteristics. Thus, it might make no difference what population one sampled in a study of critical flicker fusion frequency. For other behavioral phenomena, however, especially those beyond the reflex level, one might well expect interactions with subject characteristics.

Thus, I would suggest that our "science of the behavior of sophomores," to the extent that it has discovered reliable behavioral phenomena, is just as valid as a science of behavior based on the sampling of any other population one could tap. The generalizability of any particular finding, however, may be limited by interaction with behavioral characteristics peculiar to any population to which one is attempting to generalize. But this would be true no matter what population one sampled in the original research.[1]

A similar argument has been made for the predominant use of males in both animal and human experimentation. To the extent that most of our knowledge base is generated from males, we have a psychology of male behavior and not human behavior. It is an empirical question whether males and females respond similarly or differently. In many cases they respond differently, so it is inappropriate to assume that the data obtained from male studies would apply equally to females.

The important issue in external validity is the extent to which the experimental results can generalize across different persons, settings, and times; hence, the problem boils down to a test of interactions. Since population validity refers to the ability to generalize across persons, it is threatened by a selection by treatment interaction. If a selection by treatment inter-

---

[1]From "External Validity of the Use of Real People as Subjects" by W. Oakes, 1972, *American Psychologist, 27*, pp. 961–962. Copyright 1972 by the American Psychological Association. Reprinted by permission of the author and publisher.

action (Campbell and Stanley, 1963) exists, the experiment is externally invalid or cannot be generalized to the target population. This means that the particular sample of subjects selected for use in a study may respond differently to the experimental treatment condition than would another sample of subjects with different characteristics. Rosenthal and Rosnow (1975) have summarized the wealth of literature showing that volunteer and nonvolunteer subjects respond differently to many experimental treatment conditions. Kendler and Kendler (1959) found that kindergarten children did not achieve results consistent with those obtained from college students. They suggest that perhaps a maturational variable exists, causing subjects at different maturational levels to respond differently to selected tasks. Such research indicates that a selection by treatment interaction limits the extent to which the results of a study can be generalized beyond the groups used to establish the initial relationship.

ECOLOGICAL VALIDITY

Ecological validity
The extent to which the results of a study can be generalized across settings or environmental conditions

Ecological validity refers to the ability to generalize the results of the study across settings or from one set of environmental conditions to another. For example, the environmental setting of an experiment may require a specific arrangement of the equipment, a particular location, or a certain type of experimenter. If the results of a laboratory experiment that requires such a setting can be generalized to other settings (such as a therapy setting or a labor relations setting), then the experiment possesses ecological validity. Consequently, ecological validity exists to the extent that the treatment effect is independent of the experimental setting. The following discussion will focus on some of the characteristics of experimental settings that can threaten ecological validity.

Multiple-Treatment Interference

Multiple-treatment interference
The sequencing effect that can occur when subjects participate in more than one treatment condition

The **multiple-treatment interference** phenomenon refers to the effect that a subject's participation in one treatment condition has on his or her participation in a second treatment condition. The multiple-treatment interference phenomenon is a sequencing effect and, therefore, impedes direct generalization of the results of the study in addition to threatening internal validity, because it is difficult to separate the effect of the particular order of conditions from the effect of the treatment conditions. Any generalization that is made regarding the results of the study is restricted to the particular sequence of conditions that was administered.

There are at least two situations in which the multiple-treatment interference phenomenon may occur. It may occur in an experiment in which the subject or subjects are required to participate in more than one treatment condition. Fox (1963) found such an effect in his investigation of reading speed as relating to typeface. In an incomplete counterbalanced design, one group of subjects first read material typed in Standard Elite and then read material typed in Gothic Elite, whereas the second group of subjects took the reverse order. When the data from the reading of the Standard Elite and Gothic Elite were combined across groups, no reliable difference relating to typeface was found. However, when the two typefaces were compared for the first reading, it was found that the subjects read the Standard Elite typeface significantly faster than they read the Gothic Elite. When the two groups switched to material typed in the other typeface, they continued to read at the average speed previously established. Thus, the reading rate subjects established when reading the first typeface transferred over to the second treatment effect and thus affected the reading rate for the second typeface. Consequently, the order of presentation of the typeface determined reading speed.

Multiple-treatment interference may also occur between experiments. The effect of participation in a previous experiment may affect the subject's response to the treatment condition in a current experiment. Underwood (1957), for example, has demonstrated that recall of serially presented adjective lists is to a great extent a function of the number of previous lists that have been learned. If a subject had previously been required to recall serial adjective lists, the current one would be more difficult to recall. Holmes (1967) and Holmes and Applebaum (1970) found that prior participation in psychological experiments affects performance in verbal conditioning experiments. However, Cook, Gruder, Hennigan, and Flay (1970) found no such effect in one of their studies. It may be that a subject's history of experimental participation affects performance only in certain types of studies. Such preliminary indicators need to be pursued to determine what types of data are affected by a subject's prior experimental history and how they are affected, so that such effects can be taken into consideration in attempts to generalize experimental data.

Investigators using albino rats in their studies also must think about the potential effects of using the same rats in more than one study. The generally accepted procedure is to use the rats in only one study. When another study is to be conducted, a new sample of albino rats must be purchased. In this way, the investigator ensures that the rats are naive in the sense of not having previously participated in a psychological study.

The Hawthorne Effect

Hawthorne effect
The influence produced by knowing that one is participating in an experiment

The **Hawthorne effect** refers to the fact that one's performance in an experiment is affected by knowledge that one is in an experiment. It is similar to the effect of being on television: once you know the camera is on you, you shift to your "television" behavior.

Bracht and Glass (1968) suggest several reasons why subjects may respond differently when they know that they are participating in an experiment. Subjects may, for example, display a high degree of compliance and diligence in performing the experimental task because they are motivated by a high regard for the aims of science and experimentation. Consequently, subjects who know they are in an experiment may agree to perform a task that they otherwise would refuse to do. For example, if you walked up to a person and asked her to do ten pushups, she might tell you where to go. But if you added that you were conducting an experiment on physical fitness, she might very well comply. To the extent that such effects alter the subject's performance, the experimental results cannot be accounted for by just the treatment effect.

Novelty, or disruption, effect
A treatment effect that occurs when the treatment condition involves something new or different

A phenomenon similar to the Hawthorne effect is a **novelty, or disruption, effect.** If the experimental treatment condition involves something new or unusual, a treatment effect may result by virtue of this fact. When the novelty or disruption diminishes, the treatment effect may disappear. Van Buskirk (1932) found that placing a red (novel) nonsense syllable in the most difficult position in a serial list of black-on-white nonsense syllables greatly facilitated its being learned originally and later recalled. Brownell (1966) also provided an example of a novelty treatment effect. He set out to compare two different instructional programs in England and Scotland.

Results of the study from the two countries conflicted, however. This conflict was attributed to the novelty effect that existed in Scotland but not in England. Whereas teachers and pupils in England were accustomed to innovation and new programs, those in Scotland were not. Consequently, the new program was enthusiastically inaugurated in Scotland, but in England it was seen as just a continuation of an established pattern. Conditions such as novelty and the Hawthorne effect limit one's ability to generalize the results of a given study to populations where such an effect does not exist.

The Experimenter Effect

Chapter 7 covers in detail the potential biasing effects that the experimenter can have on the results of the experiment. Experimenter effects are a limiting factor in generalization of results to the extent that experimenter bias is not controlled and the results of the experiment interact with the attributes or expectancies of the experimenter. In other words, if the results of the experiment are partially dependent on the experimenter's particular attributes or expectancies, the results are generalizable only to other similar situations.

The Pretesting Effect

Pretesting effect
The effect that pretesting has on the influence of the treatment condition

The **pretesting effect** refers to the influence that administering a pretest may have on the experimental treatment effect. Administering a pretest may sensitize the subject in such a way that he or she approaches the experimental treatment differently than does the subject who did not receive the pretest and thus influence external validity in addition to internal validity. If it is true that pretested subjects respond differently to the dependent variable than do unpretested subjects, the results of a study that pretests subjects can be generalized only to a pretested population. However, Lana (1969) concluded that pretesting does not have the sensitizing effect that others had previously postulated. Rosenthal and Rosnow (1975) have suggested, based on the research of Rosnow and Suls (1970), that the failure to find a pretest sensitization effect was due to the failure to distinguish between the motivational sets of volunteer and nonvolunteer subjects—volunteers being the willing and eager subjects, and nonvolunteers being the unwilling subjects. Rosnow and Suls found a pretest sensitization effect operating among volunteers and an opposite dampening effect operating among nonvolunteers. This indicates that the pretest sensitization variable may be specific to the type of subject used in the study (a selection by treatment interaction). However, it seems that these results should be replicated before a firm conclusion is drawn regarding the interaction of volunteer status and pretest sensitization.

TEMPORAL VALIDITY

Temporal validity
The extent to which the results of an experiment can be generalized across time

Temporal validity refers to the extent to which the results of an experiment can be generalized across time. Most psychological studies are conducted during one time period. Carlson (1971), for example, found that 78 percent of the studies published in the *Journal of Personality and Social Psychology* and the *Journal of Personality* were based on only a single session. Five years later, Levenson, Gray, and Ingram (1976) found that 89 percent of the articles published in the same two journals were based on a single session. If these two periodicals are representative of most of the work being conducted in psychology, we are not taking the time variable into consideration. Experimenters seem to be assuming that the results of an experiment remain invariant across time.

It has been demonstrated, however, that the results of an experiment can vary depending on the amount of time that elapses between the presentation of the independent variable and the assessment of the dependent variable. Walster (1964) asked army draftees to rate the attractiveness of ten different jobs to which they could be assigned during their two-year enlistment period. After rating the ten jobs, each person was asked to choose between two jobs that had been rated similarly and were moderately attractive. After selecting one of the two, the recruits were asked to rate the attractiveness of the two jobs once again. For one group of recruits, this second rating took place immediately after the job choice was made. For the other three groups, the choice was delayed four, fifteen and ninety minutes, respectively. Figure 14.2 illustrates the change in attractiveness of the chosen job over time. Immediately after making the choice, the draftees found the chosen job more attractive. Within four minutes, however, the recruits

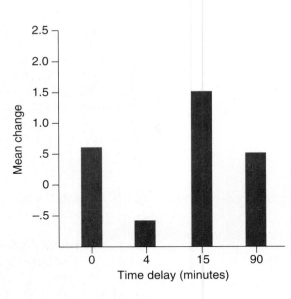

FIGURE 14.2 Illustration of how the attractiveness of a chosen job changes across time. (Based on data from "The Temporal Sequence of Post-Decision Processes" by E. Walster. In L. Festinger, *Conflict, Decision, and Dissonance.* Stanford: Stanford University Press, 1964.)

apparently experienced some degree of regret or had second thoughts about the jobs they had chosen, as they rated the job as less attractive. This regret then dissipated rather rapidly. Ratings of the chosen job reached their highest level after fifteen minutes, only to return to their original post-choice level within ninety minutes.

Experiments such as this one vividly demonstrate that experimental results can vary across time. Failure to consider the time variable can threaten the external validity of experiments. Following are some of the specific time variations identified by Willson (1981) that can threaten external validity.

Seasonal Variation

Seasonal variation
A variation in results that occurs at regular time intervals

A **seasonal variation** is a variation that appears regularly over time in parts of the population. If the automobile accident rate were plotted over time, we might find that it increased during the winter months in states such as South Dakota (Willson, 1980) because of adverse weather conditions.

Fixed-time variation
A seasonal variation that occurs at specific points in time

There are actually two forms of seasonal variation: fixed time and variable time. A **fixed-time variation** refers to a change that occurs at specific, predictable points in time, such as on holidays or weekends. For example, it is well known that traffic accidents increase during holidays and that airline traffic increases over the summer months and over holidays. In contrast, the timing of a **variable-time variation** cannot be predicted except in terms of a specific event, the timing of whose occurrence is not known. For example, we cannot know exactly when a loved one will die, but we all know that when it happens, a predictable psychological reaction will follow.

Variable-time variation
A seasonal variation that occurs after specific events

Cyclical Variation

Cyclical variation
A regular variation that occurs within people and other organisms

A **cyclical variation** is actually a form of seasonal variation. But whereas a seasonal variation appears regularly across units of the population or across people, a cyclical variation occurs within people and other organisms. Hunsicker and Mellgren (1977) demonstrated that rats apparently operate on a twelve-hour cycle, and at certain points during this cycle retention is maximized. Similarly, within humans there are cyclical variations that can interact with experimental treatment conditions. For example, pulse rate, temperature, endocrine function, and kidney function operate on a circadian (approximately twenty-four-hour) rhythm (Conroy and Mills, 1970). The many cyclical variations that exist within the organisms on which we conduct our experiments can potentially alter the influence of the experimental treatment or interact with the experimental treatment. If such an interaction takes place, the results of the experiment are generalizable only to the specific point in the cycle during which the experiment was conducted.

Personological Variation

Personological variation
A variation that occurs in the characteristics of an individual as a function of time

Personological variation refers to variation in the characteristics of individuals over time. Although it is generally assumed that the characteristics within a person are relatively stable across time, this may not be true for all traits. For example, an individual's evaluation of self and others and his or her political leanings are to a great extent dependent on the environmental stimuli to which the individual is exposed. Gergen (1973a) has noted that the variables that predicted political activism during the early years of the Vietnam War differed from those that successfully predicted political activism during later periods. This means that the factors motivating activism changed over time, much as our preference for clothing varies over the course of years. If the characteristics being investigated are subject to change, the outcome of the study will be valid only for the period during which the study is conducted and will not be generalizable across time.

RELATIONSHIP BETWEEN INTERNAL AND EXTERNAL VALIDITY

Given knowledge of the three general classes of variables that threaten external validity, it would seem logical to design experiments using a diverse sample of subjects and settings across several different time periods in order to increase external validity. The problem with such a strategy is that there tends to be an inverse relationship between internal and external validity. When external validity is increased, internal validity tends to be sacrificed; when internal validity is increased, external validity tends to suffer (Kazdin, 1980).

To gain insight into this relationship between internal and external validity, consider the following characteristics of a well-designed study. From the previous chapters, you know that a well-designed study attempts to control for the effects of all extraneous variables. The researcher selects as subjects a specific subsample of the population, such as females or sixth-grade children, in order to control for variation in different subsamples of individuals, or to create a more homogeneous sample so as to maximize the possibility of detecting a treatment effect. The experimenter conducts the experiment within the confines of a controlled laboratory setting in order to present a specific amount of the treatment condition and to eliminate the influence of extraneous variables, such as the presence of noise or weather conditions. While in the laboratory setting, the subjects receive a set of standardized instructions delivered by one experimenter or maybe by some automated device, and the study is conducted at one specific point in time. But these same features that maximize the possibility of attaining internal validity—using a restricted sample of subjects and testing them in the artificial setting of a laboratory at one specific time—limit the external validity (Kazdin, 1980) by excluding different persons, settings, and times. However, if an experimenter tried to maximize external validity by conducting the experiment on diverse groups of individuals in many settings and at different points in time, the experiment's internal validity would tend to decrease. As the number of settings or the number of types of subjects is increased, control of the extraneous variables that may influence the independent variable decreases, decreasing the likelihood that the study will identify the influence produced by the independent variable. This does not mean that external validity should be disregarded. Rather, it suggests that the first and foremost objective should be the identification of the influence of a treatment effect. Once an effect has been verified by means of well-controlled, internally valid studies, external validity can be investigated.

CAUTIONS IN EVALUATING THE EXTERNAL VALIDITY OF EXPERIMENTS

The external validity of an experiment has been defined as the extent to which the experimental results will generalize to and across different persons, settings, and times. This chapter has considered a number of the specific threats to the external validity of an experiment. Given such a listing of possible threats, one is tempted to compare the list to the characteristics of a given study and conclude that the study does not have external validity if some of the threats legitimately apply. However, such a procedure is not always valid. Let us consider the study conducted by Berkowitz and LePage (1967). These researchers investigated the influence of the mere presence of weapons on aggression as measured by the number of shocks subjects delivered to an accomplice of the experimenter. A hundred college students were randomly assigned to one of three experimental treatment conditions. Subjects were seated at a table that had on it a telegraph key either alone, with neutral objects (such as a badminton racket), or with a shotgun and a revolver. The telegraph key was connected to a shock machine. In one condition, the subject's partner was told to come up with ideas to boost the image and the record sales of a popular singer. If the subject did not like the partner's ideas, the subject was to deliver several shocks to the partner by pressing the telegraph key. Berkowitz and LePage found that subjects delivered more shocks to their partners in the condition that had the weapons on the table even though the subjects were told that these weapons were not part of the experiment but had been left by the previous researcher. The experiment suggests that the mere presence of weapons can enhance aggression. But does this experiment have external validity—can the results be applied to the world at large?

If we look at the threats to external validity, we can see that many of them apply to the Berkowitz and LePage study. The experiment did not contain a random sample of subjects or even a diverse sample of college students. The study was conducted at only one time in a college laboratory in which subjects were told that the experiment was measuring their physiological reactions to stress. The task the students were performing was one of brainstorming ideas to improve the image and record sales of a popular singer. The study would seem to be a total failure with regard to external validity, since it did not sample across people, settings, or times. However, there are several cautions of which we should be aware in assessing the external validity of experiments such as this one.

Mook (1983) has appropriately pointed out that experiments are conducted for a variety of purposes and that some of these purposes do not attempt to relate to real-life behavior. The issue of external validity is moot for such experiments. For example, we may conduct an experiment in order to determine if something *can* happen and not necessarily if it really does happen. Person–perception studies have revealed that people wearing glasses are judged as more intelligent when seen for only fifteen seconds. If these same people are viewed for five minutes, however, the glasses make no difference. The temporary effect of the glasses seems to have no meaning in real life. If the experiment had not been conducted, however, this temporary effect would not have been identified. Although it says little about real-life behavior, it does say something about humans as judges, so the experimental results are important.

A second significant point made by Mook (1983) is that the component of an experiment that is to be generalized is often the theoretical process that is being tested or the understanding that accrues from the experiment. For example, consider the difference between a study investigating the influence of a new teaching technique and a study investigating the influence of the presence of weapons. Researchers investigating a new technique for teaching basic arithmetic to seven-year-old children would like to have their results generalize to all seven-year-olds. Contrast their situation with that of Berkowitz and LePage, who found that if a weapon was present on the table a subject delivered more shocks as an indication of not liking a partner's ideas. In no way would they want to generalize the specific procedure of this experiment to other people, settings, or times. A person in a setting other than the laboratory might indeed give more shocks to a partner if a weapon were placed on the table. Such a situation, however, would seldom exist in real life, and it was not Berkowitz and LePage's intention to generalize the specific experimental procedure. Rather, it is an understanding of the impact of the presence of weapons that is to be extracted from the study, not the specific procedure.

One must also exercise caution in placing too much emphasis on the setting in which the experiment is conducted. Laboratory research has been criticized because of the artificiality of the setting, and this artificiality can impose a severe restraint on the external validity of the experiment. Because of the intuitive logic inherent in this criticism, a number of investigators have advocated that psychologists move out of the laboratory and into a real-world setting, such as a mental health clinic. The assumption seems to be that if we change the experimental setting, we will enhance external validity. For example, some people believe that if industrial psychologists conducted their research in an organizational setting, both the setting and the subjects would be more representative of the real-world population. Dipboye and Flanagan (1979) have demonstrated that such an assumption is not valid. They showed that the field research conducted in industrial organizational psychology has been performed on male, professional, technical, and managerial personnel in productive economic organizations through use of self-report inventories. Thus, in at least this one area of psychology, the assumption of greater external validity of field research has not been supported, since the field research has used a limited sample of individuals, in limited field settings, over one time period.

A last caution has to do with the assessment of the interaction between the treatment effect and any of the traits that threaten external validity. Earlier the chapter stated that assessing external validity boils down to determining whether an interaction exists between the experimental treatment and one of the characteristics that threaten external validity. If the experimental treatment effect does not interact with persons, settings, or times, then the treatment effect can be generalized. Within the field of psychology, however, most of the phenom-

Ordinal interaction
An interaction in which the rank order of the treatment effects remains the same across persons, settings, and times

Disordinal interaction
An interaction in which the rank order of the treatment effects changes as a function of persons, settings, and times

ena we investigate do interact with one of these types of characteristics. Does this mean that none of our experimental results have external validity? Such a conclusion would obviously be rash and false. When interactions do exist, it is important to distinguish between ordinal and disordinal interactions (Lindquist, 1953; Lubin, 1961). An **ordinal interaction** is one in which the rank order of the treatment effects remains the same across persons, settings, and times. Figure 14.3 illustrates an ordinal interaction of treatment by type of person. Note that the rank order of the treatments remains the same for both categories of persons, with treatment C producing the highest dependent variable scores and treatment A producing the lowest dependent variable scores.

A **disordinal interaction,** on the other hand, is one in which the rank order of the treatments changes across persons, settings, or times, as illustrated in Figure 14.4. Here, treatment A is most effective for low-ability persons, treatment C is most effective for high-ability persons, and treatment B is equally effective for both types. Consequently, the effectiveness of the condition depends on the type of person receiving the treatment.

What do these two types of interactions say about generalizations? Ordinal interactions do not limit generalizability of the results, whereas disordinal interactions do. For ordinal interactions, the one best treatment condition can be prescribed for all individuals. As Figure 14.3 illustrates, treatment condition C is superior for both types of individuals. The ordinal interaction reveals the differential effectiveness of the three treatments and so lends support to the meaningfulness of interpreting the data; however, it does not limit the ability to

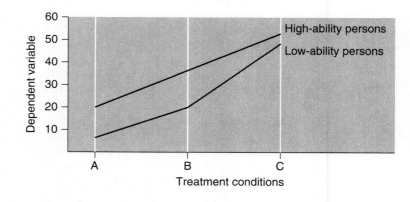

FIGURE 14.3   Ordinal interaction.

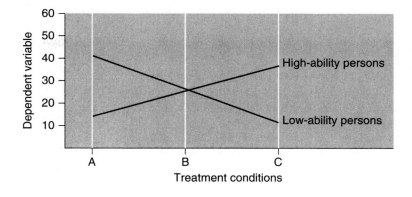

FIGURE 14.4 Disordinal interaction.

generalize the data. Therefore, if the lines of the graph do not cross, generalization of the data is not limited, despite the fact that a significant interaction exists.

If there is a statistically significant interaction and the lines of the graph cross as in Figure 14.4, a disordinal interaction exists and generalizability of the data is limited. From this figure, it is readily apparent that one of the three treatment conditions is not superior for both types of individuals; the most effective treatment depends on the type of person (high or low ability) being considered. Note that the differences between groups, settings, or times *must* be statistically significant for a true disordinal interaction to exist. For example, the scores of the high- and low-ability persons must be significantly different for treatments A and C for there to be a disordinal interaction. This is because the lines of a graph may cross as a function of random variation in response.

SUMMARY

Superimposed on the researcher's concern for conducting an internally valid experiment is a concern for understanding behavior. In addition to wanting our experiments to be internally valid, we want to produce results that will have implications for a larger population of individuals behaving in diverse settings and at different points in time. When the results of an experiment can be generalized across persons, settings, and times, the experiment is externally valid. Threats to external validity fall into three broad categories: lack of population validity, lack of ecological validity, and lack of time validity.

Population validity refers to the ability to generalize from the experimental sample of subjects to the larger population from which the sample was drawn. For population validity to exist, the experimental subjects must be randomly selected from the larger population. This is seldom the case. Most animal research has been performed on a nonrandom sample of albino rats, and most human research has been conducted on college sophomores taking an introductory psychology course. The behavioral phenomena manifested in these nonrandom samples are still real; whether the results will generalize depends on whether there is an interaction between these behavioral phenomena and subjects with characteristics different from those of subjects who participated in the study. Assessing generalizability boils down to determining if there is a subject, setting, or time characteristic by treatment interaction.

Ecological validity refers to the ability to generalize the results of the study across settings or environmental conditions. Every experiment is conducted within a specific setting. If the results of the experiment cannot be generalized to other settings, then the treatment effect is limited to the setting in which it was demonstrated. There are a number of experimental environmental conditions that can limit one's ability to generalize the results to other situations. The multiple-treatment interference phenomenon is a sequencing effect. If the experimental effect occurs only when a particular order of conditions exists, then generalization is limited to this order of conditions. The Hawthorne effect refers to the alteration that occurs in the subject's behavior when the subject knows he or she is part of an experiment. If the experimental effect is a function of this fact, then generalization is limited to similar environments. Other limiting factors include the potential influence of the experimenter, of pretesting, and of having conducted the study over only a single session. To the extent that these variables affect the results, the results can be generalized only to other environments that include these factors.

Temporal validity refers to the influence of time on the results of experiments. If an experimental result changes over time, the influence of the treatment condition is limited to the point in time at which the experiment was conducted. The primary time variables seem to be seasonal variations and cyclical variations, but characteristics of individuals may also change over time and produce a personological variation.

Although it is desirable to conduct an externally valid experiment, the first requirement is to conduct an internally valid experiment. Unfortunately, the factors a researcher must include in order to conduct an internally valid experiment tend to preclude the attainment of external validity. Thus the concern for external validity should arise only after an effect has been verified with an internally valid study.

STUDY QUESTIONS

1. What does it mean to say that a study has external validity?
2. Define population validity, and discuss the threats to population validity.
3. Define ecological validity, and discuss the threats to ecological validity.
4. Define temporal validity, and discuss the threats to temporal validity.
5. Discuss the relationship that exists between internal and external validity.
6. Identify and discuss the factors that are important in evaluating the external validity of an experiment.
7. Discuss the difference between ordinal and disordinal interaction.

KEY TERMS AND CONCEPTS

External validity
Population validity
Target population
Experimentally accessible population
Ecological validity
Multiple-treatment interference
Hawthorne effect
Novelty effect

Pretesting effect
Temporal validity
Seasonal variation
Fixed-time variation
Variable-time variation
Cyclical variation
Personological variation
Ordinal interaction
Disordinal interaction

CHAPTER 15

The Research Report

LEARNING OBJECTIVES

1. To learn to structure a research report according to APA format.
2. To develop an understanding of the type of information that must be communicated in each section of the research report.
3. To understand how a research report is to be prepared for submission to a scholarly journal.

INTRODUCTION

Throughout this book, I have presented the various steps involved in the research process and discussed in detail the intricacies of each. A thorough presentation was made to enable you to conduct a sound scientific study. As a scientist, however, you have a responsibility not only to conduct a well-designed and well-executed study but also to communicate the results of the study to the rest of the scientific community. Your study may have answered a very significant research question, but the results are of no value unless they are made public. The primary mechanism for communicating results is through professional journals. Within the field of psychology, the American Psychological Association publishes twenty-four journals and a newsletter. As Table 15.1 illustrates, these periodicals cover a wide variety of areas and provide an outlet for studies conducted within just about any field of interest. There are other journals not affiliated with the APA that also publish the results of psychological studies. In order to facilitate clear communication of research results, the APA has published a manual (American Psychological Association, 1983) that gives a standardized format for authors to follow when preparing research reports. Since many periodicals instruct their authors to prepare manuscripts according to the style specified in the APA manual, this is the format I will present for writing a research report.

Prior to preparing a report on a study that you have completed, you must ask yourself if the study is important enough to justify publication. Would others be interested in it, and, more important, would it influence their work? As a general rule you should never conduct a study you don't think is publishable. If you think the study is significant, you must decide whether it is free from flaws that would preclude drawing any causal relation between the independent and the dependent variables. For example, you must ask yourself whether you have built in the controls needed to eliminate the influence of rival hypotheses. If you can satisfy yourself with regard to these two questions, then you are justified in proceeding with the preparation of the research report.

TABLE 15.1 Journals Published by the American Psychological Association

Name of Journal	Area Covered
American Psychologist	Contains the American Psychological Association's archival documents as well as articles relating to problems and issues facing psychology, trends within the field, the relation between psychology and society, and the current status and applications of psychological knowledge.
Behavioral Neuroscience	Contains research papers, reviews, and theoretical papers on the biological basis of behavior.
Clinician's Research Digest	Highlights for clinicians selections from over fifty journals.
Contemporary Psychology	Reviews material relevant to psychology, such as books, films, and tapes.
Developmental Psychology	Publishes articles relating to human development across the life span.
Experimental and Clinical Psychopharmacology	Publishes research integrating pharmacology and behavior.
Health Psychology	Devoted to furthering an understanding of the relationship between behavioral principles and physical health or illness.
Journal of Abnormal Psychology	Publishes articles relating to the determinants, theories, and correlates of abnormal behavior.
Journal of Applied Psychology	Publishes articles that contribute to our understanding of any applied area of psychology except clinical psychology.
Journal of Comparative Psychology	Contains behavioral studies that relate to evolution, development, ecology, control, and functional significance of various species.
Journal of Consulting and Clinical Psychology	Contains research investigations pertaining to development, validity, and use of various techniques for diagnosing and treating disturbed behavior in all populations.
Journal of Counseling Psychology	Contains articles pertaining to evaluation, application, and theoretical issues surrounding counseling.
Journal of Educational Psychology	Publishes studies and theoretical papers concerned with learning and cognition.
Journal of Experimental Psychology: Animal Behavior Processes	Publishes studies relating to nonhuman learning, memory, perception, motivation, and performance.
Journal of Experimental Psychology: General	Publishes integrative articles of interest to all experimental psychologists.
Journal of Experimental Psychology: Human Perception and Performance	Focuses on perception, verbal and motor performance by humans.
Journal of Experimental Psychology: Learning, Memory, and Cognition	Contains original studies and theory on encoding, transfer, memory, and cognitive processes in human memory.
Journal of Family Psychology	Focuses on the study of family systems and processes and on problems such as marital and family abuse.
Journal of Personality and Social Psychology	Contains articles on all areas of personality and social psychology.

TABLE 15.1   *(Continued)*

Name of Journal	Area Covered
Neuropsychology	Publishes articles on the relation between the brain and human cognitive, emotional, and behavioral function.
Professional Psychology: Research and Practice	Focuses on the application of psychology.
Psychological Assessment: A Journal of Consulting and Clinical Psychology	Publishes articles concerning clinical assessment and evaluation.
Psychological Bulletin	Publishes evaluative and integrative reviews of both methodological and substantive issues.
Psychological Review	Publishes articles that make a theoretical contribution to psychology.
Psychology and Aging	Publishes articles on adult development and aging.

THE APA FORMAT

The structure of the research report is very simple and tends to follow the steps one takes in conducting a research study. To illustrate the format of the research report, an article that was published in the *Journal of Abnormal Psychology* is reproduced on the following pages, using the format required when an article is submitted for publication.[1] Adjacent to each section of the research report is an explanation of the material that should be included in that section. This explanation may include some recommendations that are not illustrated in the research report, since each study does not include all of the elements listed in the publication manual (American Psychological Association, 1983).

When reading through each section of the research report and then when writing your own report, you should keep its purpose in mind. The primary goal is to report as precisely as possible what you did, including a statement of the problem investigated, the methods used to investigate the problem, the results of your investigation, and any conclusions you may have reached. Is there any criterion you can use to determine

[1]"Characteristics of the frequent nightmare sufferer" by A. Berquier and R. Ashton, 1992, *Journal of Abnormal Psychology, 101*, pp. 246–250. Copyright 1992 by the American Psychological Association. Reprinted by permission.

whether you have clearly and explicitly reported your study? The criterion of replication is probably the most important. If another investigator could read your research report and precisely replicate your study, then chances are good that you have written a clear and complete report.

The following sample research report was prepared according to the guidelines specified in the APA publication manual. This type of research report could be submitted to a journal such as the *Journal of Consulting and Clinical Psychology,* or *Behavior Therapy.*

Title
The title should be centered on the first page of the manuscript and typed in upper- and lowercase letters. It should inform the readers about the study by concisely stating the relationship between the independent and the dependent variables. A typical title length is twelve to fifteen words.

Name of author(s)
The names of any authors who have made a substantial contribution to the study should appear immediately below the title and should be typed in upper- and lowercase letters and centered on the page. The preferred form is to list first name, middle initial, and last name, with titles and degrees omitted.

Running head
The running head is centered at the bottom of the first (title) page and is typed in all uppercase letters. It is an abbreviated title of not more than fifty characters in length, counting letters, punctuation, and spaces between words.

Nightmare

1

Characteristics of the

Frequent Nightmare Sufferer

Anne Berquier and Roderick Ashton

Running Head: NIGHTMARE SUFFERER

Nightmare

2

Abstract

Previous research has found that persons who experience frequent nightmares score highly on scales that measure psychotic symptomatology. Neurotic symptoms have also been implicated as correlates of nightmare frequency. In this study, 30 adult lifelong nightmare sufferers were compared with 30 control subjects, matched for age, sex, and socioeconomic status. Subjects were asked to record all dreams for 1 month and to complete the Minnesota Multiphasic Personality Inventory (MMPI) and the Eysenck Personality Questionnaire (EPQ). Nightmare subjects scored significantly higher on the EPQ Neuroticism scale and on MMPI clinical scales than did the control group. These scales also best discriminated between the groups in a direct discriminant analysis. The results are interpreted as a reflection of global maladjustment rather than of specific psychotic symptomatology.

Page number and short title
The page number as well as a shortened title should appear in the upper right-hand corner of all manuscript pages except those containing figures. The shortened title should consist of the first one to three words of the running head, which will allow for identification of the page of the manuscript if the pages are separated during the review process. All pages should be numbered consecutively, beginning with the title page.

Abstract
The abstract is a comprehensive summary of the contents of the research report, 100 to 150 words in length. It is typed on a separate page, with the word "Abstract" centered at the top of the page in upper- and lowercase letters and no paragraph indentation. It should include a brief statement of the problem, a summary of the method used (including a description of the subjects, instruments, or apparatus), the procedure, the results (including statistical significance levels), and any conclusions and implications.

Introduction
The research report begins with the introduction, which is not labeled because of its position in the paper. The introduction is funnel shaped in the sense that it is broad at the beginning and narrow at the end. It should begin with a very general introduction to the problem area and then start to narrow by citing the results of prior works that have been conducted in the area and that bear on the specific issue that you are investigating, leading into a statement of the variables to be investigated. In citing prior research, do not attempt to make an exhaustive review of the literature. Cite only those studies that are directly pertinent, and avoid tangential references. This pertinent literature should lead directly into your study and thereby show the continuity between what you are investigating and prior research. You should then state, preferably in question form, the purpose of your study. The introduction should give the reader the rationale for the given investigation, explaining how it fits in with, and is a logical extension of, prior research.

Characteristics of the

Frequent Nightmare Sufferer

The relation between dreams and behavior has elicited considerable interest in psychology. However, precise studies of highly arousing dreams, such as nightmares, are rather few. Until recently, theories of nightmare etiology were peripheral to overall dream theory. In many cases, nightmares were neglected as conceptually problematic to the development of coherent dream theory (Hartmann, 1984). Although sleep laboratory studies have led to a reduction in the polysemy of the term nightmare (Fischer, Byrne, & Edwards, 1968), methodological problems intrinsic to retrospective self-report measures have led to interpretive difficulties in studies to date.

The paucity of systematic studies is surprising given the prevalence of nightmares in the adult population. Nightmares are often conceptualized as a childhood development disorder (Foulkes, 1985) or as a symptom secondary to mental illness (Hefez, Metz, & Lavie, 1987). It has, however, consistently been estimated that 5%–10% of nonclinical adults experience one nightmare or more per month through their lifetime (Belicki & Belicki, 1982; Feldman & Hersen, 1967).

Recently, several studies have linked frequent nightmares to personality characteristics and, in particular, to psychopathological symptoms (Hartmann, 1984; A. Kales et al., 1980). Two sources of evidence have contributed to this theory. First, nightmares have been associated with various psychiatric disorders, such as schizophrenia (Hartmann, 1984; Stone, 1979; van der Kolk & Goldberg, 1983) and alcohol withdrawal (Cernovsky, 1985, 1986). Prepsychotic patients often report nightmares, particularly of body fragmentation and the death of the dreamer (Stone, 1979). Van der Kolk and Goldberg concluded that most chronic schizophrenics have at least one nightmare per month during periods of remission. Kestenbaum (cited in Hartmann, 1984) reported that in a longitudinal study, nightmares were a frequent symptom in children who later developed schizophrenia. Nightmares seem to occur in schizophrenics during nonpsychotic periods, although this may be an artifact of the difficulties in eliciting nightmare recall during psychotic episodes. An intermediate phase of frequent nightmares in Parkinsonian patients administered with the drug L-dopa has also led to the hypothesis that nightmares and schizophrenia are biochemically and perhaps etiologically related (Hartmann, Skoff, Russ, & Oldfield, 1978).

Second, psychopathological symptoms have been observed in nonclinical adults who reported having frequent nightmares since their childhood (Hartman, 1984; A. Kales et al., 1980). Hartmann found that a significantly high proportion of persons who experienced one nightmare per week throughout their lifetimes reported that a first- or second-degree relative suffered from schizophrenia. He also found that the nightmare group obtained significantly elevated scores on the Schizophrenia, Psychopathic Deviate, and Paranoia scales of the Minnesota Multiphasic Personality Inventory (MMPI) and concluded that the nightmare group was genetically vulnerable to schizophrenia and possibly lay on a continuum between normalcy and psychosis.

Although studies have consistently found clinically elevated symptomatology in nightmare groups, the type of symptom has varied between studies. Van der Kolk, Blitz, Buff, Sherey, and Hartmann (1984), for example, found that their subjects were clinically elevated on all scales of the MMPI. Their study, however, was conducted on psychiatric outpatients, and they were compared with posttraumatic stress disorder patients rather than with a low nightmare frequency control group.

The experience of frequent nightmares has also been correlated in nonclinical populations with the presence of such neurotic symptoms as guilt and anxiety. Nightmare frequency has been correlated with anxiety-distraction, guilt-dysphoria, and guilt-fear of failure in daydream patterns (Starker, 1974, 1984; Starker & Hasenfeld, 1976). However, there is the suggestion in at least one study that frequent nightmare groups have higher levels of hypnotizability, imagery capability, and an ability to become absorbed in daytime fantasy than do control subjects (Belicki & Belicki, 1986). Interpretation of these results is difficult, however, given another study in which only a tenuous relation between fantasy and behavior was found (Rabinowitz, 1975).

Subjects in Starker's (1974, 1984) and Starker and Hasenfeld's (1976) studies were measured on neurotic daydream patterns rather than functional social performance. Daydream patterns in nightmare subjects might also reflect intermittent daytime preoccupation with nightmare content. Conceptual similarities between daydreaming and dreaming suggest that these two measures may be assessing the same construct. An evaluation of overall daytime neuroticism conceptually independent of dream influence requires measurement of the neuroticism associated with social function.

Studies of personality characteristics in nonclinical nightmare groups have also suffered from a number of methodological and interpretative problems. A greater degree of control over the accuracy of reported nightmare frequency is needed to ensure that subjective distress does not lead to an exaggeration of nightmare frequency. Retrospective reports of nightmare frequency are inadequate, because persons who show high levels of psychopathological symptoms may exaggerate the number of nightmares they have experienced in order to obtain help for subjective distress. Studies would, therefore, reflect psychopathology alone rather than any associations between nightmare frequency and personality. In addition, subject numbers have often been small, and causal inferences have been made from correlational data. Discrepancies between studies on the presence of neurotic symptoms also suggest that independent criterion validity may contribute to the evaluation of personality characteristics.

The primary goal of this study was to assess correlates between lifelong nightmares and personality characteristics while controlling for exaggerated reports of nightmare frequency. On the basis of previous research, it was predicted that lifelong nightmare subjects would score more highly on the Schizophrenic, Psychopathic Deviate, and Paranoia scales of the MMPI than would low nightmare frequency control subjects. In addition, the nightmare subjects were expected to score more highly on scales that independently measure neuroticism and pyschoticism, the Neuroticism and Psychoticism scales of the Eysenck Personality Questionnaire (EPQ). These three scales of the MMPI and two scales of the EPQ were expected to best discriminate between the nightmare group and the control group.

Method

Subjects

Forty-three persons responded to a newspaper advertisement calling for adults who had experienced one nightmare or more per month since childhood. They were screened in a telephone interview to ensure that they were not experiencing other sleep disorders. Five women were found to be experiencing nightmares associated with

exactly how you conducted the study is necessary so that the reader can evaluate the adequacy of the research. In order to facilitate communication, the method section is typically divided into subsections: subjects; apparatus, materials, or instruments; and procedure. Deviation from this format may be necessary if the experiment is complex or a detailed description of the stimuli is called for. In such instances, additional subsections may be required to help readers find specific information.

Subjects

The subjects subsection should tell the reader who the research participants were, how many there were, their characteristics (age, sex), and how they were selected. Any other pertinent information regarding the subjects should also be included, such as how they were assigned to the experimental condition, the number of subjects that were selected for the study but did not complete it (and why), and any inducements that were given to encourage participation. If animals were

recent traumatic events, and 2 men seemed to be suffering from REM-sleep behavioral disorder. They were referred to other professionals. Four men declined to participate in the study because they expressed reluctance to reveal dream content, and a further 2 did not meet nightmare frequency and duration selection criteria.

The nightmare group therefore consisted of 30 subjects (24 women and 6 men). The mean age was 34.7 years and ages ranged from 18 to 63 years. They were compared with 30 low nightmare frequency control subjects (24 women and 6 men) with a mean age of 35.9 years and an age range from 19 to 63 years. Selection criteria for the control group was the presence of less than two nightmares per year on average. Nightmare and control subjects were also matched for socioeconomic status on Hollingshead's (1957) certification of occupations. The control subjects were recruited from friends, acquaintances, and colleagues of the experimenters.

Materials

Dream diary. A dream diary was constructed in which subjects were required to record for 1 month all dreams immediately after awakening. Each dream was rated by the subject on a scale from 1 (pleasant) to 5 (very frightening). Sub-

used, their genus, species, strain number, and supplier should be specified, in addition to their sex, age, weight, and physiological condition.

Apparatus, materials, measures, and instruments
In this subsection, the reader can learn what apparatus or materials were used. Sufficient detail should be used to enable the reader to obtain comparable equipment. Additionally, the reader should be told why the equipment was used. Any mention of commercially marketed equipment should be accompanied by the firm's name and the model number or, in the case of a measuring instrument such as an anxiety scale, a reference that will enable the reader to obtain the same scale. Custom-made equipment should be described; in the case of complex equipment, a diagram or photograph may need to be included.

Procedure
In the procedure subsection, the reader is told exactly how the study was executed, from the moment the subject and the experimenter came into contact, to the moment

jects were told only to score a dream as 5 if they were so frightened that they were awakened by it.

<u>Minnesota Multiphasic Personality Inventory.</u> The MMPI (Hathoway & McKinley, 1942) is a 566-item self-report inventory that yields three validity scales and 10 clinical scales. The Masculine-Feminine and Social Isolation scales were not used in this study.

<u>Eysenck Personality Questionnaire.</u> The EPQ (Eysenck & Eysenck, 1975) is a 90-item self-report questionnaire and is scored on one validity scale and three personality scales.

<u>Drug use questionnaire.</u> A questionnaire in which subjects were asked to rate daily use of alcohol, tobacco, caffeinated beverages and any use of prescribed and non-prescribed drugs was constructed.

<u>Procedure</u>

The order of administration of the MMPI, the EPQ, and the dream diary was varied in a 3 × 3 Latin square design. All subjects completed the drug use questionnaire at the end of the study.

Individual interviews were then conducted in which nightmare subjects were asked how long they had experienced nightmares and whether they could recall any traumatic event associated with in-their contact was terminated. Consequently, this subsection represents a step-by-step account of what both the experimenter and the subject did during the study. This section should include any instructions or stimulus conditions presented to the subjects and the responses that were required of them, as well as any control techniques used (such as randomization or counterbalancing). In other words, you are to tell the reader exactly what both you and the subjects did and how you did it.

Results

The purpose of the results section is to tell the reader exactly what data were collected, how they were analyzed, and what the outcome of the data analysis was. This section should tell what statistical tests were used. Significant values of any inferential tests (e.g., *t*-tests, *F*-tests, and chi-square measures) should be accompanied by the magnitude of the obtained value of the test, along with the accompanying degrees of freedom, probability level, and direction of the effect. In reporting and illustrating the

itial nightmare onset. All subjects were asked for a current and past psychiatric history.

Results

Dream Diary

The dream diaries yielded a mean of 5.9 dreams (range, 2–16) with a rating of 5 (very frightening) for the nightmare subjects. No dreams were given a rating of 5 by the control subjects.

The mean number of dreams recalled for the month was 23.3 (SD = 16.7) for the nightmare group and 10.9 (SD = 9.9) for the control group. An independent-groups, two-tailed $\underline{t}$ test revealed that this difference was significant, $\underline{t}(58) = 3.67$, $\underline{p} < .001$.

Interview

All of the nightmare subjects reported that they had experienced 1 or more nightmares per month since they were under 10 years of age. Four reported receiving current psychiatric care: 1 for major depressive episode, 1 for generalized anxiety disorder, and 2 for psychoactive substance abuse. A further 7 reported previous outpatient psychiatric care (as adults) but had received no definitive diagnosis. None could recall a traumatic event associated with the

direction of a significant effect (nonsignificant effects are not elaborated on for obvious reasons), you need to decide on the medium that will most clearly and economically serve your purpose. If a main effect consisting of three groups is significant, your best approach is probably to incorporate the mean scores for each of these groups into the text of the report. If the significant effect is a complex interaction, the best approach is to summarize your data by means of a figure or a table. If you do use a figure or table (a decision that you must make), be sure to tell the reader, in the text of the report, what data it depicts. Then give a sufficient explanation of the presented data to make sure that the reader interprets them correctly. In writing the results section, there are several things you should not include. Individual data are not included unless a single-subject study is conducted. Statistical formulas are not included unless the statistical test is new, unique, or in some other way not standard or commonly used.

onset of nightmares, but all indicated that self-perceived stress increased the number and intensity of nightmares. No subjects in the control group reported any past or current psychiatric history.

Drug Use Questionnaire

Table 1 shows mean daily use of alcohol, tobacco, and caffeinated beverages for each group, and Table 2 shows, for prescribed and nonprescribed drugs, the type of drug and the number of users. Independent, two-tailed t tests were performed on alcohol, tobacco, and caffeinated beverage use. None of the t tests were significant: for alcohol, $t(29) = 1.86$, $p > .05$, for tobacco, $t(29) = 1.52$, $p > .1$, and for caffeinated beverages, $t(37) = 1.47$, $p > .1$. Two subjects in the nightmare group had previously used opioid drugs and were receiving methadone treatment at the time of the study.

INSERT TABLES 1 AND 2 ABOUT HERE

Personality Scales

All subjects had valid profiles on the three validity scales of the MMPI and on the Lie scale of the EPQ. Twenty-six nightmare subjects and 2 control subjects had scores above 70 (clinically elevated) on at least one MMPI scale. On the EPQ scales, 27 nightmare subjects and 1 control subject had scores similar to psychotic groups on the Neuroticism scale (Eysenck & Eysenck, 1975). No subjects had elevated scores on the Psychoticism scale.

The nightmare group was compared with the control group by using univariate F ratios and discriminant function analysis with the Bonferroni correction for Type I error (Hertzog & Rovine, 1985). These analyses showed that differences between the groups on EPQ Neuroticism and on MMPI Hypochondriasis, Depression, Hysteria, Paranoia, Schizophrenia, and Hypomania scales were significant ($p < .01$); the nightmare group obtained higher scores on these variables than did the control group (see Table 3).

INSERT TABLE 3 ABOUT HERE

Direct discriminant analysis was performed with scores from eight of the clinical scales of the MMPI and the Psychoticism and Neuroticism scales of the EPQ as predictor variables and the two grouping variables as criterion. Four cases were identified as multivariate outliers by Mahalanobis's distance test ($p < .01$). We found that two of the outliers were subjects in the control group who scored highly on several scales and two were subjects in the nightmare group with low scores. We therefore decided to retain the outliers in order to obtain a more conservative test.

The obtained function produced a squared multiple correlation of .53, which indicates that the function accounted for 53% of the variance between the groups. The function was significant, Wilks's = .46, $\underline{X}^2(2)$ = 42.79, $\underline{p}$ < .0001. Table 3 presents the group means, standard deviations, univariate $\underline{F}$ ratios, and standardized canonical discriminant function coefficients for each group.

An examination of the coefficients in Table 3 shows that nine scales had weights above .30. The weight of .18 for the EPQ Psychoticism scale indicates that this measure contributed little to the differentiation between the groups. The direction of contribution of the scales can be inferred from the group means. The nightmare groups appeared to be higher than the control group on the EPQ Neuroticism and on the MMPI scales. One question on the MMPI directly relates to nightmares and loads onto the Schizophrenia scale, and the responses to this question were deleted and the data reanalyzed. The analysis yielded the same results as when the question was included. An initial classification correctly identified 85% of the total subject pool. Subject numbers were too small to attempt an external classification procedure.

Discussion

The purpose of the discussion section of the research report is to interpret and evaluate the results obtained, giving primary emphasis to the relationships between the results and the hypotheses of the study. It is recommended that you begin the discussion by stating whether the hypothesis of the study was or was not supported. Following this statement, you should interpret the results, telling the reader what you think they mean. In doing so, you should attempt to integrate your research findings with the results of prior research. Note that this is the only place in the research report where you are given any latitude for stating your own opinion, and even then you are limited to stating your interpretation of the

Discussion

The results indicated that the EPQ Neuroticism scale and all of the MMPI scales used in the study contributed to the overall significant difference between the nightmare group and the control group. These results lend support to the hypothesis that adults who report frequent nightmares also evince symptoms similar to psychopathological groups. Contrary to expectations, however, the nightmare subjects were not significantly higher on the EPQ Psychoticism scale. In addition, the MMPI Depression, Psychasthenia, Hypochondriasis, and Hysteria scales contributed to overall discrimination between the groups, and the nightmare group showed significantly higher levels on all four scales.

Contrary to Hartmann's (1984) results, the nightmare group in this study showed more neurotic symptoms than did the control subjects, both on the EPQ and on the MMPI. This suggests a

results and what you think the major shortcomings of the study are. When discussing the shortcomings, you should mention only the flaws that may have had a significant influence on the results obtained. You should accept a negative finding as such rather than attempting to explain it as being due to some methodological flaw (unless, as may occasionally occur, there is a very good and documented reason why a flaw did cause the negative finding).

global maladjustment rather than specific psychotic symptomatology.

The dream diaries showed that each subject in the nightmare group experienced at least two nightmares in the month under consideration. The inclusion of ordinary dreams in the diaries and the requirement that subjects record content made it unlikely that they invented nightmare frequency. Moreover, the subjects extended content far beyond the requirements of the diaries, adding extra pages to include detailed descriptions of each dream. It appears, therefore, that the subjects in this study did experience the nightmares recorded. Overall dream recall was significantly higher for the nightmare group than for the control group. This suggests that nightmare frequency may be an artifact of greater recall; however, it is also possible that increased wakening after disturbing dreams also increases recall. In addition, the strength of affect may also increase the likelihood of remembering dreams.

It is also possible that there was a subject selection bias because only nightmare sufferers willing to respond to a newspaper story were used. In addition, recruitment of the control group from persons known to the experimenters may

have led to a social desirability effect in
responses from this group. Wood and Bootzin
(1990) recently suggested that the incidence of
nightmares in the community is higher than once
thought and that the presence of nightmares is
uncorrelated with anxiety. In this respect, control subjects in the present study may represent
an extreme group. In Wood and Bootzin's study,
however, subjects were only asked to report the
number of nightmares, not overall number of
dreams or dream content, and that may have
inflated nightmare frequency. Although their subjects were told that both nightmare sufferers and
nonsufferers were of interest to the experimenters, preliminary instructions and log reports
only pertained to nightmares and thus possibly
created a bias either in eliciting or reporting
nightmare frequency. In addition, all of their
subjects who fell beyond the 95th percentile in
nightmare frequency were eliminated from their
analyses. It is possible that correlations between nightmare frequency and anxiety were artificially reduced by this strategy.

Although a selection bias may have been
operative in our study, the nightmare subjects do
represent at least a subset of frequent nightmare
sufferers who subjectively experience nightmares

as problematic. Furthermore, most subjects in this group were clinically elevated on at least one scale of the MMPI and on the Neuroticism scale of the EPQ, which indicates increased symptomatology when compared to norms within the population. Valid scores on the Lie scales for the control group also suggest that social desirability was not a strong influence on responses.

The nightmare subjects were more likely to have used nonprescribed tranquilizers in the past than were the control subjects. Although benzodiazepine withdrawal has been associated with an increase in nightmares (Cernovsky, 1985; Empson, 1989; J. Kales, Allen, Preston, Tan, & Kales, 1970), it is unlikely that drug use represented an initial causal factor as all subjects reported nightmare onset before the age of 10. It is possible that tranquilizers were used in attempts to self-medicate. Reductions in nightmare frequency have been found during benzodiazepine administration (Empson, 1989), and 2 subjects indicated that benzodiazepine had been prescribed by general practitioners specifically for frequent nightmare complaints.

Practical considerations precluded analysis of dream content. Cursory examination, however, suggested that the content of nightmares fell

broadly into two categories, dreams in which the
dreamer was consistently the victim of attack and
those in which the dreamer was consistently the
attacker. An assessment of possible differences
between these subgroups may have provided clearer
symptom differentiation, but this was not possible in this study because of the small number of
subjects (3) who fitted the second category.
Persons who experience lifelong nightmares have
been considered a homogeneous group; however,
differences in dream content may be related to
differences in daytime functioning and may also
involve different etiologies.

We cannot infer causal direction from our
data. Conceptually, it is as likely that frequent
nightmares over many years increase levels of
anxiety, paranoia, depression, and preoccupation
with bodily functions as vice versa. Similarly,
nightmares can either elicit or reflect greater
preoccupation with unusual thought patterns.
Dreams have been likened to the schizophrenic
experience in that they involve visual and auditory hallucinations without objective sensory
input (Hartmann, 1984). The strong affective
response associated with nightmares may lend
credence to the bizarre content of the dreams.

The finding that adults who report frequent nightmares have a greater ability to become absorbed in daytime fantasy and show higher imagery capability (Belicki & Belicki, 1986) also suggests that nightmare groups can experience more vivid dreams and therefore see them as more personally salient to daytime concerns. In Hartmann's (1984) research, subjects reported frequent difficulty in determining whether they were awake or still dreaming when awakening from a nightmare. It was not determined whether this represented a blurring of fantasy and reality similar to that experienced by schizophrenic patients or whether it was specific to sleep phenomena.

The high scores on the MMPI Schizophrenia scale in our study imply that the nightmare group showed higher levels of distorted perceptions and bizarre thoughts than did control subjects. However, discrepancies between clinically elevated scores and psychiatric status suggest that the behavior and functioning of the nightmare subjects was not extreme. Implications of schizophrenic ideation can be further investigated with measures developed for diagnostic purposes.

It seems therefore that the study of frequent nightmares requires greater attention in order to determine etiology and to assess symptomatic and functional correlates. Longitudinal studies on children who present with a complaint of frequent nightmares will assist in the development of nightmare etiology theory. Our finding that persons who report frequent nightmares also experience considerable distress with higher levels of anxiety, depression, somaticization, paranoia, and psychotic ideation than for low nightmare frequency control subjects needs further investigation. A rather high proportion of adults in the population report frequent nightmares, and this study provides correlates of global personality maladjustment in this group.

References
The purpose of the reference section, as you might expect, is to provide an accurate and complete list of all the references cited in the text of the report. All of the listed references must be cited in the text.

References

Belicki, K., & Belicki, D. (1982). Nightmares in a university population. Sleep Research, 11, 116–119.

Belicki, K., & Belicki, D. (1986). Predisposition for nightmares: A study of hypnotic ability, vividness of imagery, and absorption. Journal of Clinical Psychology, 42, 714–718.

Cernovsky, Z. Z. (1985). MMPI and nightmares in male alcoholics. Perceptual and Motor Skills, 61, 841–842.

Cernovsky, Z. Z. (1986). MMPI and nightmare reports in women addicted to alcohol and other drugs. Perceptual and Motor Skills, 62, 717–718.

Empson, J. (1989). Sleep and dreaming. London: Faber & Faber.

Eysenck, H. J., & Eysenck, S. B. (1975). Manual of the Eysenck Personality Questionnaire (junior and adult). Seven Oaks, United Kingdom: Hodder & Stoughton.

Feldman, M. J., & Hersen, M. (1967). Attitudes towards death in nightmare subjects. Journal of Abnormal Psychology, 72, 421–425.

Fischer, C., Byrne, J. V., & Edwards, A. (1968). MREM and REM nightmares. Psychophysiology, 5, 221-222.

Foulkes, D. (1985). Dreaming: A cognitive-psychological analysis. Hillsdale, NJ: Erlbaum.

Hartmann, E. (1984). The nightmare: The psychology and biology of terrifying dreams. New York: Basic Books.

Hartmann, E., Skoff, B., Russ, D., & Oldfield, M. (1978). The biochemistry of the nightmare: Possible involvement of dopamine. Sleep Research, 7, 186.

Hathoway, S. R., & McKinley, J. C. (1942). Minnesota Multiphasic Personality Inventory. Minneapolis: Regents of the University of Minnesota.

Hefez, A., Metz, L., & Lavie, P. (1987). Long-term effects of extreme situational stress on sleep and dreaming. American Journal of Psychiatry, 144, 344-347.

Hertzog, C., & Rovine, M. (1985). Repeated-measures analysis of variance in developmental research: Selected issues. Child Development, 56, 787-809.

Hollingshead, A. B. (1957). Two-factor index of social position. Unpublished manuscript,

Department of Sociology, Yale University, New Haven, CT.

Kales, A., Soldatas, C. R., Caldwell, A. B., Charney, D. S., Kales, J. D., Marknel, D., & Cadieue, R. (1980). Nightmares: Clinical characteristics and personality patterns. American Journal of Psychiatry, 137, 1197–1201.

Kales, J., Allen, C., Preston, A., Tan, T., & Kales, A. (1970). Changes in REM sleep and dreaming with cigarette smoking and following withdrawal. Psychophysiology, 7, 347–348.

Rabinowitz, A. (1975). Hostility measurement and its relationship to fantasy capacity. Journal of Personality Assessment, 39, 50–54.

Starker, S. (1974). Daydreaming styles and nocturnal dreaming. Journal of Abnormal Psychology, 83, 52–55.

Starker, S. (1984). Daydreams, nightmares and insomnia: The relation of waking fantasy to sleep disturbances. Imagination, Cognition and Personality, 4, 237–248.

Starker, S., & Hasenfeld, R. (1976). Daydream styles and sleep disturbance. Journal of Nervous and Mental Disease, 163, 391–400.

Stone, M. H. (1979). Dreams of fragmentation
    and of the death of the dreamer: A
    manifestation of vulnerability to psychosis.
    Psychopharmacology Bulletin, 15, 12-14.

van der Kolk, B. A., Blitz, R., Burr, W.,
    Sherey, S., & Hartmann, E. (1984). Clinical
    characteristics of traumatic and lifelong
    nightmare sufferers. American Journal of
    Psychiatry, 141, 187-190.

van der Kolk, B. A., & Goldberg, H. L.
    (1983). Aftercare of schizophrenic patients:
    Psychopharmacology and consistency of
    therapists. Hospital and Community
    Psychiatry, 34, 343-348.

Wood, J. M., & Bootzin, R. R. (1990). The
    prevalence of nightmares and their
    independence from anxiety. Journal of
    Abnormal Psychology, 99, 64-68.

Author notes
Author identification
notes appear with each
printed article and are
for the purpose of
acknowledging the basis
of a study (such as a
grant or a dissertation),
acknowledging
assistance in the
conduct of the study or
preparation of the
manuscript, specifying
the institutional
affiliation of the author,
and designating the
address of the author to
whom reprint requests
should be sent. These
notes are typed on a
separate page, with the
words "Author Notes"
centered at the top of
the page in upper- and
lowercase letters. Each
note should start with a
paragraph indentation.
The order is
acknowledgments first,
then the affiliation of
the author, and finally
the author's address.
These notes are not
numbered or cited in the
text and appear on the
title page if the report is
to be blind-reviewed, or
reviewed in the absence
of any information that
would identify the
author.

Author Notes

Both Anne Berquier and Roderick Ashton are members of the Department of Psychology, University of Queensland, St. Lucia, Queensland, Australia. Correspondence concerning this article should be addressed to Roderick Ashton, Department of Psychology, University of Queensland, St. Lucia, Queensland 4072, Australia.

Footnotes

Footnotes are numbered consecutively, with a superscript arabic numeral, in the order in which they appear in the text of the report. Most footnotes are content footnotes, containing material needed to supplement the information provided in the text. Such footnotes are typed on a separate page, with the word "Footnotes" centered in upper- and lowercase letters. The first line of each footnote is indented five spaces, and the superscript numeral of the footnote should appear in the space just preceding the beginning of the footnote. Footnotes are typed in the order in which they are mentioned in the text.

Footnotes

    There were no footnotes in this manuscript, but if there were they would be presented on this page.

TABLE 1

Alcohol, Tobacco, and Caffeine Use for Both Groups

	Nightmare group			Control group		
Drug	M	SD	n	M	SD	n
Alcohol	15.2	11.5	17	7.0	4.8	14
Tobacco	25.7	8.6	7	16.3	11.8	9
Caffeinated beverages	5.8	4.7	20	4.5	2.5	19

Note: Means are given for use per week.

TABLE 2

Drug-Taking Status for Both Groups

Drug	Nightmare group (n)	Control group (n)
Prescribed drugs		
Ventalin	2	3
Methadone	2	—
Tricyclics	1	—
Benzodiazepine	3	3
Nonprescribed drugs		
Marijuana		
Past	6	—
Current	3	3
Hallucinogens		
Current	3	1
Opioids		
Past	2	—
Tranquilizers		
Past	11	—
Current	4	3

TABLE 3

Means, Standard Deviations, Univariate F Ratios, and Standardized Discriminant Function Coefficients for Both Groups

Scale	Nightmare group		Control group		$F(1, 58)$	Discriminant function coefficient
	M	SD	M	SD		
EPQ Neuroticism	16.9	4.1	9.6	4.7	42.0**	.79
Schizophrenia	71.9	17.8	53.3	7.9	27.2**	.64
Psychasthenia	66.1	13.8	51.5	8.9	23.7**	.63
Paranoia	67.7	15.8	50.5	10.1	25.4**	.62
Psychopathic Deviate	69.1	13.3	55.7	10.7	18.7**	.59
Hypochondriasis	62.4	12.4	52.9	7.1	13.4**	.55
Hysteria	64.5	10.6	53.5	8.0	20.6**	.55
Depression	65.4	13.9	56.5	6.8	9.8*	.47
Hypomania	62.9	14.7	53.4	10.4	8.3*	.35
EPQ Psychoticism	4.8	3.4	3.6	2.7	2.4	.18

Note: EPQ = Eysenck Personality Questionnaire; all other scales are from the Minnesota Multiphasic Personality Inventory.

*$p < .01$

**$p < .001$

PREPARATION OF THE RESEARCH REPORT

In the preceding section you saw an example of the way a research report must be prepared in order to be submitted for possible publication in a psychological journal. Although the essence of the report was discussed in the marginal notes, there are still many style rules that must be considered. In order to maintain consistency across articles and ensure that articles are clear, journal editors want research reports prepared in accordance with the following principles.

Language

The language used to communicate the results of a research study should be nonsexist: that is, it should be free from bias. Sexism in journal writing usually boils down to problems of designation and evaluation. Problems of designation occur when the writer makes reference to one or more people. Perhaps the most common example is referring to people in general as *man* or *mankind.* Because such designations may imply that women are of secondary importance, they should be avoided. Problems of evaluation occur from the habitual use of familiar expressions such as *man and wife* as opposed to *husband and wife* or *man and woman.* The former may imply a difference in the freedom and activities of each individual, whereas the latter implies parallel status. The American Psychological Association (1977) has published a set of guidelines that authors can follow to eliminate sexist language; however, the procedure is rather lengthy and time consuming. Bass (1979) has identified a set of rules that will more efficiently enable you to avoid the use of sexist language in most instances:

1. Rather than using *men, man,* or *mankind* to refer to people in general, use the words *people, individuals,* or *persons.*
2. Switch from the active to the passive voice.
3. Eliminate the singular, since words such as *they, their,* and *them* are genderless. For example, rather than stating "The manager and his subordinates" you could state "Managers and their subordinates."

When writing a research report, you also have to decide whether to use a first- or third-person writing style. Some individuals prefer a first-person writing style; others believe the research report should be impersonal and written in the third person. Polyson, Levinson, and Miller (1982) found that journal editors do not agree on which type of writing style should be used. Similarly, the publication manual does not take a specific position on this issue. Rather, the emphasis is on clarity and precision in word choice. In the final analysis, it seems as though the writing style chosen should be the one that will facilitate communication of the research study.

Abbreviations

Abbreviations are to be used sparingly. Generally speaking, abbreviate only when the abbreviations are conventional and likely to be familiar to the reader (such as IQ) or when it is necessary to abbreviate to save space and avoid cumbersome repetition. In all instances, the Latin abbreviations *cf.* (compare), *e.g.* (for example), *etc.* (and so forth), *i.e.* (that is), *viz.* (namely), and *vs.* (versus, against) are to be used only in parenthetical material. The exception to this rule is the Latin abbreviation *et al.*, which can be used in the text of the manuscript. The unit of time *second* is abbreviated *s* rather than *sec.* Periods are omitted with nonmetric measurements such as *ft* and *lb.* The only exception is inch, which is abbreviated *in.* with the period. Units of time such as *day, week, month,* and *year* are never abbreviated.

Headings

Headings serve to outline the manuscript and to indicate the importance of each topic. There are five different levels of headings that can be used in an article. These are rank-ordered as follows: centered main heading in uppercase letters, centered main heading in upper- and lowercase letters, centered main heading in underlined upper- and lowercase letters, flush side heading in underlined upper- and lowercase letters, and indented paragraph heading in underlined lowercase letters and followed by a period.

Since all articles do not require all five types of headings, you should decide how many levels are needed for your article and then order them appropriately. The following illustrates the way in which you would place the headings if your article contained two, three, four, or five headings.

1. Example of the use of two headings:

Method
Subjects

2. Example of the use of three headings:

Method
Instruments
MMPI.

3. Example of the use of four headings:

Experiment 1
Method
Instruments
MMPI.

4. For five headings, use all of the headings in the order listed above.

Numbers

The general rule about expressing numbers in the text is to use words to express any number that begins a sentence as well as any number below 10. Use figures to express all other numbers. There are several exceptions to this rule, and the APA publication manual should be consulted for these exceptions. A second rule to follow in stating numbers is to use arabic and not roman numerals.

Physical Measurements

All physical measurements are to be stated in metric units. If a measurement is expressed in nonmetric units, it must be accompanied, in parentheses, by its metric equivalent.

Presentation of Statistical Results

When presenting the results of statistical tests in the text, give the symbol of the statistical test, its degrees of freedom, the value obtained from the test, and its probability value, as follows:

$$t(36) = 4.52, p < .01$$
$$F(3, 52) = 17.36, p < .01$$

When reporting a chi-square value, you should report the degrees of freedom and the sample size in parentheses as follows:

$$X^2 (6, N = 68) = 12.64, p < .05$$

Such common statistical tests are not referenced, and the formulas are not included in the text. Referencing and formulas are included only when the statistical test is new, rare, or essential to the manuscript, as when the article concerns a given statistical test.

After the results of a statistical test are reported, descriptive statistical data such as means and standard deviations must be included to clarify the meaning of a significant effect.

Tables

Tables are expensive to publish and therefore should be reserved for use only when they can convey and summarize data more economically and clearly than can a lengthy discussion. If you decide to use tables, you should number them consecutively with arabic numerals. Although a table is prepared on a separate page in the manuscript submitted for publication, it appears in the body of the published article. Therefore, you should indicate the approximate position of the table by an obvious break in the text, with instructions indicated as follows:

<u>Insert Table 1 about here</u>

In preparing the table, you can use the tables presented in the sample article as guides. Each table should have a brief title that clearly explains the data it contains. This title and the word "Table" and its number are typed flush with the left margin and at the top of the table. Each column and row of data within the table should be given a label that identifies, as briefly as possible, the data contained in that row or column. Columns within the table should be at least three spaces apart. The publication manual (American Psychological Association,

1983) lists the various types of headings that can be used in tables. When placing data in the rows and columns, carry each data point out to the same number of decimal places, and place a dash to indicate an absence of data.

When writing the manuscript, you should refer to the table somewhere in the text. This reference should tell what data are presented in the table and briefly discuss the data. When referring to a table, identify it by name, as in "the data in Table 3." Do not use a reference such as "the above table" or "the table on page 12."

Figures

Figures are also very time consuming and expensive to produce and therefore should be used only when they complement the text or eliminate a lengthy discussion of the data. Figures may consist of charts, graphs, photographs, drawings, or other similar means of representing data or pictorial concepts. In preparing a figure, you should usually use a professional drafting service because most nonprofessionals do not have the technical skill to produce an acceptable product. Computer-generated figures are acceptable if they are produced by a laser printer or another printer of similar quality. If you do not use a professional drafting service or a laser printer, you should consult the publication manual (American Psychological Association, 1983) for guidelines for the production of an acceptable figure. The publication manual also gives information on the type of figure to use in presenting certain types of data.

Once the figures have been prepared, number them consecutively with arabic numerals in the order in which they are used in the manuscript. On the back and also near the edge of the figure, write the number of the figure and a short title in pencil. Also write the word *top* on the back to designate the top of the figure. The figure's location in the body of the manuscript should be indicated by a break in the text, as follows:

<u>Insert Figure 6 about here</u>

Figure Captions

Each figure has a caption that not only provides a brief description of the contents but also serves as a title. However, these captions are not placed on the figure but are typed on a separate page with the words "Figure Captions" centered and typed in upper- and lowercase letters at the top of the page. Flush with the left margin of the page, each caption should begin with the word "Figure" and the number of the figure, both of which are underlined, followed by a period. The caption is typed on the remainder of the line. If more than one line is needed, each subsequent line also begins flush left.

Reference Citations

In the text of the research report, particularly in the introductory section, you must reference other works you have cited. The APA format is to use the author–date citation method, which involves inserting the author's surname and the publication date at the appropriate point, as follows:

> Doe (1975) investigated the . . .
>   or
> It has been demonstrated (Doe, 1975) . . .

With this information, the reader can turn to the reference list and locate complete information regarding the source. Multiple citations involving the same author are arranged in chronological order:

> Doe (1970, 1971, 1972, 1973)

Multiple citations involving different authors are arranged alphabetically, as follows:

> Several studies (Doe, 1970; Kelly, 1965; Mills, 1975) have revealed . . .

If a citation includes more than two but fewer than six authors, all authors should be cited the first time the reference is used. Subsequent citations include only the name of the first author, followed by the words "et al." and the year the article was published. If six or more authors are associated with a citation, only the surname of the first author followed by "et al." is used for all citations.

Reference List

All citations in the text of the research report must be accurately and completely cited in the reference list so that it is possible for readers to locate the works. This means that each entry should include the name of the author, year of publication, title, publishing data, and any other information necessary to identify the reference. All references are to appear in alphabetical order. Rather than elaborate on the specific style of presentation, I refer you to the reference list in the sample article, which gives many examples. However, you should be aware of one type of reference that is not illustrated. If you have cited an article that has been accepted for publication but has not as yet been published, you should reference this citation by including the author's name, followed by the words "in press" in parentheses where the date of publication would otherwise appear. This is followed by the title and the journal in which the article is to be published.

All references are to be typed on a separate page, with the word "References" centered at the top of the page in upper- and lowercase letters.

Typing

In typing the manuscript, double-space all material. The rule is to set the typewriter or word processor on double-spacing and leave it there. There should be $1\frac{1}{2}$-inch margins at the top, bottom, right, and left of every page. Words should not be divided at the end of the page, and each page should contain no more than twenty-five lines of text.

Ordering of Manuscript Pages

The pages of the manuscript should be arranged as follows:

1. *Title page.* This is a separate page (numbered page 1) and includes the title, author's name, and running head.
2. *Abstract.* This is a separate page, numbered page 2.
3. *Text of the manuscript.* The text begins on page 3 and continues on consecutive pages through the completion of the discussion section.
4. *References.* References begin on a separate page.
5. *Author identification notes.* These notes begin on a new page.
6. *Footnotes.* Footnotes also begin on a new page.

7. *Tables.* Each table should be placed on a separate page.
8. *Figure captions.* Captions should be listed together on a separate page.
9. *Figures.* Each figure should be placed on a separate page.

Writing Style

Many people have difficulty communicating by means of the written word. A clear, concise writing style is something that must be acquired. Unfortunately, many individuals conducting research have not yet mastered this skill; this becomes painfully apparent when they attempt to write the research report. Teaching the art of writing is beyond the scope of this text. However, for the student who has difficulty in writing, I recommend an excellent book by W. Strunk, Jr., and E. B. White, *The Elements of Style.* This book is a classic and has the virtue of being short. For assistance in reasoning and writing clearly, I recommend Gage's *The Shape of Reason,* and for additional assistance in preparing your research, Rosnow and Rosnow's *Writing Papers in Psychology* is excellent. Finally, some years ago H. F. Harlow published a very humorous commentary on the content and style of a research report in the *Journal of Comparative and Physiological Psychology.* (See the References for bibliographic data on all these titles.)

With respect to writing, there are a number of points I want to make that may assist you. Some people have trouble getting started. They sit down at a word processor or with a pencil and pad of paper, and the words or ideas just do not develop. In such instances you can use one of two approaches. Rosnow and Rosnow (1992) suggest that you begin with the section you feel will be easiest to write. For example, this may be the method section, because you should already know details such as the characteristics of the subjects you tested and the procedure followed in testing them. Once you have begun writing this section, you may find that other sections such as the introduction are easier to write. The other technique is to force yourself to begin writing a section even if you don't like what you are saying. This technique has the advantage of getting something down on paper and giving you something to work with and revise. It also forces you to move beyond the beginning point, which may cause the ideas to begin flowing. To use this technique you must accept the fact that your first draft is just that. Seldom if ever should you consider the first draft the final product. Rather, you should produce the first draft and then revise it. This process should continue until you are satisfied with the final product.

When you have completed the final product, you should let it rest for several days and then reread it. This rereading several days later should result in additional revisions because the time lapse should allow you to approach the paper more objectively and identify sections that need work.

In preparing the research report make sure that you avoid plagiarism. Plagiarism means that you are kidnapping another person's ideas or efforts and passing them off as your own. In several sections of the research report, particularly the introduction, you must make use of others' work. When you do so, make sure that you give them credit.

Submission of the Research Report for Publication

If you have conducted an independent research project and have completed the preparation of a research report (aside from the laboratory reports that you may have prepared in this class), you must now decide whether to submit it to a journal for possible publication. Earlier in this chapter I stated that no study should be undertaken if you do not believe it is potentially worthy of publication. But even if at the outset you believe that the study you are conducting is worthy of publication, you may change your mind once the study is completed and you have prepared the research report. Therefore, at this stage you must make a final decision whether to submit the manuscript to a journal. This final decision should be based on your judgment of the significance of the study and the extent to which rival hypotheses were controlled in the study. Frequently it is valuable to have a colleague read and provide a critique of the article before you submit it for possible publication. A colleague presents a new perspective and can evaluate the worth of the article and its potential problems more critically and objectively.

If both you, as the author, and a colleague agree that the manuscript should be submitted for publication, you must then select the journal to which you are going to submit the article. Journals vary both in the percentage of submitted manuscripts they accept and in the types of articles that they will publish. From Table 15.1 you can see that each journal published by the American Psychological Association focuses on a different subject area. You must select a journal that publishes articles on subjects similar to yours. In making this selection, you must also decide whether your manuscript makes a contribution significant enough to warrant possible publication in one of the most prestigious journals. In the field of psychology, the APA journals are generally considered the most prestigious, as well as some of the most difficult to get into. These journals typically accept only about 15 percent of the manuscripts submitted to them.

Once you have selected the appropriate journal, send the required number of copies of the manuscript to the journal editor, with a cover letter stating that you are submitting the manuscript for possible publication in that journal. At this point the control of the manuscript is out of your hands and in the hands of the journal editor. The journal editor typically sends the manuscript to several individuals who are knowledgable regarding the topic of your study, and they review the manuscript and reach a decision about its acceptability. Their comments are returned to the journal editor, who makes the final decision. This decision can be a rejection, an acceptance, or an acceptance pending approval of recommended revisions. This last is the most typical mode of acceptance. The whole process typically takes two to three months.

If you get an outright acceptance—a very rare occurrence—you can celebrate. If you get a provisional acceptance—acceptance pending approval of recommended revisions—you can evaluate the recommendations and attempt to conform to them. Once the revisions have been made, you must resubmit the manuscript, which is then reevaluated by the journal editor. The editor may elect to accept the manuscript at this point, send it out for another review, or request additional revisions. If you get a rejection, try to evaluate the reviewers' comments regarding their reasons for rejecting the manuscript. If you agree with the reviewers' comments, you may reevaluate the manuscript and decide that it really was not worthy of publication. Alternatively, you may disagree with the reviewers' comments and believe that the manuscript still warrants publication. In this case, you would find another journal that focuses on the subject matter of your study and then start the process over. As you can see, the process of getting an article published is time consuming, involves a lot of work, and is subject to the approval and recommendation of your peers. Many studies are never published. Although the procedure just outlined has its flaws, it is probably the best that can be established to ensure that only high-quality research is published.

References

Abelson, R. P., & Miller, J. C. (1967). Negative persuasion via personal insult. *Journal of Experimental Social Psychology, 3,* 321–333.

Adair, J. G. (1973). *The human subject.* Boston: Little, Brown.

Adair, J. G. (1978). Open peer commentary. *Behavioral and Brain Sciences, 3,* 386–387.

Adair, J. G., Dushenko, T. W., & Lindsay, R. C. L. (1985). Ethical regulations and their impact on research practice. *American Psychologist, 40,* 59–72.

Adair, J. G., & Spinner, B. (1981). Subjects' access to cognitive processes: Demand characteristics and verbal report. *Journal for the Theory of Social Behavior, 11,* 31–52.

Allen, K. E., Hart, B., Buell, J. S., Harris, F. R., & Wolf, M. M. (1964). Effects of social reinforcement on isolate behavior of a nursery school child. *Child Development, 35,* 511–518.

Altemeyer, R. A. (1971). Subject pool pollution and the postexperimental interview. *Journal of Experimental Research in Personality, 5,* 79–84.

Alumbaugh, R. V. (1972). Another "Malleus Maleficarum." *American Psychologist, 27,* 897–899.

American Psychological Association, ad hoc Committee on Ethical Standards in Psychological Research. (1973). *Ethi-*

cal principles in the conduct of research with human participants. Washington, DC.

American Psychological Association. (1982). *Ethical principles in the conduct of research with human participants.* Washington, DC.

American Psychological Association. (1983). *Publication manual of the American Psychological Association.* 3rd ed. Washington, DC.

American Psychological Association Publication Task Force. (1977). Guidelines for nonsexist language in APA journals. *American Psychologist, 32,* 487–494.

Aronson, E. (1961). The effect of effort on the attractiveness of rewarded and unrewarded stimuli. *Journal of Abnormal and Social Psychology, 63,* 375–380.

Aronson, E. (1966). Avoidance of inter-subject communication. *Psychological Reports, 19,* 238.

Aronson, E., & Carlsmith, J. M. (1963). Performance expectancy as a determinant of actual performance. *Journal of Abnormal and Social Psychology, 66,* 584–588.

Aronson, E., & Carlsmith, J. M. (1968). Experimentation in social psychology. In G. Lindzey & E. Aronson (Eds.), *The handbook of social psychology* (2nd ed.). Reading, MA: Addison-Wesley.

Aronson, E., & Cope, V. (1968). My enemy's enemy is my friend. *Journal of Personality and Social Psychology, 8,* 8–12.

Aronson, E., & Linder, D. (1965). Gain and loss of esteem as determinants of interpersonal attractiveness. *Journal of Experimental Social Psychology, 1,* 156–171.

Aronson, E., & Mills, J. (1959). The effect of severity of initiation on liking for a group. *Journal of Abnormal and Social Psychology, 59,* 177–181.

Asch, S. E. (1956). Studies of independence and conformity: A minority of one against a unanimous majority. *Psychological Monographs, 70* (9, Whole No. 416).

Atkeson, B. M., Calhoun, K. S., Resick, P. A., & Ellis, E. M. (1982). Victims of rape: Repeated assessment of depressive symptoms. *Journal of Consulting and Clinical Psychology, 50,* 96–102.

Babbie, E. R. (1990). *Survey research methods.* 2nd ed. Belmont, CA: Wadsworth.

Baltes, P. B., Reese, H. W., & Nesselroade, J. R. (1977). *Life-span developmental psychology: Introduction to research.* Monterey, CA: Wadsworth Publishing Co.

Bannister, D. (1966). Psychology as an exercise in paradox. *Bulletin of British Psychological Society, 19,* 21–26.

Barber, T. X. (1976). *Pitfalls in human research: Ten pivotal points.* New York: Pergamon Press.

Barber, T. X., & Silver, M. J. (1968). Fact, fiction, and the experimenter bias effect. *Psychological Bulletin Monograph, 70,* 1–29.

Barlow, D. H., Sakheim, D. K., & Beck, J. G. (1983). Anxiety increases sexual arousal. *Journal of Abnormal Psychology, 92,* 49–54.

Baron, R. A. (1973). Threatened retaliation from the victim as an inhibitor of physical aggression. *Journal of Research in Personality, 7,* 103–115.

Baron, R. A. (1976). Effects of victim's pain cues, victim's race, and level of prior instigation upon physical aggression. Unpublished manuscript, Purdue University.

Bass, B. M. (1979). Confessions of a former male chauvinist. *American Psychologist, 34,* 194–195.

Baumrind, D. (1964). Some thoughts on ethics of research: After reading Milgram's "Behavioral study of obedience." *American Psychologist, 19,* 421–431.

Baumrind, D. (1971). Principles of ethical conduct in the treatment of subjects: Reaction to the draft report of the Committee on Ethical Standards in Psychological Research. *American Psychologist, 26,* 887–896.

Baumrind, D. (1972). Reactions to the May 1972 draft report of the Ad Hoc Committee on Ethical Standards in Psychological Research. *American Psychologist, 27,* 1083–1086.

Beach, F. A. (1950). The snark was a boojum. *American Psychologist, 5,* 115–124.

Beach, F. A. (1960). Experimental investigations of species specific behavior. *American Psychologist, 15,* 1–8.

Becker, L. J., Rabinowitz, V. C., & Seligman, C. (1980). Evaluating the impact of utility company billing plans on residential energy consumption. *Evaluation and Program Planning, 3,* 159–164.

Beckman, L., & Bishop, B. R. (1970). Deception in psychological research: A reply to Seeman. *American Psychologist, 25,* 878–880.

Beecher, H. K. (1966). Pain: One mystery solved. *Science, 151,* 840–841.

Benbow, C. P., & Stanley, J. C. (1980). Sex differences in mathematical ability: Fact or artifact? *Science, 210,* 1262–1264.

Bennett, F. C., & Sherman, R. (1983). Management of childhood "hyperactivity" by primary care physicians. *Journal of Developmental and Behavioral Pediatrics, 4,* 88–93.

Bergin, A. E. (1966). Some implications of psychotherapy research for therapeutic practice. *Journal of Abnormal Psychology, 71,* 235–246.

Bergin, A. E., & Strupp, H. H. (1970). New directions in psychotherapy research. *Journal of Abnormal Psychology, 76,* 13–26.

Bergin, A. E., & Strupp, H. H. (1972). *Changing frontiers in the science of psychotherapy.* New York: Aldine-Atherton.

Berkowitz, L., & LePage, A. (1967). Weapons or aggression-eliciting stimuli. *Journal of Personality and Social Psychology, 7,* 202–207.

Berkun, M., Bialek, H. M., Kern, P. R., & Yagi, K. (1962). Experimental studies of psychological stress in man. *Psychological Monographs: General and Applied, 76*(15), 1–39.

Berscheid, E., Baron, R. S., Dermer, M., & Libman, M. (1973). Anticipating informed consent: An empirical approach. *American Psychologist, 28,* 913–925.

Bijou, S. W., Peterson, R. F., Harris, F. R., Allen, K. E., & Johnston, M. S. (1969). Methodology for experimental studies of young children in natural settings. *Psychological Record, 19,* 177–210.

Billewicz, W. Z. (1965). The efficiency of matched samples: An empirical investigation. *Biometrics, 21,* 623–644.

Boice, R. (1973). Domestication. *Psychological Bulletin, 80,* 215–230.

Bolger, H. (1965). The case study method. In B. B. Wolman (Ed.), *Handbook of Clinical Psychology* (pp. 28–39). New York: McGraw-Hill.

Boring, E. G. (1954). The nature and history of experimental control. *American Journal of Psychology, 67,* 573–589.

Box, G. E. P., & Jenkins, G. M. (1970). *Time-series analysis: Forecasting and control.* San Francisco: Holden-Day.

Box, G. E. P., & Tiao, G. L. (1965). A change in level of a non-stationary time series. *Biometrics, 52,* 181–192.

Bracht, G. H., & Glass, G. V. (1968). The external validity of experiments. *American Educational Research Journal, 5,* 437–474.

Bradley, A. W. (1978). Self-serving bias in the attribution process: A reexamination of the fact or fiction question. *Journal of Personality and Social Psychology, 36,* 56–71.

Brady, J. V. (1958). Ulcers in "executive monkeys." *Scientific American, 199,* 95–100.

Brand, M. (1976). *The nature of causation.* Urbana, IL: University of Illinois Press.

Breitling, D., Guenther, W., & Rondot, P. (1986). Motor responses measured by brain electrical activity mapping. *Behavioral Neuroscience, 100,* 104–116.

Bridgman, P. W. (1927). *The logic of modern physics.* New York: Macmillan.

Britton, B. K. (1979). Ethical and educational aspects of participation as a subject in psychology experiments. *Teaching of Psychology, 6,* 195–198.

Broden, M., Bruce, M., Mitchell, M., Carter, V., & Hall, R. V. (1970). Effects of teacher attention on attending behavior of two boys at adjacent desks. *Journal of Applied Behavior Analysis, 3,* 199–203.

Brown, R., Cayden, C. B., & Bellugi-Klima, U. (1969). The child grammar from I to III. In J. P. Hill (Ed.), *Minnesota symposia on child psychology.* Minneapolis: The University of Minnesota Press, *2,* 28–73.

Brown, W. F., Wehe, N. O., Zunker, V. G., & Haslam, W. L. (1971). Effectiveness of student-to-student counseling on the academic adjustment of potential dropouts. *Journal of Educational Psychology, 62,* 285–289.

Brownell, W. A. (1966). The evaluation of learning under dissimilar systems of instruction. In J. D. Finn (Ed.), *Introduction to research and evaluation.* Buffalo: State University of New York at Buffalo.

Campbell, D. (1957). Factors relative to the validity of experiments in social settings. *Psychological Bulletin, 54,* 297–312.

Campbell, D. T. (1969). Prospective: Artifact and control. In R. Rosenthal & R. L. Rosnow. (Eds.), *Artifact in behavioral research.* New York: Academic Press.

Campbell, D. T., & Boruch, R. F. (1975). Making the case for randomized assignments to treatments by considering the alternatives: Six ways in which quasi-experimental evaluations in compensatory education tend to underestimate effects. In C. A. Bennett & A. A. Lumsdaine (Eds.), *Evaluation and experiment: Some critical issues in assessing social programs.* New York: Academic Press.

Campbell, D. T., & Erlebacher, A. (1970). How regression artifacts in quasi-experimental evaluations can mistakenly make compensatory education look harmful. In J.

Hellmuth (Ed.), *Compensatory education: A national debate* (Volume 3, *Disadvantaged child*). New York: Brunner/Mazel.

Campbell, D. T., & Ross, H. L. (1968). The Connecticut crackdown on speeding: Time series data in quasi-experimental analysis. *Law and Society Review, 3*(1), 33–53.

Campbell, D. T., & Stanley, J. C. (1963). *Experimental and quasi-experimental designs for research.* Chicago: Rand McNally.

Campbell, K. E., & Jackson, T. T. (1979). The role of and need for replication research in social psychology. *Replications in Social Psychology, 1,* 3–14.

Caporaso, J. A. (1974). *The structure and function of European integration.* Pacific Palisades, CA: Goodyear Publishing Co.

Carlopia, J., Adair, J. G., Lindsay, R. C. L., & Spinner, B. (1983). Avoiding artifact in the search for bias: The importance of assessing subjects' perceptions of the experiment. *Journal of Personality and Social Psychology, 44,* 693–701.

Carlsmith, J. M., Collins, B. E., & Helmreich, R. L. (1966). Studies in forced compliance: I. The effect of pressure for compliance on attitude change produced by face-to-face role playing and anonymous essay writing. *Journal of Personality and Social Psychology, 4,* 1–3.

Carlson, R. (1971). Where is the person in personality research? *Psychological Bulletin, 75,* 203–219.

Carlston, D. E., & Cohen, J. L. (1980). A closer examination of subject roles. *Journal of Personality and Social Psychology, 38,* 857–870.

Christensen, L. (1968). Intrarater reliability. *The Southern Journal of Educational Research, 2,* 175–182.

Christensen, L. (1977). The negative subject: Myth, reality or a prior experimental experience effect. *Journal of Personality and Social Psychology, 35,* 392–400.

Christensen, L. (1981). Positive self-presentation: A parsimonious explanation of subject motives. *The Psychological Record, 31,* 553–571.

Christensen, L. (1982). A critique of Carlston and Cohen's examination of subject roles. *Personality and Social Psychology Bulletin, 8,* 579–582.

Christensen, L. (1988). Deception in psychological research: When is its use justified? *Personality and Social Psychology Bulletin, 14,* 664–675.

Christensen, L., Krietsch, K., White, B., & Stagner, B. (1985). The impact of diet on mood disturbance. *Journal of Abnormal Psychology, 94,* 565–579.

Christensen, L., & Redig, C. (1993). The effect of meal composition on mood. *Behavioral Neuroscience, 107,* 346–353.

Church, R. M. (1964). Systematic effect of random errors in the yoked control design. *Psychological Bulletin, 62,* 122–131.

Cialdini, R. B., Cacioppo, J. T., Bassett, R., & Miller, J. A. (1978). Low-ball procedure for producing compliance: Commitment then cost. *Journal of Personality and Social Psychology, 36,* 463–467.

Cochran, W. G., & Cox, G. M. (1957). *Experimental designs.* New York: Wiley.

Cohen, J. (1988). *Statistical power analysis for the behavioral sciences.* Hillsdale, NJ: Lawrence Erlbaum Associates.

Collins, F. L., Jr., Kuhn, I. F., Jr., & King, G. D. (1979). Variables affecting subjects' ethical ratings of proposed experiments. *Psychological Reports, 44,* 155–164.

Conover, M. R. (1985). Alleviating nuisance Canada goose problems through methiocarb-induced aversive conditions. *Journal of Wildlife Management, 49,* 631–636.

Conrad, H. S., & Jones, H. E. (1940). A second study of familial resemblances in intelligence. *39th yearbook of the National Society for the Study of Education.* Chicago: University of Chicago Press, pp. 97–141.

Conroy, R. T., & Mills, J. N. (1970). *Human circadian rhythms.* London: J. & A. Church.

Consumer Reports, 39(1974, May), p. 368.

Converse, P., & Traugott, M. (1986). Assessing the accuracy of polls and surveys. *Science, 234,* 1094–1098.

Cook, T. D., Bean, J. R., Calder, B. J., Frey, R., Krovetz, M. L., & Reisman, S. R. (1970). Demand characteristics and three conceptions of the frequently deceived subject. *Journal of Personality and Social Psychology, 14,* 185–194.

Cook, T. D., & Campbell, D. T. (1975). Experiments in field settings. In M. Dunnette (Ed.), *Handbook of industrial and organizational research.* Chicago: Rand McNally.

Cook, T. D., & Campbell, D. T. (1979). Quasi-experimentation. *Design and analysis issues for field settings.* Chicago: Rand McNally.

Cook, T. D., Gruder, C. L., Hennigan, K. M., & Flay, B. R. (1979). The history of the sleeper effect: Some logical pitfalls in accepting the null hypothesis. *Psychological Bulletin, 86,* 662–679.

Cowles, M. F. (1974). N = 35: A rule of thumb for psychological researchers. *Perceptual and Motor Skills, 38,* 1135–1138.

Cowley, G., Hager, M., Drew, L., Namuth, T., Wright, L., Murr, A., Abbott, N., & Robins, K. (1988, December 26). Of pain and progress. *Newsweek,* pp. 50–59.

Cox, D. E., & Sipprelle, C. N. (1971). Coercion in participation as a research subject. *American Psychologist, 26,* 726–728.

Cronbach, L. (1957). The two disciplines of scientific psychology. *American Psychologist, 12,* 671–684.

Cronbach, L. (1975). Beyond two disciplines of scientific psychology. *American Psychologist, 30,* 116–127.

Cronbach, L., & Furby, L. (1970). "How should we measure change"—or should we? *Psychological Bulletin, 74,* 68–80.

Culliton, B. (1990). Dingell: AIDS researcher in conflict. *Science, 248,* 676.

D'Amato, M. R. (1970). *Experimental psychology methodology: Psychophysics and learning.* New York: McGraw-Hill.

Darley, J. M., & Latané, B. (1968). Bystander intervention in emergencies: Diffusion of responsibility. *Journal of Personality and Social Psychology, 8,* 377–383.

Davis, H., & Memmott, J. (1983). Autocontingencies: Rats count to three to predict safety from shock. *Animal Learning and Behavior, 11,* 95–100.

Deese, J. (1972). *Psychology as science and art.* New York: Harcourt Brace Jovanovich.

DePaulo, B. M., Dull, W. R., Greenberg, J. M., & Swaim, G. W. (1989). Are shy people reluctant to ask for help? *Journal of Personality and Social Psychology, 56,* 834–844.

Devenport, L. D., & Devenport, J. A. (1990). The laboratory animal dilemma: A solution in our backyards. *Psychological Science, 1,* 215–216.

Dewsbury, D. A. (1990). Early interactions between animal psychologists and animal activists and the founding of the APA committee on precautions in animal experimentation. *American Psychologist, 45,* 315–327.

Diener, E., & Crandall, R. (1978). *Ethics in social and behavioral research.* Chicago: University of Chicago Press.

Dipboye, R. L., & Flanagan, M. F. (1979). Research settings in industrial and organizational psychology. Are findings in the field more generalizable than in the laboratory? *American Psychologist, 34,* 141–150.

Dorfman, D. D. (1978). The Cyril Burt question: New findings. *Science, 201,* 1177–1186.

Dukes, W. F. (1965). N = 1. *Psychological Bulletin, 64,* 74–79.

Ebbesen, E. B., & Haney, M. (1973). Flirting with death: Variables affecting risk taking at intersections. *Journal of Applied Social Psychology, 3,* 303–324.

Ebbinghaus, H. (1913). *Memory, a contribution to experimental psychology.* 1885. Translated by H. A. Ruger and C. E. Bussenius. New York: Teachers College, Columbia University.

Elkin, I., Parloff, M. B., Hadley, S. W., & Autry, J. H. (1985). NIMH treatment of depression collaborative research program. *Archives of General Psychiatry, 42,* 305–316.

Ellsworth, P. C. (1977). From abstract ideas to concrete instances: Some guidelines for choosing natural research settings. *American Psychologist, 33,* 604–615.

Ellsworth, P. C. (1978). Open peer commentary. *The Behavioral and Brain Sciences, 3,* 386–387.

Ellsworth, P. C., Carlsmith, J. A., & Henson, A. (1972). The stare as a stimulus to flight in human subjects. *Journal of Personality and Social Psychology, 21,* 302–311.

Epstein, S. (1979). The stability of behavior: I. On predicting most of the people much of the time. *Journal of Personality and Social Psychology, 37,* 1097–1126.

Epstein, S. (1981). The stability of behavior: II. Implications for psychological research. *American Psychologist, 35,* 790–806.

Epstein, T. M., Suedfeld, P., & Silverstein, S. J. (1973). The experimental contact: Subject's expectations of and reactions to some behavior of experimenters. *American Psychologist, 28,* 212–221.

Ericsson, K. A., & Simon, H. A. (1980). Verbal reports as data. *Psychological Review, 87,* 215–251.

Eysenck, H. J. (1952). The effects of psychotherapy: An evaluation. *Journal of Consulting Psychology, 16,* 319–324.

Eysenck, H. J. (1967). *The biological basis of personality.* Springfield, Ill.: Charles C Thomas.

Ferguson, G. A. (1966). *Statistical analysis in psychology and education.* New York: McGraw-Hill.

Ferster, C. B., & Skinner, B. F. (1957). *Schedules of reinforcement.* New York: Appleton-Century-Crofts.

Festinger, G. L., & Carlsmith, J. M. (1959). Cognitive consequences of forced compliance. *Journal of Abnormal and Social Psychology, 58,* 203–211.

Festinger, L. (1957). *A theory of cognitive dissonance.* Evanston, IL: Row, Peterson.

Fillenbaum, S. (1966). Prior deception and subsequent experimental performance: The faithful subject. *Journal of Personality and Social Psychology, 4,* 532–537.

Fisher, R. A. (1935). *The design of experiments.* 1st ed. London: Oliver and Boyd.

Flaherty, C. F., & Checke, S. (1982). Anticipation of incentive gain. *Animal Learning and Behavior, 10,* 177–182.

Fouts, R. S. (1973, June 1). Acquisition and testing of gestural signs in four young chimpanzees. *Science,* 978–979.

Fox, J. G. (1963). A comparison of Gothic Elite and Standard Elite type faces. *Ergonomics, 6,* 193–198.

Freedman, J. L., & Fraser, S. C. (1966). Compliance without pressure: The foot-in-the-door technique. *Journal of Personality and Social Psychology, 4,* 195–202.

Friedman, N. (1967). *The social nature of psychological research.* New York: Basic Books.

Gadlin, H., & Ingle, G. (1975). Through the one-way mirror: The limits of experimental self-reflection. *American Psychologist, 30,* 1003–1009.

Gage, J. T. 1991. *The shape of reason.* 2nd ed. New York: Macmillan.

Gage, N. L., & Cronbach, L. J. (1955). Conceptual and methodological problems in interpersonal perception. *Psychological Review, 62,* 411–422.

Gaito, J. (1958). Statistical dangers involved in counterbalancing. *Psychological Report, 4,* 463–468.

Gaito, J. (1961). Repeated measurements designs and counterbalancing. *Psychological Bulletin, 58,* 46–54.

Gallup, G. G., Jr., & Beckstead, J. W. (1988). Attitudes toward animal research. *American Psychologist, 43,* 474–476.

Gallup, G. G., & Suarez, S. D. (1985). Alternatives to the use of animals in psychological research. *American Psychologist, 40,* 1104–1111.

The Gallup opinion index: Political, social and economic trends. January 1980, Report No. 174, p. 29.

Garcia, J. (1981). Tilting at the paper mills of academe. *American Psychologist, 36,* 149–158.

Garcia, J., & Koelling, R. A. (1967). A comparison of aversions induced by X-rays, toxins, and drugs in the rat. *Radiation Research* (Suppl. 7), 439–450.

Gardner, G. T. (1978). Effects of federal human subjects regulations on data obtained in environmental stressor research. *Journal of Personality and Social Psychology, 36,* 628–634.

Gelfand, D., & Hartmann, D. (1968). Behavior therapy with children: A review and evaluation of research methodology. *Psychological Bulletin, 69,* 204–215.

Gerdes, E. P. (1979). College students' reactions to social psychological experiments involving deception. *Journal of Social Psychology, 107,* 99–110.

Gergen, K. J. (1973a). Social psychology as history. *Journal of Personality and Social Psychology, 26,* 309–320.

Gergen, K. J. (1973b). The codification of research ethics: Views of a doubting Thomas. *American Psychologist, 28,* 907–912.

Gibbons, A. (1992). Tenure: Does the old-boy network keep women from leaping over this crucial career hurdle? *Science, 255,* 1386.

Gilligan, C. (1982). *In a different voice.* Cambridge, MA: Harvard University Press.

Glass, G. (1976). Primary, secondary and meta-analysis of research. *Educational Research, 5,* 3–8.

Glass, G. V., Tiao, G. C., & Maguire, T. O. (1971). The 1900 revision of German divorce laws. *Law and Society Review, 5,* 539–562.

Glass, G. V., Willson, V. L., & Gottman, J. M. (1975). *Design and analysis of time series.* Boulder, CO: Laboratory of Educational Research Press.

Gottman, J. M., & Glass, G. V. (1978). Analysis of interrupted time-series experiments. In T. R. Kratochwill (Ed.), *Single subject research: Strategies for evaluating change.* New York: Academic Press.

Gottman, J. M., & McFall, R. M. (1972). Self-monitoring effects in a program for potential high school dropouts: A time-series analysis. *Journal of Consulting and Clinical Psychology, 39,* 273–281.

Gottman, J. M., McFall, R. M., & Barnett, J. T. (1969). Design and analysis of research using time series. *Psychological Bulletin, 72,* 299–306.

Greenough, W. T. (1991, May/June). The animal rights assertions: A researcher's perspective. *Psychological Science Agenda*, pp. 10–12.

Greenwald, A. G. (1975). Consequences of prejudice against the null hypothesis. *Psychological Bulletin, 82*, 1–20.

Grice, G. R. (1966). Dependence of empirical laws upon the source of experimental variation. *Psychological Bulletin, 66*, 488–498.

Grisso, T., Baldwin, E., Blanck, P. D., Rotheram-Borus, M. J., Schooler, N. R., & Thompson, T. (1991). Standards in research: APA's mechanism for monitoring challenges. *American Psychologist, 46*, 758–766.

Groves, R. M., & Kahn, R. L. (1979). *Surveys by telephone: A national comparison with personal interviews.* New York: Academic Press.

Hall, J. F., & Kobrick, J. L. (1952). The relationships among three measures of response strength. *Journal of Comparative and Physiological Psychology, 45*, 280–282.

Hall, R. V., & Fox, R. W. (1977). Changing-criterion designs: An alternative applied behavior analysis procedure. In C. C. Etzel, G. M. LeBlanc, & D. M. Baer (Eds.), *New developments in behavioral research: Theory, method, and application* (in honor of Sidney W. Bijou). Hillsdale, NJ: Lawrence Erlbaum Associates.

Harcum, E. R. (1990). Methodological versus empirical literature: Two views on causal acceptance of the null hypothesis. *American Psychologist, 45*, 404–405.

Hare-Mustin, R. T., & Marecek, J. (eds.). (1990). *Making a difference: Psychology and the construction of gender.* New Haven: Yale University Press.

Harlow, H. F. (1976). Fundamental principles for preparing psychology journal articles. *Journal of Comparative and Physiological Psychology, 55*, 893–896.

Harris, M. J., & Rosenthal, R. (1985). Mediation of interpersonal expectancy effects: 31 meta-analyses. *Psychological Bulletin, 97*, 363–386.

Hartmann, D. P., & Hall, R. V. (1976). A discussion of the changing criterion design. *Journal of Applied Behavior Analysis, 9*, 527–532.

Hashtroudi, S., Parker, E. S., DeLisi, L. E., & Wyatt, R. J. (1983). On elaboration and alcohol. *Journal of Verbal Learning and Verbal Behavior, 22*, 164–173.

Haslerud, G., & Meyers, S. (1958). The transfer value of given and individually derived principles. *Journal of Educational Psychology, 49,* 293–298.

Hauri, P., & Ohmstead, E. (1983). What is the moment of sleep onset for insomniacs? *Sleep, 6,* 10–15.

Hayes, S. C., & Cone, J. D. (1981). Reduction of residential consumption of electricity through simple monthly feedback. *Journal of Applied Behavior Analysis, 14,* 81–88.

Helmstadter, G. C. (1970). *Research concepts in human behavior.* New York: Appleton-Century-Crofts.

Hersen, M., & Barlow, D. H. (1976). *Single case experimental designs: Strategies for studying behavioral change.* New York: Pergamon Press.

Hersen, M., Gullick, E. L., Matherne, P. M., & Harbert, T. L. (1972). Instructions and reinforcement in the modification of a conversion reaction. *Psychological Reports, 31,* 719–722.

Hewett, F. M., Taylor, F. D., & Artuso, A. A. (1969). The Santa Monica project: Evaluation of an engineered classroom design with emotionally disturbed children. *Exceptional Children, 35,* 523–529.

Hilgartner, S. (1990). Research fraud, misconduct, and the IRB. *IRB: A Review of Human Subjects Research, 12,* 1–4.

Holden, C. (1987). NIMH finds a case of "serious misconduct." *Science, 235,* 1566–1567.

Holmes, D. S. (1967). Amount of experience in experiments as a determinant of performance in later experiments. *Journal of Personality and Social Psychology, 7,* 403–407.

Holmes, D. S. (1973). Effectiveness of debriefing after a stress-producing deception. *Journal of Research in Personality, 7,* 127–138.

Holmes, D. S. (1976a). Debriefing after psychological experiments: I. Effectiveness of postdeception dehoaxing. *American Psychologist, 31,* 858–867.

Holmes, D. S. (1976b). Debriefing after psychological experiments: II. Effectiveness of postexperimental desensitizing. *American Psychologist, 31,* 868–875.

Holmes, D. S., & Applebaum, A. S. (1970). Nature of prior experimental experience as a determinant of performance in a subsequent experiment. *Journal of Personality and Social Psychology, 14,* 195–202.

Holmes, D. S., & Bennett, D. H. (1974). Experiments to answer questions raised by the use of deception in psychological research: I. Role playing as an alternative to deception; II. Effectiveness of debriefing after a deception; III. Effect of informed consent on deception. *Journal of Personality and Social Psychology, 29,* 358–367.

Humphreys, L. (1970). *Tearoom trade.* Chicago: Aldine.

Hunsicker, J. P., & Mellgren, R. L. (1977). Multiple deficits in the retention of an appetitively motivated behavior across a 24-hr period in rats. *Animal Learning and Behavior, 5,* 14–16.

Hurley, D. (1989, July/August). Cycles of craving. *Psychology Today,* 54–58.

Hyman, R., & Berger, L. (1966). Discussion. In H. J. Eysenck (Ed.), *The effects of psychotherapy* (pp. 81–86). New York: International Science Press.

Johnson, J. M. (1972). Punishment of human behavior. *American Psychologist, 27,* 1033–1054.

Johnson, R. F. Q. (1976). The experimenter attributes effect: A methodological analysis. *Psychological Record, 26,* 67–78.

Johnson, R. W., & Adair, J. G. (1972). Experimenter expectancy vs. systematic recording errors under automated and nonautomated stimulus presentation. *Journal of Experimental Research in Personality, 6,* 88–94.

Johnson, R. W., & Ryan, B. J. (1976). Observer/recorder error as affected by different tasks and different expectancy inducements. *Journal of Experimental Research in Personality, 10,* 201–214.

Jones, J. H. (1981). *Bad blood: The Tuskegee syphilis experiment.* New York: Free Press.

Jung, J. (1971). *The experimenter's dilemma.* New York: Harper & Row.

Karhan, J. R. (1973). A behavioral and written measure of the effects of guilt and anticipated guilt on compliance for Machiavellians. Unpublished thesis, Texas A&M University.

Kavanau, J. L. (1964). Behavior: Confinement, adaptation, and compulsory regimes in laboratory studies. *Science, 143,* 490.

Kavanau, J. L. (1967). Behavior of captive whitefooted mice. *Science, 155,* 1623–1639.

Kazdin, A. E. (1973). The role of instructions and reinforcement in behavior changes in token reinforcement programs. *Journal of Educational Psychology, 64,* 63–71.

Kazdin, A. E. (1978). Methodological and interpretive problems of single-case experimental designs. *Journal of Consulting and Clinical Psychology, 46,* 629–642.

Kazdin, A. E. (1980). *Research design in clinical psychology.* New York: Harper & Row.

Kazdin, A. E., & Kopel, S. A. (1975). On resolving ambiguities of the multiple-baseline design: Problems and recommendations. *Behavior Therapy, 6,* 601–608.

Keller, E. F. (1984). Feminism and science. In S. Harding & J. F. O'Barr (Eds.), *Sex and scientific inquiry.* Chicago: The University of Chicago Press.

Kelman, H. C. (1967). Human use of human subjects. *Psychological Bulletin, 67,* 1–11.

Kelman, H. C. (1968). *A time to speak.* San Francisco: Jossey-Bass.

Kelman, H. C. (1972). The rights of the subject in social research: An analysis in terms of relative power and legitimacy. *American Psychologist, 27,* 989–1016.

Kempthorne, O. (1961). The design and analysis of experiments with some reference to educational research. In R. O. Collier, Jr. & S. M. Elam (Eds.), *Research design and analysis: Second annual Phi Delta Kappa symposium on educational research* (pp. 97–126). Bloomington, IN: Phi Delta Kappa.

Kendler, T. S., & Kendler, H. H. (1959). Reversal and nonreversal shifts in kindergarten children. *Journal of Experimental Psychology, 58,* 56–60.

Kendler, T. S., Kendler, H. H., & Learnard, B. (1962). Mediated responses to size and brightness as a function of age. *American Journal of Psychology, 75,* 571–586.

Kennedy, J. L., & Uphoff, H. F. (1939). Experiments on the nature of extrasensory perception: III. The recording error criticism of extra-chance scores. *Journal of Parapsychology, 3,* 226–245.

Kerlinger, F. (1972). Draft report of the APA Committee on Ethical Standards in Psychological Research: A critical reaction. *American Psychologist, 27,* 894–896.

Kerlinger, F. N. (1973). *Foundations of behavioral research.* New York: Holt, Rinehart and Winston.

Kerlinger, F. N., & Pedhazur, E. J. (1973). *Multiple regression in behavioral research.* New York: Holt, Rinehart and Winston.

Key, B. W. (1980). *The clam-plate orgy and other subliminal techniques for manipulating your behavior.* Englewood Cliffs, NJ: Prentice-Hall.

Kimmel, H. D., & Terrant, F. R. (1968). Bias due to individual differences in yoked control designs. *Behavior Research Methods and Instrumentation, 1,* 11–14.

Kirkhan, G. L. (1975). Doc cop. *Human Behavior, 4,* 16–23.

Klockars, C. B., & O'Connor, F. W., eds. (1979). *Deviance and decency: The ethics of research with human subjects.* London: Sage Publications.

Knight, J. A. (1984). Exploring the compromise of ethical principles in science. *Perspectives in Biology and Medicine, 27,* 432–441.

Kratochwill, T. R. (1978). Foundations of time-series research. In T. R. Kratochwill (Ed.), *Single subject research: Strategies for evaluating change.* New York: Academic Press.

Krejcie, R. V., & Morgan, D. W. (1970). Determining sample size for research activities. *Educational and Psychological Measurement, 30,* 607–610.

Kruglanski, A. W. (1976). On the paradigmatic objections to experimental psychology: A reply to Gadlin and Ingle. *American Psychologist, 31,* 655–663.

Kusche, C. A., & Greenberg, M. T. (1983). Evaluative understanding and role-taking ability: A comparison of deaf and hearing children. *Child Development, 54,* 141–147.

Lana, R. (1959). Pretest-treatment interaction effects in longitudinal studies. *Psychological Bulletin, 56,* 293–300.

Lana, R. E. (1969). Pretest sensitization. In R. Rosenthal and R. L. Rosnow (Eds.), *Artifact in behavioral research.* New York: Academic Press.

Landers, S. (1989). New animal care rules greeted with grumbles. *The APA Monitor, 20,* 1,4–5.

Lansdell, H. (1988). Laboratory animals need only humane treatment: Animals "rights" may debase human rights. *International Journal of Neuroscience, 42,* 169–178.

Lawler, E. E., III, & Hackman, J. R. (1969). Impact of employee participation in the development of pay incentive plans: A field experiment. *Journal of Applied Psychology, 53,* 467–471.

Leak, G. K. (1981). Student perception of coercion and value from participation in psychological research. *Teaching of Psychology, 8,* 147–149.

Lefkowitz, M., Blake, R. R., & Mouton, J. S. (1955). Status factors in pedestrian violation of traffic signals. *Journal of Abnormal and Social Psychology, 51,* 704–705.

Leitenberg, H. (1973). The use of single-case methodology in psychotherapy research. *Journal of Abnormal Psychology, 82,* 87–101.

Leitenberg, H., Agras, W. S., Thompson, L., & Wright, D. E. (1968). Feedback in behavior modification: An experimental analysis in two phobic cases. *Journal of Applied Behavior Analysis, 1*, 131–137.

Lenneberg, E. H. (1962). Understanding language without ability to speak: A case report. *Journal of Abnormal and Social Psychology, 65*, 419–425.

LeUnes, A., Christensen, L., & Wilkerson, D. (1975). Institutional tour effects on attitudes related to mental retardation. *American Journal of Mental Deficiency, 79*, 732–735.

Levenson, H., Gray, M., & Ingram, A. (1976). Current research methods in personality: Five years after Carlston's survey. *Personality and Social Psychology Bulletin, 2*, 158–161.

Liddle, G., & Long, D. (1958). Experimental room for slow learners. *Elementary School Journal, 59*, 143–149.

Liebert, R. M., Odem, R. D., Hill, J., & Huff, R. (1969). Effects of age and rule familiarity on the production of modeled language constructions. *Developmental Psychology, 1*, 108–112.

Lindquist, E. F. (1953). *Design and analysis of experiments in psychology and education.* Boston: Houghton Mifflin.

Lockard, R. B. (1968). The albino rat: A defensible choice or a bad habit? *American Psychologist, 23*, 734–742.

Loftus, E. F., & Fries, J. F. (1979). Informed consent may be hazardous to health. *Science, 204*, 11.

Lord, F. M. (1969). Statistical adjustments when comparing preexisting groups. *Psychological Bulletin, 72*, 336–337.

Lubin, A. (1961). The interpretation of significant interaction. *Educational and Psychological Measurement, 21*, 807–817.

Lyons, J. (1964). On the psychology of the psychological experiment. In C. Schurer (Ed.), *Cognition-theory, research, promise.* New York: Harper.

Maier, N. R. F. (1938). Experimentally produced neurotic behavior in the rat. Paper presented at the meeting of the American Association for the Advancement of Science, Richmond. In M. H. Marx and W. A. Hillix (Eds.), *Systems and theories in psychology* (1973, p. 13). New York: McGraw-Hill.

Maier, N. R. F. (1949). *Frustration: The study of behavior without a goal.* New York: McGraw-Hill.

Maier, S. F., & Laudenslager, M. (1985). Stress and health: Exploring the links. *Psychology Today, 19,* 36–43.

Marks-Kaufman, R., & Lipeles, B. J. (1982). Patterns of nutrient selection in rats orally self-administering morphine. *Nutrition and Behavior, 1,* 33–46.

Marquart, J. W. (1983). Cooptation of the kept: Maintaining control in a southern penitentiary. Unpublished dissertation, Texas A&M University.

Martin, C. J., Boersma, F. J., & Cox, D. L. (1965). A classification of associative strategies in paired-associate learning. *Psychonomic Science, 3,* 455–456.

Marx, M. H. (1963). *Theories in contemporary psychology.* New York: Macmillan.

Marx, M. H., & Hillix, W. A. (1973). *Systems and theories in psychology.* New York: McGraw-Hill.

Masling, J. (1966). Role-related behavior of the subject and psychologist and its effects upon psychological data. *Nebraska Symposium on Motivation, 14,* 67–103. Lincoln: University of Nebraska Press.

Matlin, M. W. (1993). *The psychology of women.* New York: Harcourt Brace Jovanovich.

May, W. W. (1972). On Baumrind's four commandments. *American Psychologist, 27,* 899–902.

McArdle, J. (1984). Psychological experimentation on animals: Not necessary, not valid. *The Humane Society News, 29,* 20–22.

McCabe, K. (1986, August). Who will live, who will die? *The Washingtonian,* p. 112.

McClelland, D. C. (1953). *The achievement motive.* New York: Appleton-Century-Crofts.

McCullough, J. P., Cornell, J. E., McDaniel, M. H., & Mueller, R. K. (1974). Utilization of the simultaneous treatment design to improve student behavior in a first-grade classroom. *Journal of Consulting and Clinical Psychology, 42,* 288–292.

McFall, R. M. (1970). Effects of self-monitoring on normal smoking behavior. *Journal of Consulting and Clinical Psychology, 35,* 135–142.

McGuigan, F. J. (1963). The experimenter: A neglected stimulus object. *Psychological Bulletin, 60,* 421–428.

McNemar, Q. (1946). Opinion-attitude methodology. *Psychological Bulletin, 43,* 289–374.

Mees, C. E. K. (1934). Scientific thought and social reconstruction. *Sigma Xi Quarterly, 22,* 13–24.

Mellgren, R. L., Nation, J. R., & Wrather, D. M. (1975). Magnitude of negative reinforcement and resistance to extinction. *Learning and Motivation, 6,* 253–263.

Mellgren, R. L., Seybert, J. A., & Dyck, D. G. (1978). The order of continuous, partial and nonreward trials and resistance to extinction. *Learning and Motivation, 9,* 359–371.

Meyer, R. G., & Osborne, Y. V. H. (1982). *Case Studies in Abnormal Behavior.* Boston: Allyn and Bacon.

Middlemist, R. D., Knowles, E. S., & Matter, C. F. (1976). Personal space invasions in the lavatory: Suggestive evidence for arousal. *Journal of Personality and Social Psychology, 33,* 541–546.

Milgram, S. (1963). Behavioral study of obedience. *Journal of Abnormal and Social Psychology, 67,* 371–378.

Milgram, S. (1964a). Group pressure and action against a person. *Journal of Personality and Social Psychology, 69,* 137–143.

Milgram, S. (1964b). Issues in the study of obedience: A reply to Baumrind. *American Psychologist, 19,* 848–852.

Mill, J. S. (1874). *A system of logic.* New York: Harper.

Miller, A. G. (1972). Role playing: An alternative to deception? A review of the evidence. *American Psychologist, 27,* 623–636.

Miller, N. E. (1957). Objective techniques for studying motivational effects of drugs on animals. In S. Garettini & V. Ghetti (Eds.), *Psychotropic drugs.* Amsterdam: Elsevier.

Miller, N. E. (1985). The value of behavioral research on animals. *American Psychologist, 40,* 423–440.

Mills, J. (1976). A procedure for explaining experiments involving deception. *Personality and Social Psychology Bulletin, 2,* 3–13.

Mook, D. G. (1983). In defense of external invalidity. *American Psychologist, 38,* 379–387.

Morgan, C. T., & Morgan, J. D. (1939). Auditory induction of abnormal pattern of behavior in rats. *Journal of Comparative Psychology, 27,* 505–508.

Morin, S. F., Charles, K. A., & Malyon, A. K. (1984). The psychological impact of AIDS on gay men. *American Psychologist, 39,* 1288–1293.

Morison, R. S. (1960). "Gradualness, gradualness, gradualness" (I. P. Pavlov). *American Psychologist, 15,* 187–198.

Nation, J. R., Bourgeois, A. E., Clark, D. E., & Hare, M. F. (1983). The effects of chronic cobalt exposure on behaviors and metallothionein levels in the adult rat. *Neurobehavioral Toxicology and Teratology, 9*, 9–15.

National Institutes of Health. (1985). *Guide for the care and use of laboratory animals.* NIH Publication No. 86-23. Washington, D.C.: U.S. Government Printing Office.

Neale, J. M., & Liebert, R. M. (1973). *Science and behavior: An introduction to methods of research.* Englewood Cliffs, NJ: Prentice-Hall.

Nederhof, A. J. (1985). A comparison of European and North American response patterns in mail surveys. *Journal of the Market Research Society, 27*, 55–63.

Nezu, A. M. (1986). Efficacy of a social problem-solving therapy approach for unipolar depression. *Journal of Consulting and Clinical Psychology, 54*, 196–202.

Nicewander, W. A., & Price, J. M. (1978). Dependent variable reliability and the power of significance tests. *Psychological Bulletin, 85*, 405–409.

Oakes, W. (1972). External validity and the use of real people as subjects. *American Psychologist, 27*, 959–962.

Oliver, R. L., & Berger, P. K. (1980). Advisability of pretest designs in psychological research. *Perceptual and Motor Skills, 51*, 463–471.

Orne, M. T. (1962). On the social psychology of the psychological experiment: With particular reference to demand characteristics and their implications. *American Psychologist, 17*, 776–783.

Orne, M. (1973). Communication by the total experimental situations: Why is it important, how is it evaluated, and its significance for the ecological validity of findings. In P. Pliner, L. Kramer, and T. Alloway (Eds.), *Communication and affect.* New York: Academic Press.

Ossip-Klein, D. J., Epstein, L. H., Winter, M. K., Stiller, R., Russell, P., & Dickson, B. (1983). Does switching to low tar/nicotine/carbon monoxide–yield cigarettes decrease alveolar carbon monoxide measures? A randomized trial. *Journal of Consulting and Clinical Psychology, 51*, 234–241.

Page, M. M. (1968). Modification of figure-ground perception as a function of awareness of demand characteristics. *Journal of Personality and Social Psychology, 9*, 59–66.

Page, M. M. (1969). Social psychology of a classical conditioning of attitudes experiment. *Journal of Personality and Social Psychology, 11*, 177–186.

Page, M. M., & Kahle, L. R. (1976). Demand characteristics in the satiation-deprivation effect on attitude conditioning.

Journal of Personality and Social Psychology, 33, 553–562.

Page, M. M., & Scheidt, R. J. (1971). The elusive weapons effect: Demand awareness, evaluation apprehension, and slightly sophisticated subjects. *Journal of Personality and Social Psychology, 20,* 304–318.

Page, S., & Yates, E. (1973). Attitudes of psychologists toward experimenter controls in research. *The Canadian Psychologist, 14,* 202–207.

Pappworth, M. H. (1967). *Human guinea pigs: Experimentation on man.* Boston: Beacon Press.

Pardes, H., West, A., & Pincus, H. A. (1991). Physicians and the animal-rights movement. *The New England Journal of Medicine, 324,* 1640–1643.

Paul, G. L. (1969). Behavior modification research: Design and tactics. In C. M. Franks (Ed.), *Behavior therapy appraisal and status.* New York: McGraw-Hill.

Pavlov, I. P. (1928). *Lecture on conditioned reflexes.* Translated by W. H. Gantt. New York: International.

Pellegrini, R. J. (1972). Ethics and identity: A note on the call to conscience. *American Psychologist, 27,* 896–897.

Pfungst, O. (1965). *Clever Hans (the horse of Mr. Von Osten): A contribution to experimental, animal, and human psychology.* Translated by C. L. Rahn. New York: Holt, Rinehart and Winston. (Originally published in 1911.)

Pihl, R. D., Zacchia, C., & Zeichner, A. (1981). Follow-up analysis of the use of deception and aversive contingencies in psychological experiments. *Psychological Reports, 48,* 927–930.

Plutchik, R. (1974). *Foundations of experimental research.* New York: Harper.

Polanyi, M. (1963). The potential theory of absorption. *Science, 141,* 1010–1013.

Polyson, J., Levinson, M., & Miller, H. (1982). Writing styles: A survey of psychology journal editors. *American Psychologist, 37,* 335–338.

Popper, K. R. (1968). *The logic of scientific discovery.* London: Hutchinson and Co.

Poulton, E. C., & Freeman, P. R. (1966). Unwanted asymmetrical transfer effects with balanced experimental designs. *Psychological Bulletin, 66,* 1–8.

Pribram, K. H. (1971). *Languages of the brain: Experimental paradoxes and principles in neuropsychology.* Englewood Cliffs, NJ: Prentice-Hall.

Quattrochi-Tubin, S., & Jason, L. A. (1980). Enhancing social interactions and activity among the elderly through stimulus control. *Journal of Applied Behavior Analysis, 13,* 159–163.

Reed, J. G., & Baxter, P. M. (Eds.). (1991). *Library use: A handbook for psychology* (2nd ed.). Washington, DC: American Psychology Association.

Resnick, J. H., & Schwartz, T. (1973). Ethical standards as an independent variable in psychological research. *American Psychologist, 28,* 134–139.

Rich, C. L. (1977). Is random digit dialing really necessary? *Journal of Marketing Research, 14,* 300–305.

Richter, C. P. (1959). Rats, man, and the welfare state. *American Psychologist, 14,* 18–28.

Ring, K., Wallston, K., & Corey, M. (1970). Mode of debriefing as a factor affecting reaction to a Milgram-type obedience experiment: An ethical inquiry. *Representative Research in Social Psychology, 1,* 67–88.

Risley, T. R., & Wolf, M. M. (1972). Strategies for analyzing behavioral change over time. In J. R. Nesselroade & H. W. Reese (Eds.), *Life-span developmental psychology: Methodological issues.* New York: Academic Press.

Ritchie, E., & Phares, E. J. (1969). Attitude change as a function of internal-external control and communicator status. *Journal of Personality, 37,* 429–443.

Roberson, M. T., & Sundstrom, E. (1990). Questionnaire design, return rates, and response favorableness in an employee attitude questionnaire. *Journal of Applied Psychology, 75,* 354–357.

Robinson, J., & Cohen, L. (1954). Individual bias in psychological reports. *Journal of Clinical Psychology, 10,* 333–336.

Rogers, T. F. (1976). Interviews by telephone and in person: Quality of responses and field performance. *Public Opinion Quarterly, 40,* 51–65.

Rosenberg, M. J. (1969). The conditions and consequences of evaluation apprehension. In R. Rosenthal and R. L. Rosnow (Eds.), *Artifact in behavioral research.* New York: Academic Press.

Rosenberg, M. J. (1980). Experimenter expectancy, evaluation apprehension, and the diffusion of methodological angst. *The Behavioral and Brain Sciences, 3,* 472–474.

Rosenthal, R. (1966). *Experimenter effects in behavioral research.* New York: Appleton-Century-Crofts.

Rosenthal, R. (1969). Interpersonal expectations: Effects of the experimenter's hypothesis. In R. Rosenthal and R. L. Rosnow. (Eds.), *Artifact in behavioral research.* New York: Academic Press.

Rosenthal, R. (1976). *Experimenter effects in behavioral research.* 2nd ed. New York: Irvington.

Rosenthal, R. (1978). How often are our numbers wrong? *American Psychologist, 33,* 1005–1007.

Rosenthal, R. (1980). Replicability and experimenter influence: Experimenter effects in behavioral research. *Parapsychology Review, 11,* 5–11.

Rosenthal, R., & Fode, K. L. (1963). The effect of experimenter bias on the performance of the albino rat. *Behavioral Science, 8,* 183–189.

Rosenthal, R., Persinger, G. W., Vikan-Kline, L., & Mulry, R. C. (1963). The role of the research assistant in the mediation of experimenter bias. *Journal of Personality, 31,* 313–335.

Rosenthal, R., & Rosnow, R. L. (1975). *The volunteer subject.* New York: Wiley.

Rosenthal, R., & Rubin, D. B. (1978). Interpersonal expectancy effects: The first 345 studies. *The Behavioral and Brain Sciences, 3,* 377–415.

Rosnow, R. L., & Rosnow, M. (1992). *Writing papers in psychology* (2nd ed.). New York: Wiley.

Rosnow, R. L., & Suls, J. M. (1970). Reactive effects of pretesting in attitude research. *Journal of Personality and Social Psychology, 15,* 338–343.

Rotton, J., & Kelly, I. W. (1985). Much ado about the full moon: A meta-analysis of lunar–lunacy research. *Psychological Bulletin, 97,* 286–306.

Rugg, E. A. (1975). Ethical judgments of social research involving experimental deception. Unpublished doctoral dissertation, George Peabody College for Teachers, Nashville, Tenn.

Rumenik, D. K., Capasso, D. R., & Hendrick, C. (1977). Experimenter sex effects in behavioral research. *Psychological Bulletin, 84,* 852–877.

Ryan, J. P., & Isaacson, R. L. (1983). Intraaccumbens injections of ACTH induce excessive grooming in rats. *Physiological Psychology, 11,* 54–58.

Sanders, G. S., & Simmons, W. L. (1983). Use of hypnosis to enhance eyewitness accuracy: Does it work? *Journal of Applied Psychology, 68,* 70–77.

Saxe, L. (1991). Thoughts of an applied social psychologist. *American Psychologist, 46,* 409–415.

Schachter, S., & Singer, J. E. (1962). Cognitive, social and physiological determinants of emotional state. *Psychological Review, 69,* 379–399.

Schafer, R., & Murphy, G. (1943). The role of autism in visual figure–ground relationship. *Journal of Experimental Psychology, 32,* 335–343.

Schenk, S., Lacelle, G., Gorman, K., & Amit, Z. (1987). Cocaine self-administration in rats influenced by environmental conditions: Implications for the etiology of drug abuse. *Neuroscience Letters, 81,* 227–231.

Schoenthaler, S. J. (1983). The Los Angeles probation department diet–behavior program: An empirical analysis of six institutional settings. *International Journal of Biosocial Research, 5,* 88–98.

Scholtz, J. A. (1973). Defense styles in suicide attempters. *Journal of Consulting and Clinical Psychology, 41,* 70–73.

Sears, D. O. (1986). College sophomores in the laboratory: Influences of a narrow data base on social psychology's view of human nature. *Journal of Personality and Social Psychology, 51,* 515–530.

Sears, R. R., Whiting, J. W. M., Nowlis, V., & Sears, P. S. (1953). Some child-rearing antecedents of aggression and dependence in young children. *Genetic Psychology Monographs, 47,* 135–234.

Seeman, J. (1969). Deception in psychological research. *American Psychologist, 24,* 1025–1028.

Selltiz, C., Jahoda, M., Deutsch, M., & Cook, S. W. (1959). *Research methods in social relations.* New York: Holt.

Sharpe, D., Adair, J. G., & Roese, N. J. (1992). Twenty years of deception research: A decline in subjects' trust? *Personality and Social Psychology Bulletin, 18,* 585–590.

Shine, L. C., II. (1975). Five research steps designed to integrate the single-subject and multi-subject approaches to experimental research. *Canadian Psychological Review, 16,* 179–183.

Shuell, T. J. (1981). Distribution of practice and retroactive inhibition in free-recall learning. *Psychological Record, 31,* 589–598.

Sidman, M. (1960). *Tactics of scientific research.* New York: Basic Books.

Sidowski, J. B., & Lockard, R. B. (1966). Some preliminary considerations in research. In J. B. Sidowski (Ed.), *Experimen-*

tal methods and instrumentation in psychology. New York: McGraw-Hill.

Sieber, J. E. (1982). Deception in social research: I. Kinds of deception and the wrongs they may involve. *IRB: A Review of Human Subjects Research, 4*(9), 1–5.

Sieber, J. E. (1983a). Deception in social research: II. Evaluating the potential for wrong. *IRB: A Review of Human Subjects Research, 5*(1), 1–5.

Sieber, J. E. (1983b). Deception in social research: III. The nature and limits of debriefing. *IRB: A Review of Human Subjects Research, 5*(3), 1–4.

Sieber, J. E., & Stanley, B. (1988). Ethical and professional dimensions of socially sensitive research. *American Psychologist, 43,* 49–55.

Sigall, H., Aronson, E., & Van Hoose, T. (1970). The cooperative subject: Myth or reality. *Journal of Experimental Social Psychology, 6,* 1–10.

Silverman, I. (1974). The experimenter: A (still) neglected stimulus object. *The Canadian Psychologist, 15,* 258–270.

Skinner, B. F. (1948). *Walden two.* New York: Macmillan.

Skinner, B. F. (1953). *Science and human behavior.* New York: Macmillan.

Skinner, B. F. (1956). A case history in scientific method. *American Psychologist, 11,* 221–223.

Skinner, B. F. (1971). *Beyond freedom and dignity.* New York: Knopf.

Small-sample techniques. (1960). *The NEA Research Bulletin, 36,* 99–104.

Smith, R. E. (1969). The other side of the coin. *Contemporary Psychology, 14,* 628–630.

Smith, S. S., & Richardson, D. (1983). Amelioration of deception and harm in psychological research: The important role of debriefing. *Journal of Personality and Social Psychology, 44,* 1075–1082.

Solomon, R. (1949). An extension of control group design. *Psychological Bulletin, 44,* 137–150.

Sperry, R. W. (1968). Hemisphere deconnection and unity in conscious awareness. *American Psychologist, 23,* 723–733.

Spring, B., Chiodo, J., Harden, M., Bourgeois, M. J., & Lutherer, L. (1989). Psychobiological effects of carbohydrates. *Journal of Clinical Psychiatry, 50* (Suppl.), 27–33.

Steele, C. M., & Southwick, L. (1985). Alcohol and social behavior: I. The psychology of drunken excess. *Journal of Personality and Social Psychology, 48,* 18–34.

Steinberg, L., & Dornbusch, S. M. (1991). Negative correlates of part-time employment during adolescence:
Replication and elaboration. *Developmental Psychology,
27,* 303–313.

Sternglass, E., & Bell, S. (1979). Fallout and the decline of
Scholastic Aptitude scores. Paper presented at the annual
meeting of the American Psychological Association, New
York.

Stevens, S. S. (1939). Psychology and the science of science.
Psychological Bulletin, 36, 221–263.

Streiner, D. L. (1990). Sample size and power in psychiatric
research. *Canadian Journal of Psychiatry, 35,* 616–620.

Stratton, G. M. (1897). Vision without inversion of the retinal
image. *Psychological Review, 4,* 341–360, 463–481.

Strunk, W., Jr., & White, E. B. (1979). *The elements of style.* 3rd
ed. New York: Macmillan.

Sudman, S., & Bradburn, N. M. (1982). *Asking questions: A
practical guide to questionnaire design.* San Francisco:
Jossey Bass.

Sulik, K. K., Johnston, M. C., & Webb, M. A. (1981). Fetal alcohol syndrome: Embryogenesis in a mouse model. *Science, 214,* 936–938.

Sullivan, D. S., & Deiker, T. E. (1973). Subject–experimenter
perception of ethical issues in human research. *American
Psychologist, 28,* 587–591.

Sulzer-Azaroff, B., & Consuelo de Santamaria, M. (1980). Industrial safety hazard reduction through performance
feedback. *Journal of Applied Behavior Analysis, 13,* 287–
295.

Sutcliffe, J. P. (1972). On the role of "instructions to the subject" in psychological experiments. *American Psychologist, 27,* 755–758.

Swann, W. B., Jr., Wenzlaff, R. M., Krull, D. S., & Pelham,
B. W. (1992). Allure of negative feedback: Self-verification
strivings among depressed persons. *Journal of Abnormal
Psychology, 101,* 293–306.

Taffel, C. (1955). Anxiety and the conditioning of verbal behavior. *Journal of Abnormal and Social Psychology, 51,* 496–
501.

Tedeschi, J. T., Schlenker, B. R., & Bonoma, T. V. (1971). Cognitive dissonance: Private ratiocination or public specta-
cle. *American Psychologist, 26,* 685–695.

Tesch, F. E. (1977). Debriefing research participants: Though this be method there is madness to it. *Journal of Personality and Social Psychology, 35,* 217–224.

Thorne, S. B., & Himelstein, P. (1984). The role of suggestion in the perception of satanic messages in rock-and-roll recordings. *Journal of Psychology, 116,* 245–248.

Tunnell, G. B. (1977). Three dimensions of naturalness: An expanded definition of field research. *Psychological Bulletin, 84,* 426–437.

Turner, L. H., & Solomon, R. L. (1962). Human traumatic avoidance learning: Theory and experiments on the operant–respondent distinction of failures to learn. *Psychological Monographs, 76*(Whole No. 559), 1–32.

Underwood, B. J. (1957). Interference and forgetting. *Psychological Review, 64,* 49–60.

Underwood, B. J. (1959). Verbal learning in the educative process. *Harvard Educational Review, 29,* 107–117.

Unger, R., & Crawford, M. (1992). *Women and gender.* New York: McGraw-Hill.

U.S. Department of Agriculture. (1989, August 21). Animal welfare: Final rules. *Federal Register.*

U.S. Department of Agriculture. (1990, July 16). Animal welfare: Guinea pigs, hamsters and rabbits. *Federal Register.*

U.S. Department of Agriculture. (1991, February 15). Animal welfare: Standards; final rule. *Federal Register.*

U.S. Department of Health and Human Services. (1989, September 15). Proposed guidelines for policies on conflict of interest. *NIH Guide for Grants and Contracts, 18,* 1–5.

Van Buskirk, W. L. (1932). An experimental study of vividness in learning and retention. *Journal of Experimental Psychology, 15,* 563–573.

Vernon, H. M., Bedford, T., & Wyatt, S. (1924). *Two studies of rest pauses in industry.* Medical Research Council, Industrial Fatigue Research Board Report No. 25. London: His Majesty's Stationery Office.

Veroff, J., Douvan, E., & Kulka, R. A. (1981). *The inner American.* New York: Basic Books.

Videbeck, R., & Bates, H. D. (1966). Verbal conditioning by a simulated experimenter. *Psychological Record, 16,* 145–152.

Vinacke, E. (1954). Deceiving experimental subjects. *American Psychologist, 9,* 155.

Vokey, J. R., & Read, D. (1985). Subliminal messages: Between the devil and the media. *American Psychologist, 40,* 1231–1239.

Wade, E. A., & Blier, M. J. (1974). Learning and retention of verbal lists: Serial anticipation and serial discrimination. *Journal of Experimental Psychology, 103,* 732–739.

Walker, H. M., & Buckley, N. K. (1968). The use of positive reinforcement in conditioning attending behavior. *Journal of Applied Behavior Analysis, 1,* 245–250.

Walster, E. (1964). The temporal sequence of post-decision processes. In L. Festinger (Ed.), *Conflict, decision, and dissonance.* Stanford: Stanford University Press.

Webb, E. J., Campbell, D. T., Schwartz, R. D., & Sechrest, L. (1966). *Unobstructive measures: Nonreactive research in the social sciences.* Chicago: Rand McNally.

Wellman, P. J., Malpas, P. B., & Witkler, K. C. (1981). Conditioned taste aversion of unconditioned suppression of water intake induced by phenylpropanolamine in rats. *Physiological Psychology, 9,* 203–207.

West, S. G., Gunn, S. P., & Chernicky, P. (1975). Ubiquitous Watergate: An attributional analysis. *Journal of Personality and Social Psychology, 32,* 55–65.

Willson, V. L. (1980). Estimating changes in accident statistics due to reporting requirement changes. *Journal of Safety Research, 12,* 36–42.

Willson, V. L. (1981). Time and the external validity of experiments. *Evaluation and Program Planning, 4,* 229–238.

Wilson, D., & Donnerstein, E. (1976). Legal and ethical aspects of nonreactive social psychological research: An excursion into the public mind. *American Psychologist, 31,* 765–773.

Woodworth, R. S., & Sheehan, M. R. (1964). *Contemporary schools of psychology.* (3rd ed.). New York: Ronald Press.

Woolf, P. K. (1988). Deception in science. In American Association for the Advancement of Science and American Bar Association Conference of Lawyers and Scientists, *Project of Scientific Fraud and Misconduct: Report on Workshop Number One.* Washington, DC: AAAS.

Wyer, R. S., Jr., Dion, K. L., & Ellsworth, P. C. (1978). An editorial. *Journal of Experimental Social Psychology, 14,* 141–147.

Yoburn, B. C., Cohen, P. S., & Campagnoni, F. R. (1981). The role of intermittent food in the induction of attack in pigeons. *Journal of the Experimental Analysis of Behavior, 36,* 101–117.

Zaidel, D., & Sperry, R. W. (1974). Memory impairment after commissurotomy in man. *Brain, 97,* 263–272.

Zimbardo, P. G., Cohen, A. R., Weisenberg, M., Dworkin, L., & Firestone, I. (1966). Control of pain motivation by cognitive dissonance. *Science, 151,* 217–219.

Zimney, G. H. (1961). *Method in experimental psychology.* New York: Ronald Press.

Index